Artificial Intelligence and Machine Learning in the Travel Industry

Ben Vinod
Editor

Artificial Intelligence and Machine Learning in the Travel Industry

Simplifying Complex Decision Making

Previously published in *Revenue and Pricing Management* Special Issue "Artificial Intelligence and Machine Learning in the Travel Industry" Volume 20, Issue 3, March 2021

palgrave
macmillan

Editor
Ben Vinod
Grapevine, TX, USA

Spin-off from Journal: "Artificial Intelligence and Machine Learning in the Travel Industry" Volume 20, Issue 3, March 2021

ISBN 978-3-031-25455-0

This Palgrave Macmillan imprint is published by the registered company Springer Nature Switzerland AG
The registered company address is: Gewerbestrasse 11, 6330 Cham, Switzerland

Contents

Journal of Revenue and Pricing Management (2021) 20:211–212
https://doi.org/10.1057/s41272-021-00307-0

EDITORIAL

Special issue on artificial intelligence/machine learning in travel

B. Vinod[1]

Published online: 13 March 2021

Over the past decade, Artificial Intelligence has proved invaluable in a range of industry verticals such as automotive and assembly, life sciences, retail, oil and gas, and travel. The leading sectors adopting AI rapidly are Financial Services, Automotive and Assembly, High Tech and Telecommunications. Travel has been slow in adoption, but the opportunity for generating incremental value for AI over other analytics is extremely high (Chui et al. 2018).

In September 2019, Ian Yeoman and I discussed creating a special issue for the Journal of Revenue and Pricing Management on Artificial Intelligence in Travel. Information from airlines and vendors on AI in travel has been sporadic, usually discussed at industry conferences. Yet it was abundantly clear to me based on my interactions with travel suppliers, software vendors, OTAs and GDSs that they were leveraging core concepts in Artificial Intelligence and Machine Learning to create new value propositions or improve on existing applications related to travel. This was an opportunity to showcase in a single issue the breadth and scope of what individuals in these organizations were focused on with applications and business process.

The research papers…

An excellent contribution from Rodrigo Acuna-Agost, Eoin Thomas and Alix Lh'eritier from Amadeus, who propose a new method to estimate price elasticity for deep learning-based choice models with an excellent set of references. The insights they provide are particularly relevant for airline offers based on customer segment and context.

Ahmed Abdelghany and Ching-Wen Huang from Embry-Riddle University and Khaled Abdelghany from Southern Methodist University propose a novel reinforcement learning approach to calibrate itinerary choice models and measure schedule profitability.

Melvin Woodley from Sabre solves the attribution problem of associating sales or revenue to individual marketing efforts. Companies invest broadly across different platforms such as Google or Facebook, and it is important to understand the marginal impact of a specific marketing program or initiative on revenue. He uses a novel approach by casting the well-known Koyck distributed lag model in state space form to analyze the effectiveness of each marketing channel and subsequent allocation of marketing budgets.

The paper by Ravi Kumar, Wei Wang, Ahmed Simrin, Sivarama Krishnan Arunachalam and Bhaskar Rao Guntreddy and Darius Walczak on competitive revenue management models is a collaboration between PROS and Etihad Airways. Their paper proposes a demand model that captures realistic competitive dynamics by considering two types of customer behaviors: airline's loyal customers who prefer to buy from the airline even if their price is not the lowest in the market and fully flexible customers who buy the lowest fare in the market. They develop a Bayesian machine learning-based demand forecasting methodology for these models in both class-based and class-free settings that explicitly considers competitive market information.

Norbert Remenyi and Xiaodong Luo from Sabre discuss practical limitations of the choice-based demand models found in the literature to estimate demand from sales transaction data. They propose modifications and extensions under partial availability and extend the Expected Maximization (EM) algorithm for nonhomogeneous product sets. The data preprocessing and solution techniques are useful for practitioners.

The practice papers…

The paper on recommender systems by Amine Dadoun, Michael Defoin Platel, Thomas Fiig, Corinne Landra and Raphael Troncy from Amadeus highlight the central role of recommender systems to create personalized offers and its growing importance with IATA's New Distribution Capability messaging standard. It is a well-researched paper and truly relevant to the future of airline retailing.

Michael Byrd from Yum! and Ross Darrow from Charter and Go make the case for contextual bandits, a reinforcement learning technique for personalizing offers in retailing. They provide insights into the use of Thompson sampling, a

✉ B. Vinod
benvinod@yahoo.com

[1] Southlake, USA

Chapter 1 was originally published as Vinod, B. Journal of Revenue and Pricing Management (2021) 20:211–212. https://doi.org/10.1057/s41272-021-00307-0.

popular exploration heuristic and how they can be deployed. They discuss the step improvement that can be achieved with contextual bandits, despite greater computational complexity incurred when contextual features are included in the model.

Tomasz Szymanski from Nordea Bank and Ross Darrow from Charter and Go discuss the important topic of shelf placement on agency storefronts. While airlines focus on offer creation, the GDS desktop must display non-homogenous content that is addressed in this paper. A shelf product assortment method is proposed for categorizing airline offers into utility levels, thus facilitating the itinerary selection process for travelers.

Jian Wang from Realpage outlines a practical application of reinforcement learning used to determine reference rents for apartments. He demonstrates how the new approach outperforms the traditional rules-based approach.

Shriguru Nayak, Nitin Gautam and Sergey Shebalov from Sabre apply machine learning models to estimate market size and market share from competitive future schedules and augmented data sources, a key component for developing airline schedules. They also discuss how revenue management practices can be improved with access to data from network planning.

The paper by Rimo Das, Harshinder Chaddha and Somnath Banerjee from LodgIQ focuses on forecasting market demand considering seasonality and market events. They examine a variety of machine learning techniques that the data were calibrated upon and report on the accuracy of the forecasts.

In January 2018, an AI initiative was established at Sabre to identify industry-relevant problems suitable for AI-based solutions, raise internal awareness and accelerate adoption. This initiative also led to the creation and distribution of an internal AI newsletter, quarterly town halls to monitor progress and discuss use cases that I was responsible for. My contribution to the special issue reflects this initiative and steps taken to solve a range of problems in travel.

Foremost on the minds of corporations as they leverage AI for competitive advantage is how to scale AI across the organization. Deborah Leff and Kenneth Lim from IBM draw upon their extensive experience working with many companies to provide insights into the various organizational barriers to scale AI, the importance of executive sponsorship and recommend best practices. This paper is a "must read" for anyone who is a practitioner of AI.

The futures article…

Ross Darrow's future's article is thought provoking. "The Future of AI is the Market" paints a picture of how the future travel distribution landscape will be influenced by interactions in the marketplace and less on targeted one-off solutions.

I would like to take this opportunity to thank all the anonymous referees I reached out to over the past few months to provide feedback on the submitted papers. This special issue would not have been possible without your feedback and requests for revisions.

Reference

Chui, M., R. Chung, N. Henke, S. Malhotra, J. Manyika, M. Miremadi, and P. Nel. 2018. Notes from the AI Frontier: Applications and Value of Deep Learning. McKinsey.com, April 2018.

B. Vinod serves as Chief Scientist and Senior Vice President at Sabre (2008–2020). Before rejoining Sabre in 2004, he was Vice President at Sabre Airline Solutions, responsible for Pricing and Yield Management.

Journal of Revenue and Pricing Management (2021) 20:213–226
https://doi.org/10.1057/s41272-021-00308-z

RESEARCH ARTICLE

Price elasticity estimation for deep learning-based choice models: an application to air itinerary choices

Rodrigo Acuna-Agost[1] · Eoin Thomas[1] · Alix Lhéritier[1]

Received: 10 May 2020 / Accepted: 4 September 2020 / Published online: 22 March 2021

Abstract
One of the most popular approaches to model choices in the airline industry is the multinomial logit (MNL) model and its variations because it has key properties for businesses: acceptable accuracy and high interpretability. On the other hand, recent research has proven the interest of considering choice models based on deep neural networks as these provide better out-of-sample predictive power. However, these models typically lack direct business interpretability. One useful way to get insights for consumer behavior is by estimating and studying the price elasticity in different choice situations. In this research, we present a new methodology to estimate price elasticity from Deep Learning-based choice models. The approach leverages the automatic differentiation capabilities of deep learning libraries. We test our approach on data extracted from a global distribution system (GDS) on European market data. The results show clear differences in price elasticity between leisure and business trips. Overall, the demand for trips is price elastic for leisure and inelastic for the business segment. Moreover, the approach is flexible enough to study elasticity on different dimensions, showing that the demand for business trips could become highly elastic in some contexts like departures during weekends, international destinations, or when the reservation is done with enough anticipation. All these insights are of a particular interest for travel providers (e.g., airlines) to better adapt their offer, not only to the segment but also to the context.

Keywords Price elasticity · Discrete choice modeling · Deep learning · Interpretability · Automatic differentiation · Travel industry

Introduction

Discrete choice models describe the decision-making process when choosing among a set of distinct alternatives by defining a probability distribution on them. These methods have been employed both to better understand the factors leading to decisions and to predict individual decisions. Historically, the use of discrete choice models has been extensively investigated in relation to travel choices and travel offer pricing (Garrow 2016). For example, analysis of discrete choice models learned from survey data has been used to predict and inform new modes of transport such as new rail lines (McFadden 1974). Nowadays, choice models are integral parts of many revenue management systems, performing tasks such as demand modeling and assortment optimisation (Strauss et al. 2018). As more and more purchasing is made online, large volumes of purchasing choices are tracked and stored, from which choice models can be trained to select among competing offers. In these applications, the predictive power of the discrete choice model is directly correlated to conversion rates and thus revenue. In such scenarios, predictive power is valued more highly than the interpretability of the model.

One of the oldest, and still widely used, approach is the multinomial logit (MNL) model that assumes that the choice probability is proportional to the *utility* of each alternative (Luce 1959; Block and Marschak 1960), which is equivalently to satisfying Luce's axiom also known as *independence of irrelevant alternatives* (IIA) (Duncan Luce 1977). In McFadden (1973), the utility is defined as a function of observable attributes of the alternatives and some random independent and identically distributed component, which implies that a) all the decision makers behave similarly and b) the utility of the alternatives is independent.

✉ Rodrigo Acuna-Agost
rodrigo.acunaagost@amadeus.com

[1] AI, Research and Technology Department, Amadeus S.A.S., 485 Route du Pin Montard, BP 69, 06902 Sophia Antipolis Cedex, France

Chapter 2 was originally published as Acuna-Agost, R., Thomas, E. & Lhéritier, A. Journal of Revenue and Pricing Management (2021) 20:213–226. https://doi.org/10.1057/s41272-021-00308-z.

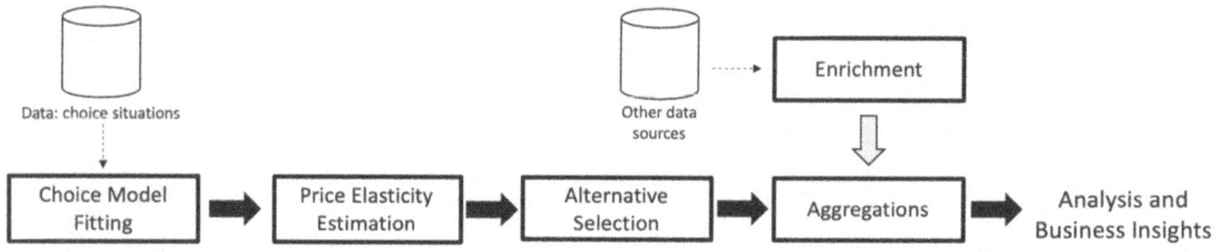

Fig. 1 The proposed methodology to extract elasticity estimates from choice models. Note that the enrichment process is optional, and here it is used to describe additional attributes of the data that are added for analysis, but not considered as features by the choice model

Natural evolutions of the MNL model deal with these limitations. For example, the mixed-MNL model allows grouping the decision makers by segments (Hayden Boyd and Mellman 1980) and, similarly, the nested logit model allows grouping alternatives by nests (McFadden 1980). More recently, (Lhéritier et al. 2019) uses a machine learning approach to allow an arbitrarily complex utility function that can also depend on the decision maker's attributes.

Deep learning is a branch of machine learning that uses artificial neural networks to model functions via an arbitrary number of composed transformations allowing to achieve high performance in a large variety of tasks (see, e.g., Goodfellow et al. 2016). Recently, some works used such approach to build more flexible choice models. In Mottini and Acuna-Agost (2017), the authors propose a sequence transformation approach to define the choice probability allowing to condition on the decision maker's attributes. More recently, Lhéritier (2020) uses a deep learning approach to parameterize the flexible class of Pairwise Choice Markov Chains that allows to escape traditional choice-theoretic assumptions such as IIA, stochastic transitivity and regularity (Ragain and Ugander 2016).

For the application of air itinerary choice prediction, there is a high interest in better understanding the choices of travelers. This application of choice models can have important impacts on revenue from using the most accurate models, but also by being able to get actionable insights that travel providers like airlines or travel agencies could exploit to improve their business metrics and offers.

The high predictive performance of neural networks comes at the cost of a difficult interpretation of the models, which has sparked research into complementary techniques to some shed light on how inputs attributes influence the outputs of the model. Various methods for performing feature importance have been proposed (Molnar 2019), as well as more specific methods to explain individual decisions. Local interpretable model explanations (LIME) (Ribeiro et al. 2016) can be used to explain a specific decision, by building a linear model from samples close in the feature space to the target sample. Shapley values have also been proposed as a method inspired by coalitional game theory (Lundberg and Lee 2017). A prediction can be explained by assuming that each feature value of the instance is a player in a game where the prediction is the payout. Shapley values determine how to fairly distribute the payout among the features. However, both LIME and Shapley values are relatively expensive to compute, as they require a sampling of data based on the decision to be explained.

Furthermore, the aim of this study is to extract meaningful economic information, such as price *elasticity*[1] from complex choice models, which are not provided by generic model interpretation techniques. The proposed methodology is summarized in Fig. 1. The first step is to estimate the choice probabilities by the choice prediction model, then elasticities are extracted for all the alternatives and choice situations. In order to give meaningful interpretation a filtering step is needed to discard all non-relevant alternatives. Finally, the analysis is based on different aggregations using a representative value (e.g., the median) per several dimensions (e.g., customer segments and other features of interest depending on the application).

The structure of the paper is as follows. We initially introduce the economic concept of elasticity and give an overview of how it can be estimated for deep learning models. We then present the motivating application of air itinerary choice modelling. This is followed by the numerical analysis of elasticities for flight choices, and finally, we present future research directions and conclusions.

Elasticity estimation from neural network-based choice models

In this section, we first introduce the economic concept of elasticity and then we show how it can be computed on complex deep neural network-based choice models.

[1] Elasticity relates to the relative change of one variable (e.g., demand) to the relative change in another variable (e.g., price).

Elasticity

Elasticity is an economic measure of how sensitive is a variable (e.g., demand) with respect to another one (e.g., price). More formally, the elasticity of x with respect to p at reference values x_0 and p_0 is defined as

$$\epsilon_p^x = \frac{\partial x}{\partial p}\frac{p_0}{x_0}. \quad (1)$$

A particular case is when x corresponds to the quantity of demand of a given good with a price p. In that case ϵ_p^x represents the price elasticity of demand, a concept of special importance in economics that will be leveraged in this work. It should be noted that this value corresponds approximatively to the percentage change in the demand that is caused by a 1% change in the price of that good.

Strictly speaking, elasticity can be either positive or negative. However, in most cases elasticity is negative: when price increases, the demand decreases. However, economic theory provides two particular exceptions of goods that defy common sense: Giffen and Veblen goods. On the one hand, *Giffen goods* are associated to the case of an inferior good (i.e., a good whose demand decreases when consumer income rises) where a negative income effect induced by the price change is strong enough to overcome a potential substitution effect (Spiegel 1994). This was studied for the first time in 1815, when it was reported that a rise in the price of bread corn, beyond a certain threshold, tended to increase the consumption of it, as a consequence that people could not afford more expensive substitutes (e.g., meat) (Heijman and Mouche 2011). On the other hand, *Veblen goods* concerns normal goods (i.e., demand increases when consumer income rises) that are particularly expensive and exclusive. On those products, a higher price may make them desirable as a status symbol, for example as observed in some luxury good and services (Veblen 1899).

In terms of magnitude of elasticity, economists usually look at their absolute value, classifying them into three groups:

- Elastic reaction, if ϵ_p^x is larger than 1. Examples: luxury items, vacations, high-end electronics, and generally goods with many substitutes.
- Inelastic reaction, if ϵ_p^x is smaller than 1. Examples: food, medicine, in general goods with vital importance and few or no substitutes.
- Isoelastic reaction, if ϵ_p^x is equal to 1.

For an in-depth treatment, the reader is referred to, e.g., Dorman (2014) and Kolmar (2017).

Compared to other industries (e.g., retail), the study of price elasticity of demand in the air transportation has remained relatively unexplored in the literature. Nevertheless we can highlight some of works: Jung and Fujii (1976), Ghoshal (1981), Brons et al. (2002), Castelli et al. (2003), Njegovan (2006), Richard (2009), Schiff and Becken (2011), Granados et al. (2012), Granados et al. (2012), and Morlotti et al. (2017).

Overall, these previous studies give similar results when it comes to order of magnitudes of price elasticities, varying from inelastic values (− 0.3) to more elastic relations (− 2.0) depending on their context (different periods and/or markets).

It should be noted that most previous works do not report elasticities aggregated by relevant dimensions such as the trip characteristics (e.g., elasticity per day of the week). It has been suggested that the price elasticity might vary according to the nature of the travel (Brons et al. 2002; Oum et al. 1992; Morlotti et al. 2017) and the presence of substitute modes (Brons et al. 2002). For example, Granados et al. (2012) reports a price elasticity of − 1.03, with difference depending on the channel (online vs. offline) and different market segments (business vs. leisure). The authors find that the values range from inelastic − 0.34 for business trip booked offline to a more elastic value of − 1.56 for online channel for the leisure segment.

The contribution of this paper is the ability to provide this type of insights from any available dimensions for Neural Network-based choice models.

Neural Networks

Deep feedforward networks are made of a series of *layers* consisting of linear combinations of its inputs followed by a non-linear *activation function*. The coefficients and the intercept of each linear combination are the free parameters of the model, and their number define therefore its modeling *capacity*. More complex *architectures* can be defined by allowing other combinations of inputs or, for example, feedback loops as in *recurrent neural networks*. Thanks to their versatility, deep networks can be used for a large variety of tasks, ranging from classical ones like classification or regression to more complex ones like reinforcement learning (e.g., Bondoux et al. 2020). Some architectures are designed to reduce the dimension of the inputs in order to learn good representations of the features for the task under consideration, allowing to process complex inputs like images or text. The parameters of the linear combinations are typically fit by minimizing a *cost function* on some *training dataset* using some variant of the *gradient descent* algorithm, i.e., gradients of the cost function are taken with respect to the parameters and these are moved in the negative gradient direction using some step size. For an in-depth treatment of deep neural networks, see e.g., Goodfellow et al. (2016).

State-of-the-art deep network-based choice models

Alternatives and individuals making choices can be described by a set of features that can be then used to condition the choice probability given by a choice model. A linear-in-parameters MNL assumes that the *representative utility* of each alternative is given by a linear combination of its features (McFadden 1973). These simple MNL models can be represented by a shallow neural network that uses only one linear combination of the attributes of each alternative and a *softmax* function applied to each linear combination. More sophisticated deep neural network-based models have been proposed in the literature. In Hruschka et al. (2001), a multi-layer extension has been proposed, allowing to learn non-linear MNL models. In Mottini and Acuna-Agost (2017), the authors propose a recurrent neural network architecture with an *attention mechanism* that learns to point, within a sequence of alternatives, to the chosen one.

PCMC-Net (Lhéritier 2020) is a deep network that parameterizes Pairwise Choice Markov Chains from the alternatives' and the individuals' features by combining a series of modules performing representation learning, pairwise combination, standard feedforward processing and linear system solving to finally obtain the choice distribution. PCMC-Net exhibits excellent predictive performance on a complex airline itinerary choice dataset where the alternatives are strongly dependent on an individual-specific query and some features, like price, can vary with time. In order to understand which kind of behavioral properties are captured by these complex deep networks, elasticities are of particular interest. The probabilities that are provided by a choice model can be interpreted as market shares at equilibrium and therefore can be used as demand quantities in Eq. 1.

Automatic differentiation-based estimation

When using neural network-based choice models, probabilities can be easily differentiated with respect to any of the inputs using the *automatic differentiation* mechanism of modern neural network libraries [e.g., PyTorch, see Paszke et al. (2019) or Tensorflow, see Abadi et al. (2016)]. An automatic differentiation system converts a program specifying operations (e.g., a neural network) into a computational graph that represents it in terms of a composition of primitive operations that have specified routines for computing derivatives. Then, the derivative of some node (e.g., corresponding to the cost function or the probability given by a neural network) with respect to some leaf node of the graph (e.g., corresponding to parameters or inputs of the neural network) on some given point x can be obtained by applying basic differentiation rules and the chain rule to propagate the applied values through the graph (see Fig. 2 for an example). It is different from numerical and symbolic differentiation, and is both efficient (linear in the cost of computing the value that is being differentiated) and numerically stable. See, e.g., Baydin et al. (2017) for an in-depth treatment.

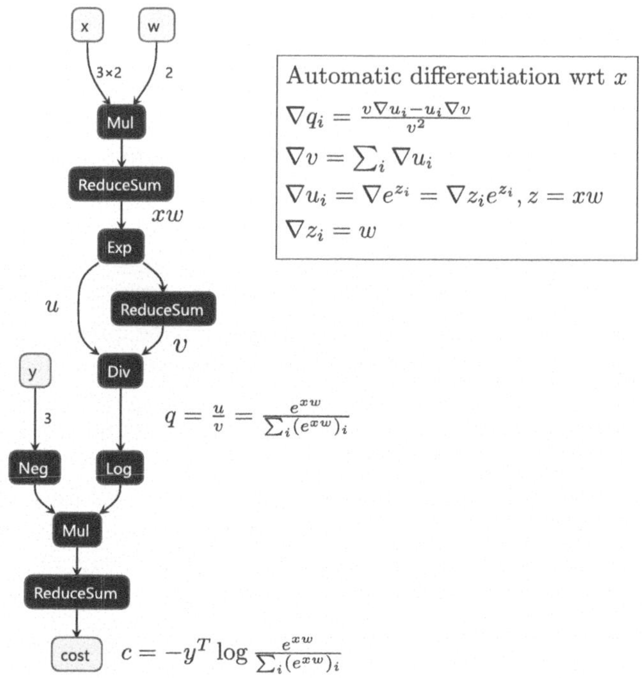

Fig. 2 Computational graph obtained from PyTorch for a linear-in-parameters MNL for three alternatives with two features represented in a 3×2 matrix x. The column vector w corresponds to the parameters of the model and the column vector y represents the actual choice with a 1 in the corresponding position and 0 elsewhere. Automatic differentiation allows to obtain derivatives of the cost function c with respect to w in order to fit the model and derivatives of the probability vector q with respect to inputs x in order to compute elasticities. q_i denotes ith component of the vector q

The derivative of the choice probability with respect to some input usually depends on the point x whose coordinates are the alternatives' and the individual's features of a given choice situation. In order to interpret the predictions given by a model on some given dataset using elasticities, derivatives can be computed on it and summarized by taking e.g., the median, as shown in the numerical results of this paper.

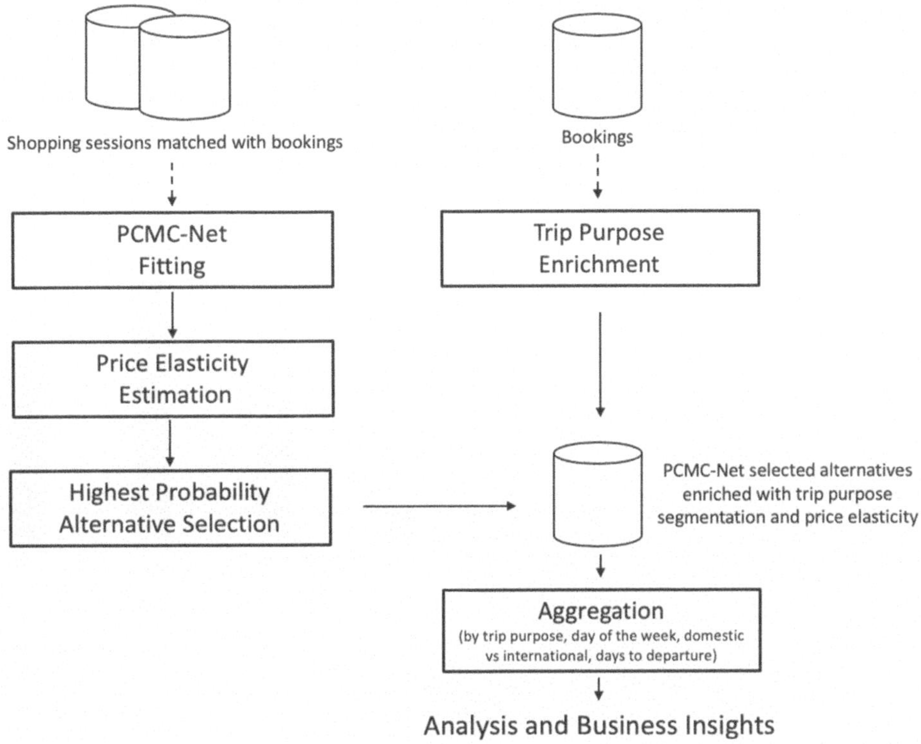

Fig. 3 The methodology applied to air itinerary choice. Price elasticities are extracted from the PCMC-Net choice model. For analysis, only the alternative with the largest probability of selection is selected in each choice session. Trip purpose segmentation is applied based on a training set consisting of bookings

Airline itinerary choice dataset with price elasticity estimation and trip purpose segmentation

Nowadays travelers have higher expectations and choice than in the past. This is mostly driven by the experience they already have in other industries, in particular online retail. Some examples of the new standards are: relevant and timely recommendations, fully customized products and services, transparent pricing, modern search, shopping cart functionalities on different channels (e.g., mobile and desktop), and high flexibility (e.g., be able to cancel subscriptions at any time or to send back products and being fully reimbursed). Many, if not all, of these improvements observed in retail have been boosted by leveraging data and newer algorithms thanks to machine learning, and naturally the travel industry is following this trend too.

Figure 3 summarizes the application of the proposed methodology for getting elasticities and business insights from air itinerary choice models. The remainder of this section presents the methodology to obtain this enriched dataset of chosen alternatives with elasticity and trip purpose estimates. The aggregations and business insights are presented in the results section of this paper.

Price elasticity estimation from airline itinerary choices

A passenger name record (PNR) contains relevant data regarding travel bookings, such as flight information of each segment of a journey and information about the individual, as well as information about ancillary services and special service requests. In order to obtain a full choice set,

Table 1 Features of the airline itinerary choice dataset

	Type	Feature	Range/cardinality
Individual	Cat.	Origin/destination	97
		Search office	11
	Num.	Departure weekday	[0, 6]
		Stay Saturday	[0, 1]
		Continental trip	[0, 1]
		Domestic trip	[0, 1]
		Days to departure	[0, 343]
Alternative	Cat.	Airline (of first flight)	63
	Num.	Price	[77.15, 16,781.50]
		Stay duration (min)	[121, 434,000]
		Trip duration (min)	[105, 4314]
		Number connections	[2, 6]
		Number airlines	[1, 4]
		Outbound departure time (in s)	[0, 84,000]
		Outbound arrival time (in s)	[0, 84,000]

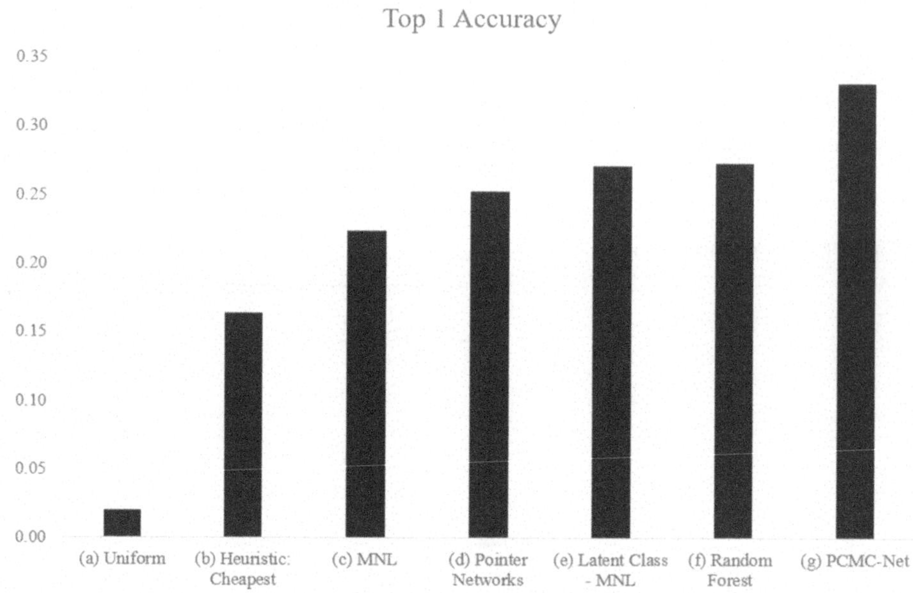

Fig. 4 Performance of different heuristics and choice models on the airline itinerary choice problem (Lhéritier 2020; Lhéritier et al. 2019; Mottini and Acuna-Agost 2017)

data from PNRs are matched with search log activity, which shows all available options presented to the traveller prior to booking.

In this experiment, the dataset from Mottini and Acuna-Agost (2017) consisting of flight bookings sessions on a set of European origins and destinations is used. Each choice session contains up to 50 different proposed itineraries, one of which has been booked by the customer. There are 815,559 distinct alternatives among which 84% are singletons and 99% are observed at most seven times. In total, there are 33,951 choice sessions of which 27160 were used for training and 6791 for testing. The dataset has a total of 13 features, both numerical and categorical, corresponding to individuals and alternatives, as shown in Table 1.

Choice models are important in helping to select, highlight and rank different offers. Several methods have previously been suggested, the performance of which is shown in Fig. 4 on a common training and test dataset. The metric used here is the Top 1 accuracy, which measures the percentage of sessions for which the most probable alternative identified by the model is indeed chosen by the user. Each choice set contains 50 alternatives, thus a uniform sampling method achieves 2% Top 1 accuracy. Simple heuristics such as selecting the cheapest offer can be used to give a baseline performance for the problem. As can be seen, non-linear methods such as Deep Pointer networks, Latent Class MNL and Random Forests give better performance than the linear MNL model. However, PCMC-net results in the best Top 1 accuracy of all the methods tested. For more details on each method, the reader is referred to Lhéritier (2020), Lhéritier et al. (2019) and Mottini and Acuna-Agost (2017).

Elasticity can be calculated for any reference value and for any alternative in the choice set. We are particularly interested in price elasticity, that is, the elasticity of the demand, estimated by the probability of being chosen, with respect to the price. Moreover, we consider only one alternative per choice situation. For each choice situation, we take the alternative with the largest probability of selection as estimated by the PCMC-Net model. It should be noted that this alternative does not necessarily match the alternative that was chosen by the consumer. This decision was taken as our main goal is to give explainability to the neural network-based model, rather than understanding individual choices.

Trip purpose segmentation: motivation

The first step to better modeling the traveller decision-making process is to understand the reason why the traveler would like to travel, which we will refer to hereafter as the *trip purpose*. If travel providers could get this information accurately at shopping time (i.e., before the booking), they could greatly improve the shopping experience: offer the best product, at the best price, at the best moment, to the targeted customer.

Business trips are driven by convenience and usually subject to companies' travel policies (i.e., the passenger do not pay for this trip, but her company). It is also the case that sometimes the passenger has a less active role in the decision of the trip as the task is delegated to travel arrangers such as travel agencies or assistants. The authors in Teichert et al. (2008) confirm that people traveling for business have a strong correlation with these attributes: efficiency, punctuality, and flexibility.

On the other hand, leisure trips are driven mostly by price (Teichert et al. 2008). Although in some cases, the passengers are sufficiently wealthy that they may give more importance to comfort and efficiency. In practice, many trips are not easily classified on exactly one of these two segments as they can be both at the same time: *bleisure* trips (i.e., extending a business trip for leisure activities) (Vivion 2016).

As an important question for the industry, it is not surprising that the problem has been addressed previously (Teichert et al. 2008; Chatterjee et al. 2020; Tahanisaz and shokuhyar 2020; Martinez-Garcia and Royo-Vela 2010; Jin-Long 2017; Dresner 2006; Vinod 2008). Most of the previous work is based on stated preference surveys (Dresner 2006; Martinez-Garcia and Royo-Vela 2010; Tahanisaz and shokuhyar 2020) i.e., asking current or potential travelers about their preferences and the reason of the trips. This kind of approach brings a lot of flexibility in terms of the type of questions as for example the analysts could even ask about hypothetical scenarios. It is well known that stated preference data present a series of inconveniences (Abdullah et al. 2011), for example, their incapacity to capture accurately all the market and personal limitations that occurs in the real world. Another limitation of these works based on surveys is that the conclusions are drawn based on relatively small amount of data [around 3000 in Dresner (2006), 300 in Tahanisaz and shokuhyar (2020), 808 in Martinez-Garcia and Royo-Vela (2010), and 5800 in Teichert et al. (2008)].

Therefore, to adequately discuss the price elasticities obtained by the choice model, these should be done in the context of business and leisure trip independently. However, this information is not available from the choice dataset, and as such must be inferred.

Trip purpose segmentation: dataset and experimental protocol

In this section, we present an analysis of business vs leisure segmentation performed over a set of labeled bookings from a larger set of unlabeled data. The bookings correspond to indirect bookings made by customers at traditional travel agencies, online travel agencies and travel management companies (among others) which are then processed by a GDS. We can consider the dataset as being partially labeled, as most offices are identified as belonging to particular market segments which deal almost exclusively with either business travel or leisure travel.

The dataset used in this trip purpose model is a sample of indirect bookings from the European market for the full year 2019. A balanced dataset is obtained by sampling 200,000 bookings made from offices tagged as "Retail- small medium enterprises" which are labeled as leisure travel, and sampling another 200,000 bookings made from offices tagged as "Global Travel Management Companies" which are labeled as business travel. From this set of 400,000 bookings, 40,000 random samples are held out as the test set for the classification task.

For each booking there are 48 features available corresponding to all non-sensitive attributes of the trip. These relate to the origin, destination, route, carrier, various aspects linked to the time and duration of travel along with aspects of the booking such as the number of passengers in the booking, the number of days prior to boarding that the trip was booked, etc. A comparison of different models based on various feature subsets is provided in Appendix A.

In order to apply the segmentation of bookings to the choice dataset, only features common to both the bookings and choice datasets are selected. These are the origin and destination, international/domestic, stay duration, days to departure, the day of the week for the booking, outbound flight and return flight and a stay Saturday feature.

A gradient boosting machine model (Friedman 2001) is used as a classifier, with grid search and early stopping selecting a maximum depth of 6 for 46 trees. On the hold-out test set, the model obtains 84.20% *accuracy*.[2] Feature importance analysis suggests that the stay duration, destination and origin airports, stay Saturday, return day of week and days to departure are the most important features, respectively. Note that this segmentation is not used as an input to the choice model, but only to segment the dataset for analysis purposes (see Fig. 3).

Analysis and business insights

In this section, we analyze and discuss the elasticities obtained by the approach presented previously using different aggregations.

In order to aggregate the estimated elasticities, we use the median since it is a measure of central tendency robust with respect to outliers and skewed data (see Fig. 5). Another important element to remark is that the analyses are based on the additive inverse of elasticity because the price elasticities are usually negative. This transformation helps to construct readable charts following the convention of economists that are interested on the absolute values (magnitudes) instead of the real number.

Overall elasticity The median value for whole data is − 1.73 which can be classified as *elastic*. Note that this value is in the same order of magnitude to the values published in previous works.

Elasticity by trip purpose In order to understand better this value, we analyze the two main customer segments in

[2] Accuracy: ratio of number of correct predictions to the total number of input samples.

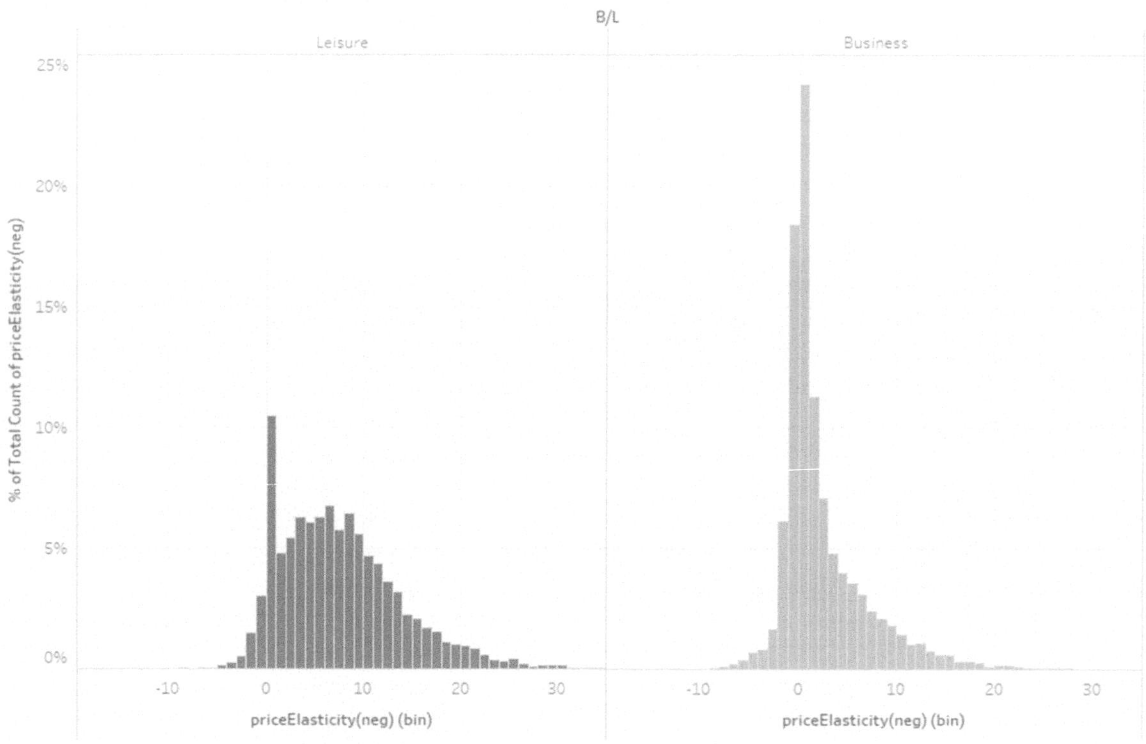

Fig. 5 Distribution of elasticity values for both segments: Leisure (left) and Business (right). The distributions are not symmetric, and in both segments there is a peak of observations near to zero. Note (a) we analyze one alternative per choice situation, the one with the highest probability, (b) all charts (and this one in particular) present the additive inverse of the elasticity that is usually negative in its original form. It should be also noted the presence of few negative values correspond to rare observations. These choices could be explained by Veblen or Giffen behaviors of some consumers in certain circumstances

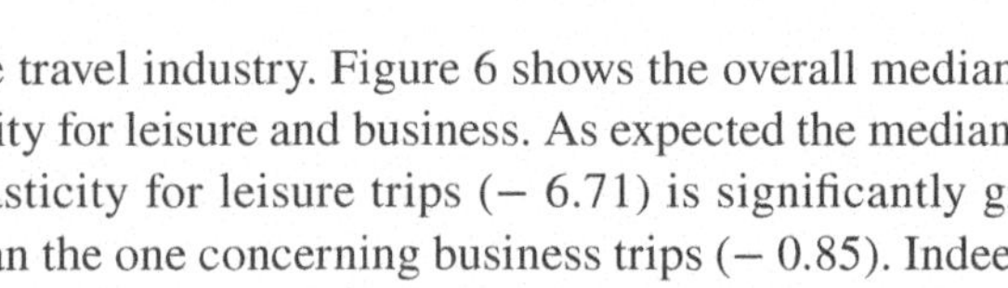

Fig. 6 Overall median negative elasticity for both segments: Business and Leisure

the travel industry. Figure 6 shows the overall median elasticity for leisure and business. As expected the median price elasticity for leisure trips (− 6.71) is significantly greater than the one concerning business trips (− 0.85). Indeed, the price elasticity for leisure trips is considered to be *elastic,* while the one for business is *inelastic*. The interpretation of this value is that if the price is changed 1%, we should expect a change in the probability of being chosen of approximately 6.71% for a consumer looking for leisure trips (and 0.85% for business trip, respectively).

Elasticity by day of the week Figure 7 shows median elasticities on both segments per day of the week (DOW). The results confirm the difference between both segments. More interesting are the values on different DOW. Consumers related to leisure trips seem to have a more regular price sensitivity, while in business trips we see a significant difference between working days and weekends. These results are consistent with previous research reporting − 0.64 for working days and − 1.05 for weekends (see Morlotti et al. 2017).

Elasticity by domestic/international trip Figure 8 shows the median elasticity on both segments split by domestic and international trips. Similarly to the previous chart, leisure trips exhibit less variance in price elasticity, while, for business trips, there is a significant difference between domestic and international trips. This could be explained by travel policies in place in most companies, where it is common to put price restrictions as a function of the distance of the flights. Note that this particular dataset is related to European markets, where international trips tend to be more expensive than domestic trips.

Elasticity by days to departure Another interesting analysis is presented in Fig. 9. The chart shows the median elasticity as a function of the trip advance purchase, also known as days to departure (DTD). It is interesting to remark, as expected, that the price elasticity increases if the consumer has more time to decide before the departure of the trip. Another interesting insight is that business and leisure trips seem to converge to similar values for large DTD. On the other hand, it is clear that for both segments, trips happening

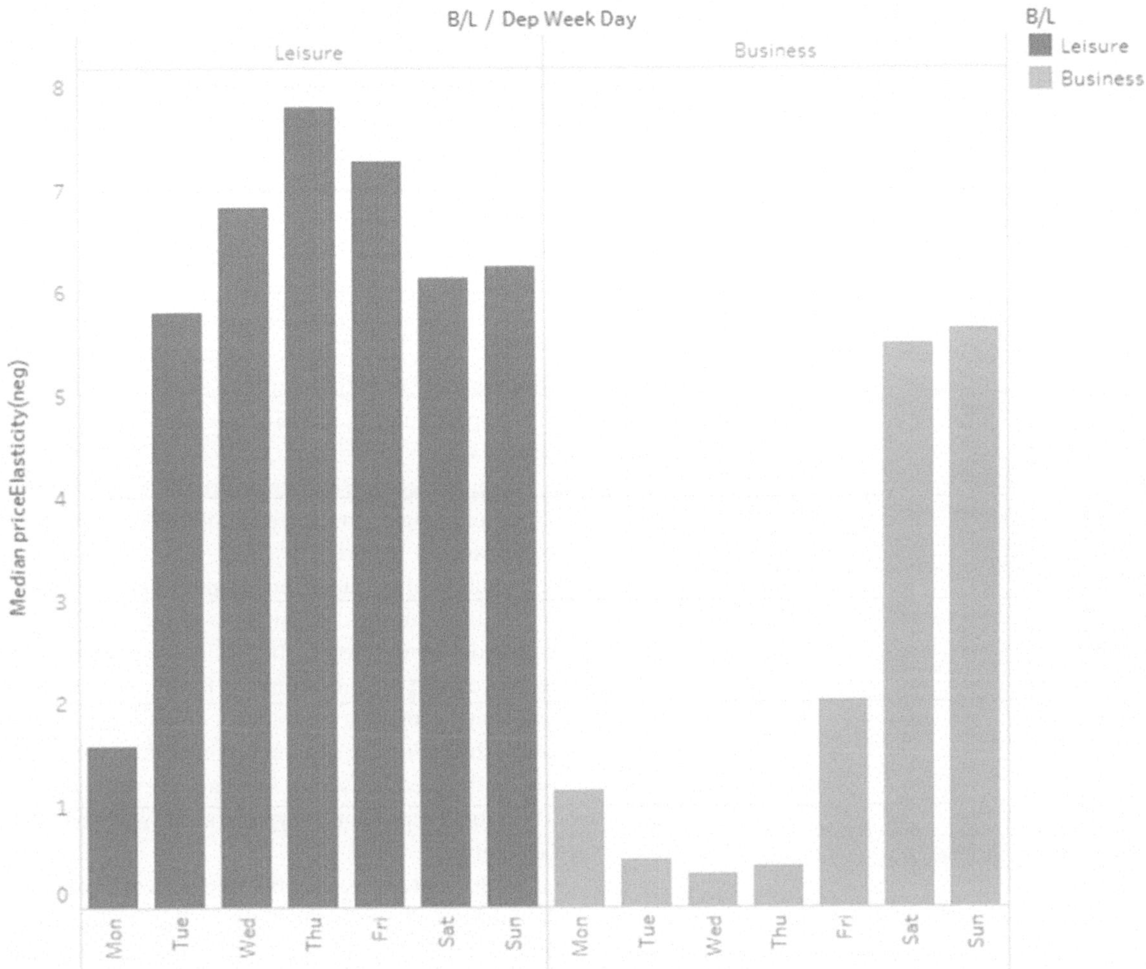

Fig. 7 Median elasticity (neg) for both segments (Business and Leisure) on different departure days of the week

in the near future (short-term) show very low price elasticities, probably explained as trips booked for closer dates are particularly related to urgent matters, in those cases the price is less relevant than other aspects like the schedule or the total trip duration. This is consistent with previous research where elasticity was reported in the interval $\epsilon \in \{-2.0, -0.5\}$ for DTD $\in \{2, 21\}$ days (see Morlotti et al. 2017). Our results extend the previous results showing that the absolute value continues to increase to larger values until 100 days approximately. The chart also shows an increase on the dispersion of values for larger DTD, which is explained mainly by the number of observations used to calculate the median values (represented by the darkness of the line).

Elasticity by stay duration An important criterion in price elasticity is the stay duration, especially when weekend stays are factored into the analysis. In Fig. 10, the price elasticity is shown as a function of stay duration in orange for trips where the traveler stayed an entire Saturday and in blue for when the traveler did not stay at destination a full Saturday. Note that the blue curve rises in a linear fashion, indicating that stay duration is directly proportional to price elasticity for trips which return during the same week as departure (this also includes trips leaving Sunday and returning prior to the following Saturday). For trips including Saturday stays (in orange), all trips are highly price elastic, but it does appear that for stay durations between 1 and 3 nights are less price elastic. These short trips always contain a Saturday night stay, and therefore often correspond to weekend getaways and possibly city breaks. Such trips are often to geographically closer destinations, which are associated with cheaper prices overall, and thus price may be a less important factor than for longer duration trips which can be associated with a higher overall budget. Furthermore, for such short trips, other factors such as time of arrival and departure may be more important to the travelers in

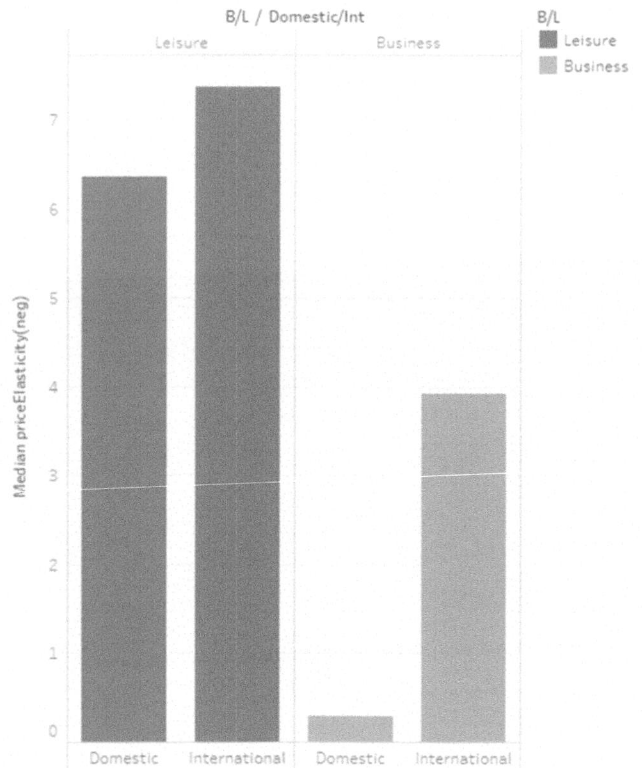

Fig. 8 Median elasticity (neg) for both segments (Business and Leisure) on domestic and international trips

order to maximise their time at destination, thus reducing the importance of price.

Conclusions

In this paper, we have proposed a new methodology to extract price elasticity from deep learning-based choice models. The approach leverages the efficiency of automatic differentiation capabilities of deep learning libraries. With this capability, we were able to estimate price elasticities on all the data points (choice set). As a consequence, the approach was flexible enough to permit deeper analysis in any dimension available in the data. With this in mind, we focused on understanding price elasticity on the classical segments in the industry (business and leisure) and the features/context that may affect their elasticity.

Regarding numerical experimentation, the elasticities extracted from PCMC-Net suggest that the demand for air travel for the business and leisure segments differs significantly. While business trips are price inelastic, leisure trips are estimated to be highly price elastic. The presented methodology allowed to explore the elasticity on other dimensions of interest as well. For example, we found that, depending on different variables, business trips can be inelastic or not. Some cases where we observed elastic reaction on business trips was on international trips, departures during weekends, and reservations done with enough anticipation (i.e., more than 30 days). This could be explained by travel policies in place on most companies. All these insights could be of the particular interest on pricing strategies and other initiatives done by travel providers to better adapt their offer to the market conditions.

We suggest the following ideas that can inform future research: (a) use elasticity as a new manner to segment passengers and (b) estimate other economic metrics of interest, in particular the willingness to pay (WTP) for certain features.

With respect to segmentation, the most common way to group passengers in the travel industry has been to separate business and leisure trips. These two segments allow differentiation of products, but assume a clear difference in the price elasticities and willingness to pay of the segments. As concluded in this work, the business segment may become highly elastic in some context. In that regard, we believe a new way to segment customers could be leveraged from this work, thus allowing segmentation of travel requests depending on the price elasticity that can be estimated in real time. This would represent a new opportunity for dynamic pricing/packaging.

With regard to economic metrics, we have presented the calculation of price elasticity in this work. Nevertheless, there are other economic metrics that are important in the industry. One of the them is the willingness to pay, usually defined as the maximum amount of money a consumer is willing to hand over to buy a product or service (Lu and Shon 2012; Carlos Martín et al. 2008; Chang and Sun 2012; Carlsson 1999; Tsamboulas and Nikoleris 2008; Merkert and Beck 2017). It should be noted that a way to estimate WTP is by looking at the derivative of a feature with respect to the price, something that could be explored using the approach presented in this paper.

Appendix: A trip purpose segmentation

To better understand the value of the business vs leisure segmentation proposed in this study, we showcase and compare three variations of the model. In the first approach, all features available in the dataset are used to generate the best performing model and provide some analysis to understand what characteristics of travel are important to differentiate different types of travel. In order to better understand the relationship between different features of travel, we also showcase a simplified model based on a decision tree, with minimal features and complexity, which can be fully interpreted. Finally, we will also construct a model which can

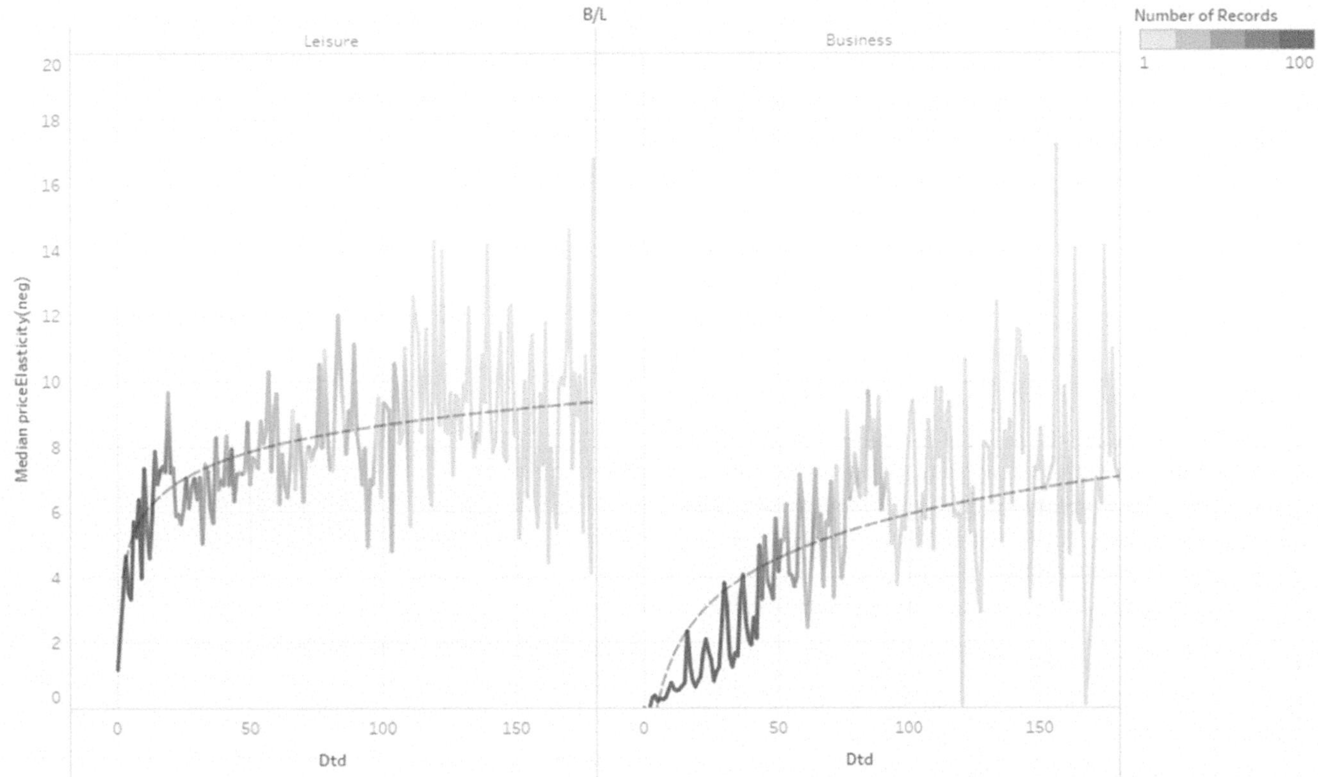

Fig. 9 Median elasticity (neg) for both segments: leisure (left) and business (right) as a function of different advance purchase days (a.k.a. days to departure)

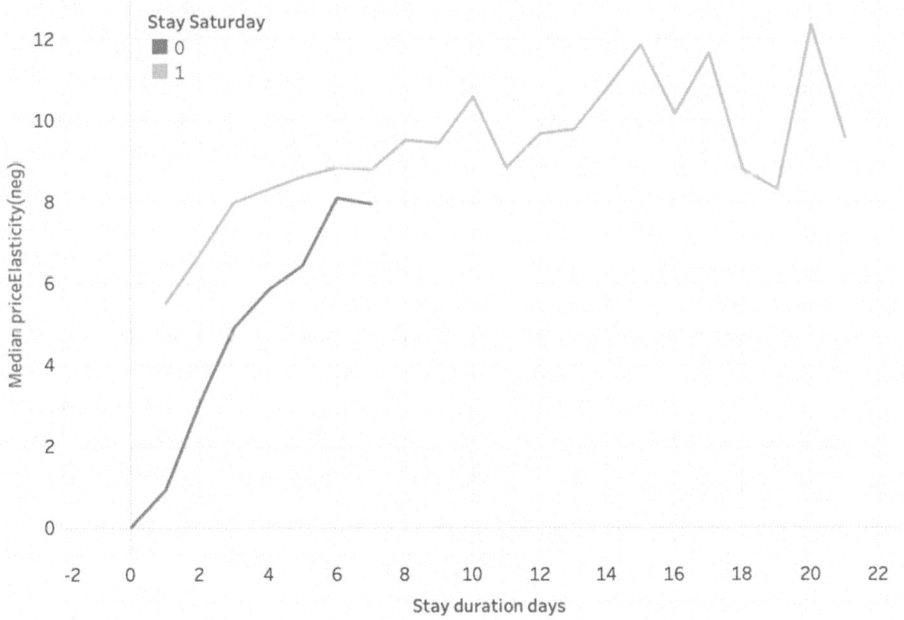

Fig. 10 Median elasticity (neg) as a function of stay duration, for trips containing a Saturday stay or not

be applied to a choice dataset used in the main article. This choice data profile segmentation model is effectively a compromise in performance, due to only some features overlapping between the bookings dataset and the choice dataset.

Gradient boosting models have proven particularly adept at classification, here the H2O.ai library is used to train the models and determine the feature importance (Candel and Malohlava 2020). The training phase uses a hold-out

Table 2 Trip purpose experiments and results on hold-out test set

Model	Model type	No. of features	Test accuracy (%)
Full feature set	H2O GBM	48	86.5
Choice feature set	H2O GBM	9	84.2
Reduced feature set	Scikit learn DT	9	80.8

Note that the H2O package was used to train the gradient boosting machines in order to obtain best performance, whereas Scikit-learn was used to train a decision tree that could be visualized and interpreted

validation set to estimate the depth of the individual trees via grid search and the number of trees via a stopping criterion. In contrast, the interpretable model is trained with scikit learn (Pedregosa et al. 2011) and consists of a single decision tree which was manually tuned to give an acceptable tradeoff between human interpretability and performance on the test set.

For the interpretable model features, these are selected to represent attributes that are not overly dependent on the market, e.g., carrier and destination country, but rather those that are representative of general travel aspects, such as stay duration and number of passengers in a booking. Specifically, the features are limited to the number of passengers, distance of trip, international, stay duration, days to departure, day of week of outbound and return flights as well as booking date, and a feature indicating whether the travelers stayed Saturday night during their travel.

The classification accuracy for the hold-out test set is given in Table 2. The full feature set results are given in order to set an upper bound estimate of the performance given no constraints for the classification task. The accuracy of 86.5% suggests that generally the trip purpose can be determined on this dataset, but there are bookings that are misclassified. The features ranked highest in terms of feature importance for this model are the carrier, destination country, whether the passengers stayed Saturday night at destination, the number of passengers in the booking and the advance purchase time prior to the flight.

In comparison, despite only accessing nine of the features available, the choice feature set obtains 84.2%. Feature importance analysis suggests that the stay duration, destination and origin airports, stay Saturday, return day of week and days to departure are the most important features, respectively. This model is applied to the flight choice dataset in order to provide additional context to the analysis in the main article.

For the interpretable model, we use a single decision tree with limited complexity. The complexity of the tree is constrained during training by limiting the number of splits performed in the tree. This can be done by setting the maximum

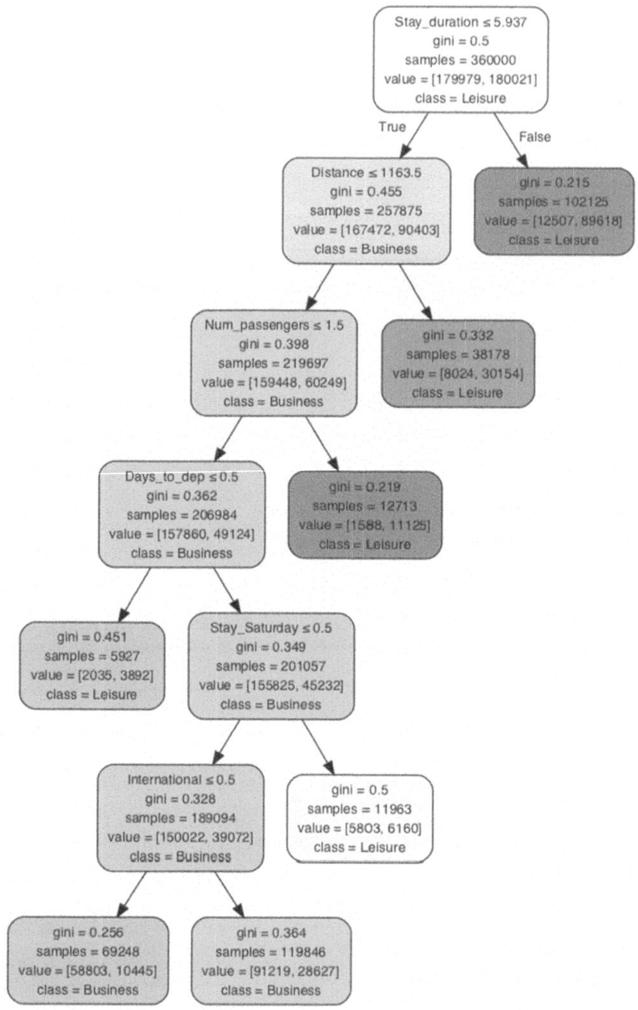

Fig. 11 Interpretable decision tree for business vs leisure prediction. Nodes that are split show the feature and threshold used for the split. All nodes show the gini value, where lower values indicate purer nodes, the total number of training samples at the node, the number of training samples from each class at the node [business, leisure], and the resulting class label. The color is based on the class label, and the intensity is a function of gini value. (Color figure online)

depth of any branch, or by only allowing splits when sufficient data are present before or after the split (often referred to as minimum samples in split or minimum samples in leaf, respectively). Testing over different choices of parameters resulted in the choice of restricting the model complexity using minimum samples before a split. This resulted in the best accuracy on the hold-out test set for any model with under 10 splits, and also produced a model for which increasing complexity did not lead to large improvements in accuracy over the test set, relative to other choices available to restrict complexity.

This interpretable tree results in 80.8% accuracy on the same hold-out test set as used in the previous experiment.

We note that the best single tree performance obtained with more complex trees was 82.4% accuracy.

The resulting model is visualised in Fig. 11. It can be seen that the first split reflects stay duration, indicating that travelers who stayed for more than 5.9 days at destination were primarily leisure travelers (87.7). Travelers who stayed less than 5.9 days, but whose destination was more than 1163 km from the origin were also predominantly leisure travelers. For short duration and short distance trips, the main differentiator appears to be the number of passengers in the booking, with solo travelers being more likely to be business. For most other conditions in the tree, we see the output is mostly business or for a small number of samples leisure but with low certainty. Based on the lowest Gini value, we can conclude that bookings will be classified as business for short durations (< 5.9 days), short distances (< 1163 km), solo travelers, booking before the day of travel, not staying Saturday and flying domestically.

References

Abadi, Martín, Paul Barham, Jianmin Chen, Zhifeng Chen, Andy Davis, Jeffrey Dean, Matthieu Devin, Sanjay Ghemawat, Geoffrey Irving, and Michael Isard, et al. 2016. Tensorflow: A system for large-scale machine learning. In *12th {USENIX} Symposium on operating systems design and implementation ({OSDI} 16)*, 265–283.

Abdullah, Sabah, Anil Markandya, and Paulo A.L.D. Nunes. 2011. Introduction to economic valuation methods. In *Research tools in natural resource and environmental economics*, 143–187. Singapore: World Scientific.

Baydin, Atılım Günes, Barak A. Pearlmutter, Alexey Andreyevich Radul, and Jeffrey Mark Siskind. 2017. Automatic differentiation in machine learning: A survey. *The Journal of Machine Learning Research* 18 (1): 5595–5637.

Block, H.D., and Jacob Marschak. 1960. Random orderings and stochastic theories of response. *Contributions to Probability and Statistics* 2: 97–132.

Bondoux, Nicolas, Anh Quan Nguyen, Thomas Fiig, and Rodrigo Acuna-Agost. 2020. Reinforcement learning applied to airline revenue management. *Journal of Revenue and Pricing Management* 19: 332–348.

Brons, Martijn, Eric Pels, Peter Nijkamp, and Piet Rietveld. 2002. Price elasticities of demand for passenger air travel: A meta-analysis. *Journal of Air Transport Management* 8 (3): 165–175.

Candel, A., and M. Malohlava. 2020. Gradient boosted models. R package version 3 (0.4).

Carlos Martín, Juan, Concepción Román, and Raquel Espino. 2008. Willingness to pay for airline service quality. *Transport Reviews* 28 (2): 199–217.

Carlsson, Fredrik. 1999. Private vs. business and rail vs. air passengers: willingness to pay for transport attributes. Working Papers in Economics No. 14.

Castelli, Lorenzo Walter Ukovich, and Raffaele Pesenti. 2003. An airline-based multilevel analysis of airfare elasticity for passenger demand. In *Air Transport Research Society (ATRS) world conference*

Chang, Li-Yen, and Pei-Yu Sun. 2012. Stated-choice analysis of willingness to pay for low cost carrier services. *Journal of Air Transport Management* 20: 15–17.

Chatterjee, Sujoy, Nicolas Pasquier, Simon Nanty, and Maria A. Zuluaga. 2020. Multi-objective consensus clustering framework for flight search recommendation. arxiv:2002.10241.

Dorman, Peter. 2014. *Microeconomics—A fresh start*. Cham: Springer.

Dresner, Martin. 2006. Leisure versus business passengers: Similarities, differences, and implications. *Journal of Air Transport Management*, 12 (1): 28–32. Leisure Traffic and Tourism: New Strategies for Airlines, Airports and the Travel Trade.

Duncan Luce, R. 1977. The choice axiom after twenty years. *Journal of Mathematical Psychology* 15 (3): 215–233.

Friedman, Jerome H. 2001. Greedy function approximation: A gradient boosting machine. *Annals of Statistics* 29 (5): 1189–1232.

Garrow, Laurie A 2016. *Discrete choice modelling and air travel demand: Theory and applications*. London: Routledge.

Ghoshal, Animesh. 1981. Price elasticity of demand for air passenger service: Some additional evidence. *Transportation Journal* 20 (4): 93–96.

Goodfellow, Ian, Yoshua Bengio, and Aaron Courville. 2016. *Deep learning*. Cambridge: MIT.

Granados, Nelson, Alok Gupta, and Robert J. Kauffman. 2012. Online and offline demand and price elasticities: Evidence from the air travel industry. *Information Systems Research* 23 (1): 164–181.

Granados, Nelson, Robert J. Kauffman, Hsiangchu Lai, and Huang-chi Lin. 2012. À la carte pricing and price elasticity of demand in air travel. *Decision Support Systems* 53 (2): 381–394.

Hayden Boyd, J., and Robert E. Mellman. 1980. The effect of fuel economy standards on the us automotive market: An hedonic demand analysis. *Transportation Research Part A: General* 14 (5–6): 367–378.

Heijman, Wim, and Pierre Mouche. 2011. *New insights into the theory of Giffen goods*, vol. 655. Berlin: Springer.

Hruschka, Harald Werner Fettes, and Markus Probst. 2001. Analyzing purchase data by a neural net extension of the multinomial logit model. In *International conference on artificial neural networks*, 790–795. Cham: Springer.

Jin-Long, Lu. 2017. Segmentation of passengers using full-service and low-cost carriers - evidence from taiwan. *Journal of Air Transport Management* 62: 204–216.

Jung, J.M. and E.T. Fujii. 1976. The price elasticity of demand for air travel: some new evidence. *Journal of Transport Economics and Policy* 10: 257–262.

Kolmar, Martin. 2017. *Principles of microeconomics*. Cham: Springer.

Lhéritier, Alix. 2020. PCMC-Net: Feature-based pairwise choice markov chains. In *International conference on learning representations (ICLR 2020)*, Addis Ababa, Ethiopia, 26–30 April 2020.

Lhéritier, Alix, Michael Bocamazo, Thierry Delahaye, and Rodrigo Acuna-Agost. 2019. Airline itinerary choice modeling using machine learning. *Journal of Choice Modelling* 31: 198–209.

Luce, R. Duncan. 1959. *Individual choice behavior: A theoretical analysis*. New York: Wiley.

Lundberg, Scott M., and Su-In Lee. 2017. A unified approach to interpreting model predictions. In *Advances in neural information processing systems*, 4765–4774.

Lu, Jin-Long , and Zhang Yi Shon. 2012. Exploring airline passengers' willingness to pay for carbon offsets. *Transportation Research Part D: Transport and Environment* 17 (2): 124–128.

Martinez-Garcia, Esther, and Marcelo Royo-Vela. 2010. Segmentation of low-cost flights users at secondary airports. *Journal of Air Transport Management* 16 (4): 234–237.

McFadden, Daniel, et al. 1973. *Conditional logit analysis of qualitative choice behavior*. Berkeley: Institute of Urban and Regional Development, University of California.
McFadden, Daniel. 1974. The measurement of urban travel demand. *Journal of public economics* 3 (4): 303–328.
McFadden, Daniel. 1980. Econometric models for probabilistic choice among products. *Journal of Business* 53 (3): S13–S29.
Merkert, Rico, and Matthew Beck. 2017. Value of travel time savings and willingness to pay for regional aviation. *Transportation Research Part A: Policy and Practice* 96: 29–42.
Molnar, Christoph. 2019. *Interpretable machine learning*. https://christophm.github.io/interpretable-ml-book/.
Morlotti, Chiara, Mattia Cattaneo, Paolo Malighetti, and Renato Redondi. 2017. Multi-dimensional price elasticity for leisure and business destinations in the low-cost air transport market: Evidence from easyjet. *Tourism Management* 61: 23–34.
Mottini, Alejandro, and Rodrigo Acuna-Agost. 2017. Deep choice model using pointer networks for airline itinerary prediction. In *Proceedings of the 23rd ACM SIGKDD international conference on knowledge discovery and data mining*, 1575–1583.
Njegovan, Nenad. 2006. Elasticities of demand for leisure air travel: A system modelling approach. *Journal of Air Transport Management* 12 (1): 33–39.
Oum, Tae Hoon, William G. Waters, and Jong-Say. Yong. 1992. Concepts of price elasticities of transport demand and recent empirical estimates: An interpretative survey. *Journal of Transport Economics and Policy* 26 (2): 139–154.
Paszke, Adam, Sam Gross, Francisco Massa, Adam Lerer, James Bradbury, Gregory Chanan, Trevor Killeen, Zeming Lin, Natalia Gimelshein, Luca Antiga, et al. 2019. Pytorch: An imperative style, high-performance deep learning library. In *Advances in neural information processing systems*, 8024–8035. Red Hook: Curran Associates.
Pedregosa, Fabian, Gaël. Varoquaux, Alexandre Gramfort, Vincent Michel, Bertrand Thirion, Olivier Grisel, Mathieu Blondel, Peter Prettenhofer, Ron Weiss, Vincent Dubourg, et al. 2011. Scikit-learn: Machine learning in python. *The Journal of Machine Learning Research* 12: 2825–2830.
Ragain, Stephen, and Johan Ugander. 2016. Pairwise choice markov chains. In *Advances in neural information processing systems*, 3198–3206.
Ribeiro, Marco Tulio, Sameer Singh, and Carlos Guestrin. 2016. Why should I trust you? Explaining the predictions of any classifier. In *Proceedings of the 22nd ACM SIGKDD international conference on knowledge discovery and data mining*, 1135–1144.
Richard, David B. 2009. The changing price elasticity of demand for domestic airline travel. In *50th Annual transportation research Forum*, Portland, Oregon.
Schiff, Aaron, and Susanne Becken. 2011. Demand elasticity estimates for new zealand tourism. *Tourism Management* 32 (3): 564–575.
Spiegel, Uriel. 1994. The case of a "giffen good". *The Journal of Economic Education* 25 (2): 137–147.
Strauss, Arne K., Robert Klein, and Claudius Steinhardt. 2018. A review of choice-based revenue management: Theory and methods. *European Journal of Operational Research* 271 (2): 375–387.
Tahanisaz, Sahar, and Sajjad shokuhyar. 2020. Evaluation of passenger satisfaction with service quality: A consecutive method applied to the airline industry. *Journal of Air Transport Management* 83: 101764.
Teichert, Thorsten, Edlira Shehu, and Iwan von Wartburg. 2008. Customer segmentation revisited: The case of the airline industry. *Transportation Research Part A: Policy and Practice* 42 (1): 227–242.
Tsamboulas, Dimitrios A., and Anastasios Nikoleris. 2008. Passengers' willingness to pay for airport ground access time savings. *Transportation Research Part A: Policy and Practice* 42 (10): 1274–1282.
Veblen, Thorstein. 1899. *The Theory of the Leisure Class*. New York: The Macmillan Company.
Vinod, B. 2008. The continuing evolution: Customer-centric revenue management. *Journal of Revenue Pricing Management* 7: 27–39.
Vivion, Nick. 2016. Bleisure travel: The benefit of mixing business travel with leisure. Online (visited on March 2, 2020). https://www.sabre.com/insights/ble-travel-the-benefits-of-mixing-business-travel-with-leisure.

Rodrigo Acuna-Agost serves as the Head of Artificial Intelligence, Research and Technology team at Amadeus. Before joining Amadeus in 2010, he worked on applications of Operations Research and AI for different industries, including manufacturing and travel.

Eoin Thomas is with Amadeus since 2014, and currently part of the AI, Research and Technology team, exploring how machine learning can help the travel industry. Prior to this, he completed his Ph.D. in University College Cork, Ireland and a postdoctoral contract with Inria, France in the domain of biomedical signal processing.

Alix Lhéritier is with Amadeus since 2015, working on machine learning research and applications for the travel industry. He was previously with Inria, France as a doctoral student and received his doctorate in computer science from the Université Nice Sophia Antipolis.

Journal of Revenue and Pricing Management (2021) 20:227–247
https://doi.org/10.1057/s41272-021-00309-y

RESEARCH ARTICLE

An integrated reinforced learning and network competition analysis for calibrating airline itinerary choice models with constrained demand

Ahmed Abdelghany[1] · Khaled Abdelghany[2] · Ching-Wen Huang[1]

Received: 10 April 2020 / Accepted: 19 September 2020 / Published online: 15 March 2021

Abstract
This paper presents a novel methodology to develop itinerary choice models (ICM) for air travelers that addresses the limitations of the traditional utility-maximization approach. The methodology integrates a reinforcement learning algorithm and an airline network competition analysis model. The reinforcement learning algorithm searches for the values of parameters of the itinerary choice model while considering maximizing a reward function. The reward function is measured as the negative of the difference between the estimated and observed system metrics. The airline network competition analysis model is used to calculate the estimated system metrics. It is a simulation model that represents passenger-itinerary assignment. It captures the demand–supply interactions at the network level while considering the competition among all airlines. An ICM system is calibrated using the developed framework considering the global airline network, which includes more than 500,000 airport pairs. Validating the model against ground truth data shows that the developed model adequately captures the travelers' itinerary choice behavior and replicates the competition pattern among airlines.

Keywords Airlines · Competition · Itinerary choice · Reinforced learning · Simulation · Market shares

Introduction

Profitability forecasting models (PFM) are crucial for airline strategic planning. They are integral part of many processes including schedule profitability prediction, competition evaluation, new route development, fleet planning and assignment, airline merger and acquisition, and pricing scenario evaluation, to name a few (Abdelghany and Abdelghany 2016 and 2018). In particular, PFM predicts the profitability of an airline schedule by determining how travelers of the different origin–destination (OD) pairs are assigned to available itineraries offered by the competing airlines. A key component of the PFM is an itinerary choice model (ICM), which represents travelers' behavior related to evaluating and choosing among itineraries scheduled in their corresponding OD pairs. The ICM captures how travelers evaluate the trade-off among itineraries' characteristics such as travel time, fare price, number of connections, itinerary circuity, connection duration, departure time, equipment types, and carrier reputation, to determine their best option while considering heterogeneity in the travelers' choice preferences.

The problem of estimating and forecasting airline demand and market share has been widely studied in the literature. Most of these studies have the objective of estimating the air-travel demand assignment to the different itineraries as a function of their attractiveness. Examples of early studies include Anderson and Kraus (1981), Ippolito (1981), Nason (1981), and Abrahams (1983). Many subsequent studies adopted the utility-maximization discrete choice modeling approach, using the maximum likelihood estimation (MLE) technique for model calibration. Examples of these studies include Nako (1992), Ghobrial and Soliman

✉ Ahmed Abdelghany
abdel776@erau.edu

Khaled Abdelghany
khaled@lyle.smu.edu

Ching-Wen Huang
HUANGC2@my.erau.edu

1 David O'Maley College of Business, Embry-Riddle Aeronautical University, 1 Aerospace Blvd, Daytona Beach, FL 32114, USA

2 Department of Civil and Environmental Engineering, Southern Methodist University, P.O. Box 750335, Dallas, TX 75275-0340, USA

Chapter 3 was originally published as Abdelghany, A., Abdelghany, K & Huang, C. -W. Journal of Revenue and Pricing Management (2021) 20:227–247. https://doi.org/10.1057/s41272-021-00309-y.

(1992), Proussaloglou and Koppelman (1995), Yoo and Ashford (1996), Algers and Beser (1997), Corsi et al. (1997), Proussaloglou and Koppelman (1999), Suzuki et al. (2001), Coldren et al. (2003), Coldren and Koppelman (2005), Warburg et al. (2006), Carrier (2008), Hess et al. (2013), and Delahaye et al. (2017).

While ICM that is based on the theory of utility-maximization is widely adopted (Ben-Akiva and Lerman (1985)), the theory has several basic assumptions that are evidently violated by the itinerary choice problem, which could significantly impact the interpretation and accuracy of these models. First, it is assumed that travelers of an airport pair make itinerary choice considering the same set of itineraries (i.e., fixed choice set). In reality, the itinerary choice set is dynamic and requires continuous update. If all seats of a flight are sold out, all itineraries that include this sold-out flight should be eliminated from the choice set of all travelers making subsequent bookings. Thus, future travelers will have a smaller set of itineraries to choose from, resulting in different itinerary choice sets across travelers of the same airport pair. Second, the utility-maximization choice theory requires that options in the choice set of travelers to be fully independent (i.e., mutually exclusive). However, this requirement could be violated as connecting itineraries between most airport pairs in a typical airline network could share one or more flights. Having a common fight among itineraries in a traveler's choice set makes them not fully independent. Third, most developed itinerary choice models use number of passenger bookings of the itineraries as an indication of their level of attractiveness. However, number of bookings of any itinerary is censored by the available seat capacity making it a biased measure for its level of attraction (Zeni 2001; Vulcano et al. 2012; Nikseresht and Ziarati, 2017). For instance, an attractive itinerary with a limited seat capacity will always have less bookings. Thus, using the capacity-constrained number of bookings could be misleading for estimating the relative attractiveness of itineraries in a traveler's choice set. Finally, a utility-maximization ICM only estimates choice probabilities at the itinerary level. The statistical significance of the model is only considered at this level. Predicting the demand at the flight level, which is needed for most studies, requires aggregating the choice probabilities of all itineraries that include each flight. The accuracy of a flight's predicted demand could be questionable considering that the flights' seat capacities are ignored while estimating the itinerary-level choice probabilities.

Considering these violations and their possible adverse impact on the fidelity of PFM, there are increasing calls in the airline network planning community to revisit methodologies adopted for developing ICM. This paper contributes to the literature by introducing a novel methodology to develop ICM that addresses the limitation associated with adopting the utility-maximization theory to model the air-travel itinerary choice behavior. The methodology integrates a reinforcement learning (RL) algorithm and an airline network competition analysis model. The RL algorithm searches for the values of parameters of the itinerary choice model while considering maximizing a reward function. The reward function is measured as the negative of the difference between the estimated and observed system metrics. The airline network competition analysis (NCA) model is used to calculate the estimated system metrics. It is a simulation model that represents passenger-itinerary assignment. It captures the demand–supply interactions at the network level, while considering the competition among all airlines.

This paper is organized as follows. "Overall framework" section presents the overall framework. The application of the framework considering the worldwide airline services is presented in "Model application" section. "Results" section gives the model results as well as the results of several case studies to illustrate its accuracy and interpretation. Finally, "Summary and conclusions" section provides concluding comments and suggestions for research extensions.

Overall framework

This section describes the overall framework used for developing an ICM that addresses the limitations of adopting the utility-maximization theory for modeling itinerary choice behavior. The proposed ICM model overcomes these limitations by explicitly capturing the impact of the airlines' seat capacity constraints on the travelers' itinerary choice behavior. Figure 1 illustrates the different components of the framework, which consists of a RL algorithm that integrates a full-scale airline network competition analysis model. The RL search algorithm is developed to estimate the parameters of the itinerary choice models. The algorithm seeks to maximize a reward function, which represents the negative of the deviation between the actual and estimated airline network metrics. The purpose of the full-scale airline network competition analysis model is to estimate the network metrics at each iteration, which is passed to the calculations of the reward function that facilitates the learning mechanism in the RL algorithm. The following subsections explain these components, which include input data, the structure of itinerary choice models, the NCA model, and the parameter estimation of the ICM using the reinforced learning algorithm.

Input data

The framework requires four main input data elements for all airlines competing in the markets under study. These data elements include (1) the OD passenger demand matrix, (2) the fare data, (3) the flight schedule data (i.e., timetable),

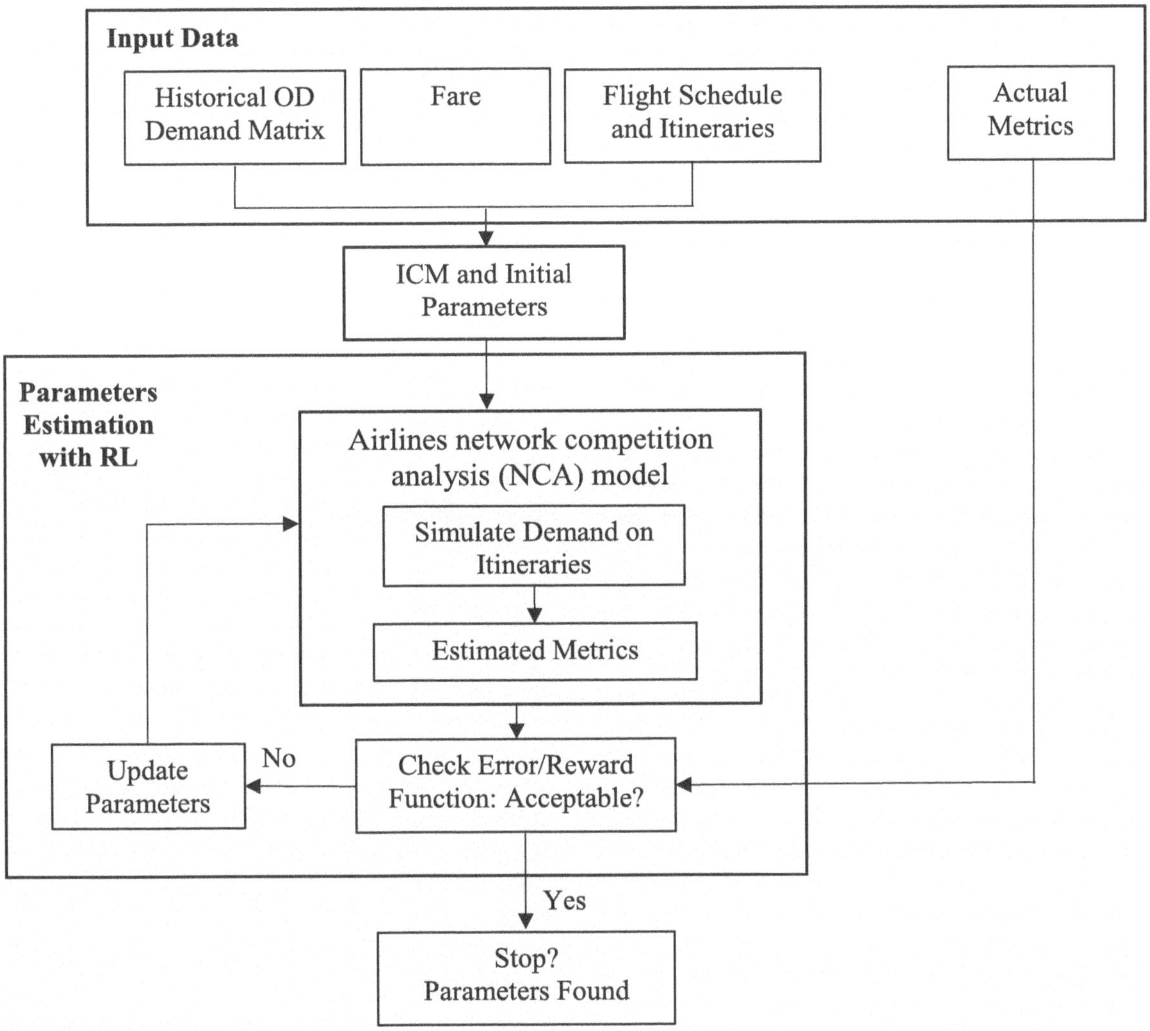

Fig. 1 The different components of the methodology

and (4) the available actual system metrics. The OD demand matrix data give the number of passengers traveling by all airlines for each airport pair. These data are usually reported monthly for the entire month and also for the average day of the month, which is known as passengers per day each way (PPDEW). The second input data element is the fare data which are widely available in the form of the average fare for each airport pair calculated based on all itineraries (e.g., non-stop and connecting) serving that airport pair. Both these data elements are available for each past month of the year through several commercial vendors (e.g., Diio by Cirium 2020; Sabre 2020), which compile and validate the data from different data sources.

The flight schedule data give the list of scheduled flights (timetable) for all airlines. Each flight is reported in terms of its main attributes including origin airport, destination airport, departure time, arrival time, fleet type, seat capacity, code-share information, day of week, and flight aeronautical distance. Historical flight schedule data are available for each month from several vendors (Sabre 2020). These vendors also provide the latest flight schedule data for near future months as published by the airlines. The last data element includes the actual system metrics. These data are used primarily to calculate the reward function of the RL algorithm. The reward function is to benchmark the model estimation results against the ground truth in the successive iterations of the RL-based calibration process. It should be mentioned that the framework is flexible to adopt any type of metrics for benchmarking and calibration including disaggregate, aggregate, or both.

For the purpose of this study and because they are widely available, two aggregate metrics are used in the current implementation of the framework, which are (a) the airline leg/route data and (b) the market share data at the airline-stop level. These data are also available as part of the Sabre Data & Analytics Market Intelligence 6.3 platform

(Sabre 2020). The airline leg data include the average load factor and the total PPDEW for each operating airline for each airport pair. The load factor is a measure of the average flight occupancy measured as the percentage of sold capacity. For example, in August 2018, airport pair John F. Kennedy International–London Heathrow (JFK-LHR) was served by four operating airlines, which are British Airways (BA), Virgin Atlantic Airways (VS), American Airlines (AA), and Delta Airlines (DL). These four airlines had an average monthly load factor of 82.1%, 78.0%, 74.1%, and 76.9%, respectively (Sabre 2020). The PPDEW for these four airlines were 1812, 1268, 836, and 359, respectively. The market share data at the airline-stop level include the PPDEW for each service. For example, in August 2018, for airport pair Daytona Beach International–John F. Kennedy International (DAB-JFK), there were 122 passengers flew using JetBlue Airways (B6), four passengers flew by Delta Air Lines (DL) via Hartsfield–Jackson International airport (ATL), and one passenger flew by American Airlines (AA) via Douglas International Airport (CLT). The market shares of the three services were 95.7%, 3.4%, and 0.9%, respectively (Sabre 2020). Unfortunately, these data are not available by the time of day. Of course, if time-dependent version of these data becomes available, it can be incorporated to better explain the travelers' itinerary choice behavior with respect to their preferences to departure/arrival times.

Itinerary choice models

The itinerary choice problem can be generally described as a statistical classification problem, which determines to which of a set of itineraries a new traveler will belong (i.e., choose). To categorize this new traveler choice observation, a training dataset is used which contains historical observations. The individual traveler observations are analyzed into a set of quantifiable properties, also known as explanatory variables or attributes. These explanatory variables, hopefully, can explain the membership of each traveler choice observation to an itinerary. Many classification algorithms adopt a linear function to combine these explanatory variables. The linear function assigns a score to itinerary in the choice set by combining the attribute vector of a traveler observation with a vector of parameters (weight) using a dot product. As such, the score can be viewed as the utility associated with traveler i choosing itinerary k. Algorithms with this basic setup are known as linear classifiers. Classification algorithms vary in terms of the procedure for determining (training) the optimal parameters (coefficients) and the way that the score is interpreted. For the purpose of this study, a probabilistic function is used to represent the relationship between the choice of an itinerary and its attributes and the attributes of the other competing itineraries. In its general form, the probability that individual i chooses itinerary k is expressed as given in (1).

$$P_{ik} = S(x_{ik}, x_{il \neq k}, S_i, \beta) \forall i \; \forall k \tag{1}$$

where P_{ik} is the probability that traveler i choses itinerary k, x_{ik} a vector of attributes of itinerary k in the choice set of traveler i, $x_{il \neq k}$ a vector of attributes of the other itineraries (l other than k) in the choice set of traveler i, S_i a vector describing the characteristics/preferences of traveler i, β a set of parameters that describes the effects of the variables on probabilities, which are estimated statistically, and $S(.)$ is a score (utility) function, which measures the attractiveness of traveler i to itinerary k.

Classification of airport pairs

The travelers' itinerary choice behavior could significantly vary across airport pairs (Coldren and Koppelman 2005). For example, the travelers' itinerary choice behavior in short-haul markets could be different from medium or long-haul markets. This behavior is also expected to be different in airport pairs with only connecting itineraries, where the connection duration and itinerary circuity might be the main attributes governing the travelers' choices. As such, in this study, the different airport pairs are classified into homogeneous classes and an itinerary choice classifier model is developed for each class. Three main criteria are used to classify airport pairs, including (1) the number of stops in the best itinerary scheduled by any of the airlines in the airport pair, (2) the distance (aeronautical miles) between the airport pair, and (3) the number of daily flight departures in the airport pair. It worth mentioning that future work is underway to investigate other/additional classification criterion such as region, time zone, and type of market (domestic/international). The main goal is to generate homogenous classes with minimum variation among the members of each class. Figure 2 illustrates the airport pairs' classification considering these three criteria, resulting in thirteen different classes of airport pairs. Accordingly, there will be thirteen ICM to be calibrated. As shown in the figure, airport pairs are first classified into non-stop, single-stop, and two-stops or more. Airport pairs served by non-stop and single-stop are further classified into three classes based on the aeronautical miles between the airport pair. These classes include distances less than or equal 2000 miles, 2000–4000 miles, and greater than 4000 miles. Finally, based on the number of daily flight departures, airport pairs are classified into three classes including greater than or equal 15 daily flights, 5–15 daily flights, and less than 5 daily flights. This classification is only considered for non-stop airport pairs. The reason for not further classifying the one-stop and the two-stop classes

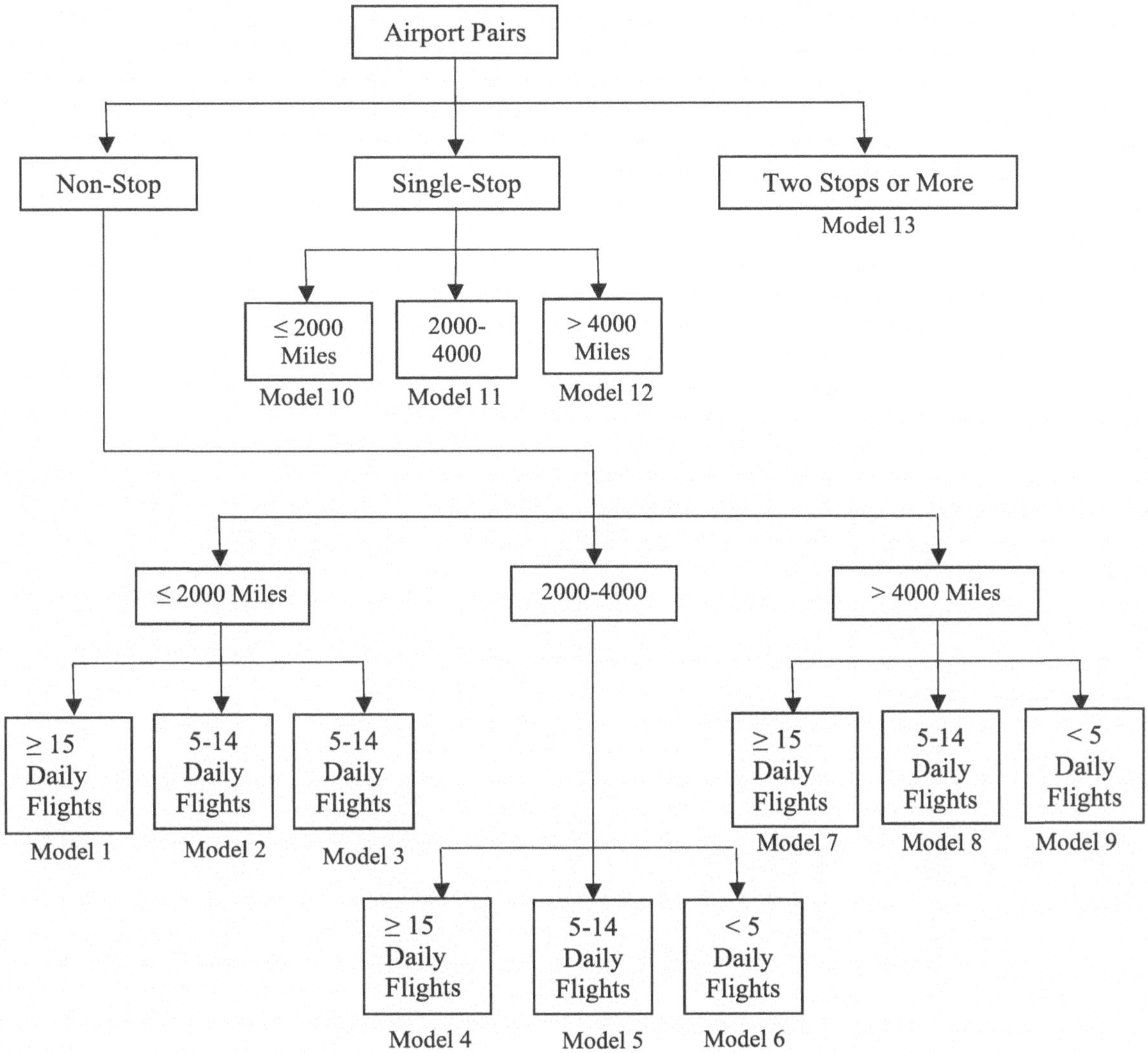

Fig. 2 The proposed classification of the different airport pairs

is to ensure that a representative sample of airport pairs exist in all classes.

Explanatory variables

Several explanatory variables are suggested in the literature to explain the travelers' itinerary choice behavior in markets with different characteristics (Coldren et al. 2003; Coldren and Koppelman 2005). These variables could be categorized into three main classes: itinerary characteristics, traveler attributes, and trip attributes. Examples of variables that describe the itinerary characteristics include trip travel time, circuity, connection quality, airline presence and brand, and departure/arrival times. Variables related to traveler attributes include, for example, income, gender, and age. Variables related to trip attributes include trip purpose (business or leisure), trip type (domestic or international), length of stay, day of week, etc. It would be ideal if information on these variables is available for all demand in all airport pairs, and for all bookings. Typically, variables that describe the itinerary characteristics are widely available compared to travelers and trip attributes. Accordingly, this study is focusing on using these variables in model specification.

Table 1 gives the list of explanatory variables considered for the models developed in this study. The table also gives the definition of each variable. As shown in the table, five main variables are considered which are the itinerary level of service, type of the itinerary, connection quality, time of day, and airline presence and brand. No variables are considered for travelers or trip attributes mainly because such data are not available for the purpose of this study. Including these variables would be a recommendation for further research. The itinerary level of service is a dummy variable that indicates the number of connections/stops per itinerary.

Table 1 The explanatory variables of the ICM

Variable	Definition
Level of service	
Less than non-stop	A dummy variable, which is equal to one if the itinerary has at least one stop in non-stop airport pairs, and zero otherwise
Less than one-stop	A dummy variable, which is equal to one if the itinerary has at least two stops in one-stop airport pairs, and zero otherwise
Itinerary type	
Code share	A dummy variable, which is equal to one if the itinerary is a code share or interline itinerary, and zero otherwise
Operated by regional	A dummy variable, which is equal to one if any flights of the itinerary is operated by a regional airline, and zero otherwise
Connection quality	
Distance ratio	A ratio that measures the itinerary aeronautical distance compared to the aeronautical distance between the airport pair of the itinerary
Connection duration	Total connection duration, if any (in hours)
Time of day	
0:00–4:59 AM	A dummy variable, which is equal to one if the itinerary's departure time is between 0:00–4:59 AM, and zero otherwise
5:00–5:59 AM	A dummy variable, which is equal to one if the itinerary's departure time is between 5:00–5:59 AM, and zero otherwise
6:00–6:59 AM	A dummy variable, which is equal to one if the itinerary's departure time is between 6:00–6:59 AM, and zero otherwise
7:00–7:59 AM	A dummy variable, which is equal to one if the itinerary's departure time is between 7:00–7:59 AM, and zero otherwise
8:00–8:59 AM	A dummy variable, which is equal to one if the itinerary's departure time is between 8:00–8:59 AM, and zero otherwise
9:00–9:59 AM	A dummy variable, which is equal to one if the itinerary's departure time is between 9:00–9:59 AM, and zero otherwise
10:00–10:59 AM	A dummy variable, which is equal to one if the itinerary's departure time is between 10:00–10:59 AM, and zero otherwise
11:00–11:59 AM	A dummy variable, which is equal to one if the itinerary's departure time is between 11:00–11:59 AM, and zero otherwise
12:00–12:59 PM	A dummy variable, which is equal to one if the itinerary's departure time is between 12:00–12:59 PM, and zero otherwise
1:00–1:59 PM	A dummy variable, which is equal to one if the itinerary's departure time is between 1:00–1:59 PM, and zero otherwise
2:00–2:59 PM	A dummy variable, which is equal to one if the itinerary's departure time is between 2:00–2:59 PM, and zero otherwise
3:00–3:59 PM	A dummy variable, which is equal to one if the itinerary's departure time is between 3:00–3:59 PM, and zero otherwise
4:00–4:59 PM	A dummy variable, which is equal to one if the itinerary's departure time is between 4:00–4:59 PM, and zero otherwise
5:00–5:59 PM	A dummy variable, which is equal to one if the itinerary's departure time is between 5:00–5:59 PM, and zero otherwise
6:00–6:59 PM	A dummy variable, which is equal to one if the itinerary's departure time is between 6:00–6:59 PM, and zero otherwise
7:00–7:59 PM	A dummy variable, which is equal to one if the itinerary's departure time is between 7:00–7:59 PM, and zero otherwise
8:00–8:59 PM	A dummy variable, which is equal to one if the itinerary's departure time is between 8:00–8:59 PM, and zero otherwise
9:00–9:59 PM	A dummy variable, which is equal to one if the itinerary's departure time is between 9:00–9:59 PM, and zero otherwise
10:00–11:59 PM	A dummy variable, which is equal to one if the itinerary's departure time is between 10:00–11:59 PM, and zero otherwise
Airline presence and brand	
Fare ratio	A ratio that measures the itinerary fare compared to the average fare in the airport pair

Table 1 (continued)

Variable	Definition
Airline presence	A ratio that measures the presence of the marketing airline of the itinerary in the airport pair. It is calculated by dividing the non-stop seats of the airline by the total non-stop seats in the airport pair
Airline name	A dummy variable, which is equal to one if the itinerary is marketed by this airline, and zero otherwise

For Non-Stop airport pairs, a dummy variable is considered to indicate connecting itineraries. The Less than Non-Stop variable is an indicator to represent an itinerary that has at least one connection. For the Single-Stop airport pairs, a variable is considered to indicate itineraries with two stops or more. The Less than One-Stop variable is when the itinerary has at least two connections.

The itinerary type is represented by two dummy variables. The first variable is the Code Share dummy variable, which represents the case when the itinerary is a code share (or interline). The second variable is Operated by Regional, which represents itineraries that have at least one flight operated by a regional/affiliated airline. Regional airlines are airlines that operate regional aircraft to provide air service to small communities. They are usually contracting with a major airline, operating under their brand name, to deliver passengers to the airline's hub from surrounding small airports.

The Connection Quality is measured by the Distance Ratio and the Connection Duration. The Distance Ratio measures the itinerary aeronautical distance compared to the direct aeronautical distance between the airport pair of the itinerary. The Connection Duration is the total connection duration in hours (i.e., ground time between flights) of the itinerary.

The Time of Day variable represents the departure hour of the itinerary. This variable is included to control for travelers' assignment to the different departure times. This variable is only included in Model 1 (i.e., airport pairs that are short haul and have at least 15 daily departures). It is not included when the number departures are less than 15 because in this case some departure times are not represented. The departure time variables are exceptionally important, but they must be modeled correctly. They are included in the short-haul market with at least 15 daily departures because this is the only case that they can be correctly modeled. In this case, when there are at least 15 departures per day, it is expected that most departure time intervals are represented in the choice set and travelers are selecting among all options. In other words, the choice reflects the actual preference of the traveler. On the contrary, for the other airport pairs in other clusters, which have a few departure times, not all the departure time intervals throughout the day will be represented in the choice set. This implies that when a traveler chooses an itinerary, his/her choice does not necessarily reflect his/her true preference because this traveler is captive to choose only from the available intervals. For example, assume there is an airport pair that has only two departures in the morning (e.g., 7:00 AM and 10:00 AM). In this case, all travelers are selecting either the 7:00 AM or 10:00 AM departure. When these data are used to calibrate a model, the parameters of the departure time intervals in the afternoon, for example, will indicate that the afternoon departures are not preferred. This is not correct. The reality is that these afternoon intervals are not available and not represented in the choice set.

Also, in the short-haul market, the departure and arrival times are correlated. For example, an itinerary that has a morning departure will most likely have a morning arrival due to the short trip time. In other airport pairs in other clusters (i.e., medium- and long-haul clusters), time zones might make the departure time variable to have a biased impact. For example, in the United States and for a cluster that includes the transcontinental routes, a midnight departure from the west coast to the east coast (red eye flight) will arrive in the morning of the second day. However, a midnight departure from the east coast to the west coast will arrive in the middle of the night. This example shows that the midnight departure interval has different impact depending on the location of the origin and destination and their time zones. Accordingly, the Time of Day variable is not included in models of medium- and long-haul airport pairs. Many of these airport pairs are expected to have significant time-zone differences and in this case the departure time preference will not be meaningful in the model. To correct for this factor, an additional clustering of the airport pairs based on their location (time zone) is needed. While this addition is expected to improve the model accuracy, it is expected to increase the computational burden of the model because of the new parameters. We recommend considering other specifications with additional clusters and variables in future research.

Finally, the Airline Presence and Brand are represented in terms of three variables, which are the Fare Ratio, Airline Presence, and Airline Name. The Fare Ratio measures the itinerary fare compared to the average fare in the airport pair. For instance, low-cost airlines are expected to have low Fare Ratio compared to full-service airlines. The Airline Presence measures the presence of the marketing airline of the itinerary in the airport pair. It is calculated by dividing the non-stop seats of the airline by the total non-stop seats in the airport pair. This variable is included only for airport pairs that have non-stop service. The Airline Name is a variable to control for any difference between airlines that are not captured by the other explanatory variables (e.g., brand, business model, network structure, ticket distribution strategy). This variable is included as a global variable among all the thirteen models, as it is believed that the characteristics that this variable might represent do not change by the cluster of the airport pairs. A summary of the set of variables considered for each of the thirteen models is given in Table 2.

The airline network competition analysis (NCA) model

As mentioned above, a simulation-based NCA model is adopted in the framework (Abdelghany and Abdelghany 2007 and 2008). The main purpose of the NCA model is to estimate the values of several system metrics based on the parameters of the itinerary choice models that are estimated at each iteration. The core component of the NCA model is the ICM system that replicates the passengers' itinerary choice behavior in the different airport pairs as a function of the attractiveness of the itineraries scheduled by the host and competing airlines. Thus, the NCA model captures competition among the airlines and estimates their market shares and corresponding revenues. It also tracks the demand spill and recapture among the different itineraries. In the proposed methodology, the NCA model is used at each iteration to evaluate the ability of the impeded ICM system to replicate the ground truth system metrics (e.g., route load factor and market shares). This section provides a brief description of the NCA model.

The overall structure of the model is depicted in Fig. 3. As shown in the figure, the input to the NCA model includes the origin–destination demand matrix, the fare data, and the flight schedule, which are defined above. Given the flight schedule data, the itinerary builder module is activated to determine all feasible itineraries for all airport pairs in the network. The module examines a set of criteria to identify infeasible itineraries and eliminate them from the passengers' choice set. These criteria include the itinerary's maximum number of connections, minimum and maximum passenger connection time between flights, and the ratio between the itinerary's total mileage and the direct distance between the airport pair. User-specified thresholds are used to determine the acceptable values of these criteria. The itinerary builder module also considers the interline and code-share agreements between airlines. The output of the itinerary builder module is the set of feasible itineraries for all airport pairs scheduled by the different competing airlines. The goal is to be able to generate all feasible itineraries that are considered by the historical demand. Each

Table 2 Correspondence between the explanatory variables and the different models

Variable	Model 1	Model 2	Model 3	Model 4	Model 5	Model 6	Model 7	Model 8	Model 9	Model 10	Model 11	Model 12	Model 13
Level of service													
Less than non-stop	×	×	×	×	×	×	×	×	×				
Less than one stop										×	×	×	
Itinerary type													
Code share	×	×	×	×	×	×	×	×	×	×	×	×	×
Operated by regional	×	×	×	×	×	×	×	×	×	×	×	×	×
Connection quality													
Distance ratio	×	×	×	×	×	×	×	×	×	×	×	×	×
Connection duration:	×	×	×	×	×	×	×	×	×	×	×	×	×
Time of day	×												
Airline presence and brand													
Fare ratio	×	×	×	×	×	×	×	×	×	×	×	×	×
Airline presence	×	×	×	×	×	×	×	×	×				
Airline name	×												

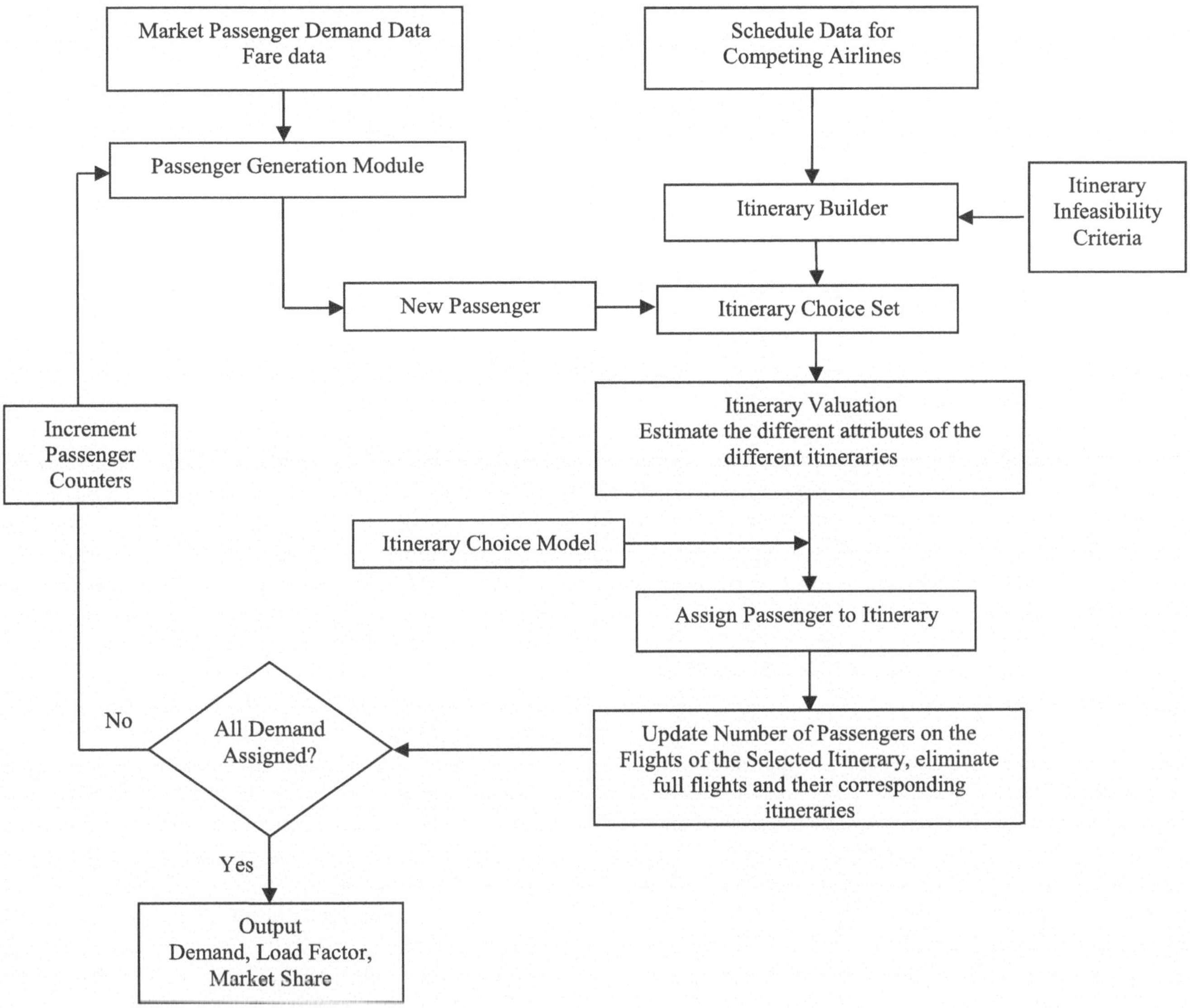

Fig. 3 The main structure of the airline network competition analysis model

itinerary is described in terms of its constituting flights along with a set of attributes including the marketing/operating airlines, the number of stops, departure and arrival times, connection quality, fare, aircraft type, and travel distance. Given the demand data for each airport pair, passengers are randomly generated to emulate the booking process in the different airport pairs. The random demand loading process does not assign priority to any of the airport pairs, and hence it prevents bias in the network demand projection in favor of any specific market. The list of feasible itineraries in the choice set is defined for each traveler. This choice set includes itineraries scheduled by all competing airlines that have seat availability, when the traveler is generated. Each generated passenger evaluates the set of itineraries available in her/his airport pair and selects one itinerary for her/his travel. The itinerary choice behavior is represented by the ICM described above. Score values that reflect the travelers' attractiveness to each itinerary are computed using this ICM. Here, a linear function is assumed to estimate the score for each possible alternative by combining the attribute vector of the alternative and a vector of parameters (weight), using a dot product. In the first iteration, a set of random (or assumed) initial values are given to the parameters of the score function. As discussed later, the values of these parameters are adjusted iteratively to minimize the difference between the difference between the ground truth values and the corresponding estimated metrics that are calculated by the NCA model. Once the score of each alternative in the choice set is calculated for each itinerary, a probabilistic function is used to calculate the probability of choosing each itinerary. According to this function, the probability p_i that an average passenger in airport pair $\{o, d\}$ chooses itinerary

k from itineraries in the choice set $\Pi_{\{o,d\}}$ is given in (2). Without loss of generality, any probabilistic function can be used in the application. The equation in (2) is used for its simplicity and it has been widely considered for ICM development (Coldren et al. 2003; Coldren and Koppelman 2005).

$$p_{ik} = e^{Score_k} / \sum_{j \in \Pi_{\{o,d\}}} e^{Score_j} \quad \forall k \in \Pi_{\{o,d\}}. \tag{2}$$

Once an itinerary is chosen, the assignment module is activated to update the number of passengers of the chosen itinerary and its corresponding flights. At each assignment step, the model tracks seat availability and eliminates itinerary when any of its flights becomes full. The process terminates when the entire demand is assigned. At termination, the model generates an array of output statistics (metrics) including number of passengers assigned to each itinerary and flight for all airlines. For the purpose of this study, the airline leg/route load factor and the demand share for each itinerary defined at the airline-stop level estimated by the NCA model are compared against their corresponding ground truth values.

Parameters estimation using reinforced learning algorithm

The objective is to estimate the values of the parameters of the ICM (described in "Itinerary choice models" section), which is incorporated in the NCA model described above. The optimal values of these parameters minimize the difference between the metrics estimated by the NCA model and their corresponding ground truth values. This problem can be generally described in the form of a non-linear mathematical program with side constraints that define the feasible range of each parameter in the ICM model. The feasible range for each parameter is a function of the seat capacity available in the different markets. The feasible range for each parameter is also dependent on those set for other parameters. A Reinforced Learning (RL) algorithm is adopted to solve this optimization problem (Sutton and Bart 2018). RL is an area of machine learning concerned with how software agents take actions in an environment in order to maximize some notion of cumulative reward. RL is mimicking the problem of learning from interaction to achieve a goal. The learner (decision-maker) is called the agent. The agent interacts with an environment, where the agent selects an action and the environment responds to those agent's action and presents the corresponding new state of the environment and a reward. The reward is a special numerical value that the agent tries to maximize. The reward signal is a way of communicating to the agent of what need to be achieved. The process is repeated until the reward cannot be further maximized. The agent and the environment interact at each of a sequence of discrete time steps$t = 0, 1, 2, 3, \ldots,$ where at each time step$t$, the agent receives some representation of the environment's state,s_t, and on that basis selects an action,$a_t \in \mathcal{A}(s_t)$, where $\mathcal{A}(s_t)$ is the set of actions available in states_t. In the next time step, as a consequence of its action, the agent receives a numerical reward, r_{t+1}, and finds itself in a new state,s_{t+1}. At each time step, the agent implements a mapping from states to probabilities of selecting each possible action. This mapping is called the agent's policy and is denotedπ_t, where $\pi_t(s, a)$ is the probability that $a_t = a$ if$s_t = s$. Reinforcement learning methods specify how the agent changes its policy as a result of its experience. Informally, the agent's goal is to maximize the total amount of reward it receives, which is the cumulative reward in the long run (Szepesvári 2010; Busoniu et al. 2010; and Wiering and Van Otterlo 2012).

The above described framework is an abstract and flexible to be applied to the problem of estimating the parameters of the ICM. For instance, in this problem, the agent represents the mechanism that is selected to update the values of the parameters of the ICM. The actions represent the decision to change the values of the parameters to a different value. The environment is represented by the NCA model, which updates the state of the system defined in the form of a passenger-itinerary assignment pattern as a result of an action of changing the values of the parameters of the ICM model. The NCA model also provides the information needed for the computation of the reward. The reward is calculated as the difference between the estimated and the ground truth metrics. The time steps refer to arbitrary successive stages of updating the values of the ICM's parameters.

The successful implementation of the RL algorithm requires a mechanism to map the reward obtained in each iteration with the value given for each parameter in the ICM. A probabilistic mapping mechanism is considered. This mapping mechanism is in the form of a probability distribution for each model parameter. Each distribution gives the probability that a certain value assigned to this model parameter is resulting in a positive reward for the system. Figure 4 shows an example of the proposed probability mapping mechanism for a hypothetical parameter. In Fig. 4a, the x-axis represents the different possible values of the parameter (i.e., feasible range of the parameter) and the y-axis gives the percentage of times a specific value is associated with a positive reward from the environment. As shown in the figure, this value is given at four different hypothetical stages (e.g., $t = 0; t = t_{1000}; t = t_{2000}; and t = t_{5000}$). At $t = 0$, the initial knowledge of the agent is that each parameter value has 50% chance of providing a positive reward. Assume that at $t = 1$, a specific value of the parameter is selected and used in the ICM. Based on this update, the reward is calculated. If the reward is positive, the probability that this specific value of the parameter is having a positive reward

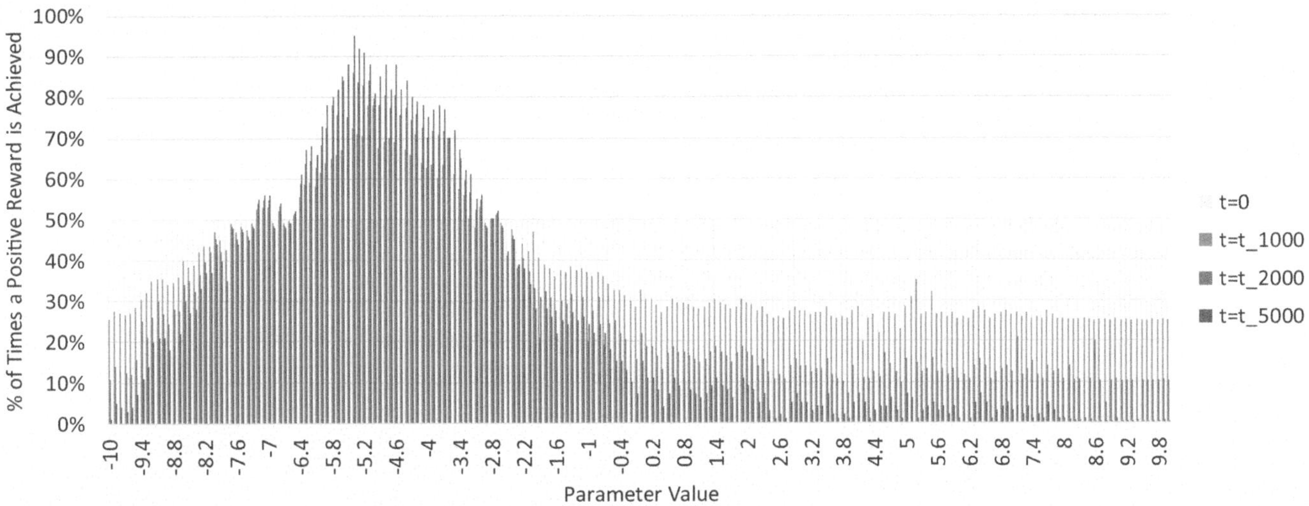

(a) Mapping distribution before normalization

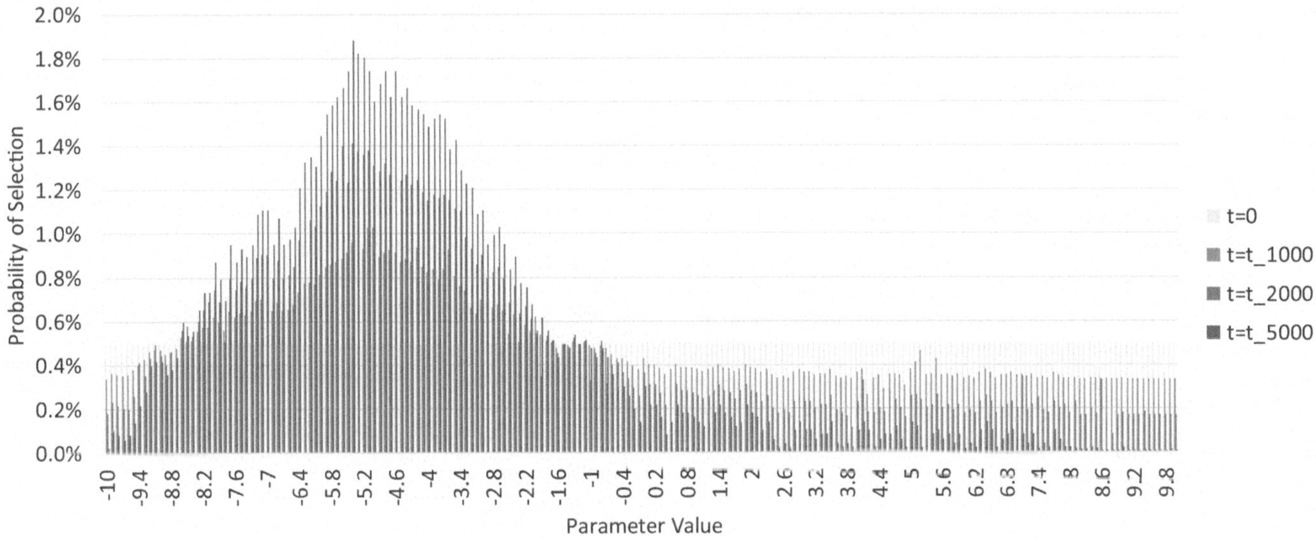

(b) Mapping distribution after normalization

Fig. 4 An example of the proposed probability mapping mechanism for a hypothetical parameter. **a** Mapping distribution before normalization, **b** mapping distribution after normalization

is updated to 2/3 (67%). On the other hand, if the reward of the environment is negative, the probability that this specific value of the parameter is having a positive reward is updated to 1/3 (33%). The probabilities of being associated with a positive reward are updated in the consecutive stages for all possible parameter values. The agent eventually converges to a specific value for the parameter, at which the environment will tend to always provide a positive reward compared to other values. Based on the data accumulated on the rewards at each stage, a reward-mapping probability distribution is developed for the parameter values. Figure 4b shows the probability distribution after normalizing the data given in Fig. 4a. As shown in the figure, the parameter's values that are associated with having positive reward are more likely to be selected in future iterations. In addition, the probability distribution tends to converge to prioritize the selection of the optimal parameter value(s). It should be noted that if the distribution in Fig. 4b does not converge to a specific value(s), this indicates that the variable associated with this parameter is not statistically significant and need to be excluded from the model.

Model application

We present the application of the framework for the global airline network considering a weekly flight schedule data. For this purpose, the global flight schedule data of the first week of July 2019 are considered in the analysis. Each flight is defined in terms of its origin and destination airports, operating airline, departure and arrival times, equipment type, seat capacity, day of week, flight aeronautical distance and duration. For flights that are operated by regional airlines, the marketing carrier of the flight is given. Each flight is also defined by its restriction code, which indicates how interline itineraries between airlines can be generated. In addition, if any of the flights is a code-share flight, the marketing airline(s) of this flight and the associated code-share flight number are given. The weekly flight schedule data include about 770,000 flights, operated by about 720 airlines in 3700 airports worldwide. For the OD demand matrix, the average PPDEW for the month of July 2019 is considered. The OD matrix includes about 573,000 city pairs worldwide, with a total of about 11.8 Million passengers per day. For each airport pair, the average total fare and the aeronautical distance between the two airports are also given. The OD demand matrix data are also available at a disaggregate level, which reveal how demand is assigned to the different airlines and connecting airports, if any. For each itinerary, the PPDEW and the total fare are given. This total fare is used as an approximation of the fare of the itineraries generated by the itinerary builder module of the NCA model to calculate the Fare Ratio variable defined above. Finally, the leg (route) for July 2019 is also considered. The leg data provide an aggregation of the performance metrics for non-stop service between airport pairs. Each leg is defined by the origin airport, destination airport, and the operating airline. For each leg, the average load factor is given. The leg data are given for about 119,000 records worldwide.

As mentioned above, the objective is to minimize the difference between estimated and ground truth metrics. Two main metrics are used: (1) the leg (route) load factor and (2) the market share for the itinerary defined at the airline-stop level. For this purpose, a weighted error function, E, is considered as the weighted sum of the error in the airline flight load factor, E_{LF}, and the market share, E_{Share}, as given in (3).

$$E = \omega_{\text{LF}}.E_{\text{LF}} + \omega_{\text{Share}}.E_{\text{Share}}, \tag{3}$$

where ω_{LF} and ω_{Share} are user-define weights for the error in the airline route load factor and the error in the itinerary market share in the airport pair ($0 \leq \omega_{\text{LF}}, \omega_{\text{Share}} \leq 1$), respectively. In this implementation, both ω_{LF} and ω_{Share} are assumed to be equal to 0.50, which means that both errors are weighted equally. The values of the error terms E_{LF} and E_{Share} are calculated as given in (4) and (5), respectively. Each error term is calculated as the sum of the squared error between the estimated and measured value weighted by the observed demand level of the flight/itinerary. Thus, flight legs/itineraries with high demand have more contribution to the error terms.

$$E_{\text{LF}} = \sqrt{\sum_{i=1}^{N} Dep_i.(LF_i^{est} - LF_i^{act})^2} \tag{4}$$

$$E_{\text{Share}} = \sqrt{\sum_{j=1}^{M} PPDEW_j.(\text{share}_j^{est} - \text{Share}_j^{act})^2}, \tag{5}$$

where N is the set of all legs (routes), indexed by i, M the set of all itineraries defined at the airline-stop level, indexed by j, Dep_i the weekly departures per airline route $i\forall N$, LF_i^{est} the estimated load factor per airline route $i\forall N$, LF_i^{act} the actual load factor per airline route $i\forall N$, $PPDEW_j$ the number of passengers per day each way for itinerary $j\forall M$, $Share_j^{est}$ the estimated market share for itinerary $j\forall M$, and $Share_j^{act}$ is the actual market share for itinerary $j\forall M$.

Results

This section presents the results related to calibrating the ICM model for the global airline network. These results include three main elements: the convergence pattern of the RL algorithm, the results on the error distribution for the different metrics, and the estimated values of the ICM parameters and their relative significance on the itinerary choice behavior.

Model convergence

Figure 5 shows the convergence pattern of the RL algorithm. The figure gives the convergence of the error terms defined above including E, E_{LF}, E_{Share}. The first iteration starts by assigning random values to the parameters of the ICM. As shown in the figure, the total error, E, was reduced from 17.1 to 11.6%. The model converges at 13.1 and 10.1% for E_{LF} and E_{Share}, respectively. While the estimated ICM contributed significantly in reducing the difference between the estimated and the measured metrics, the remaining amount of error could be contributed to several factors. For example, one can revisit the structure of the ICM in terms of its explanatory variables and/or the clustering scheme considered for the airport pairs. Enriching the explanatory variables could help enhance the model. Also, a finer disaggregation of the airport pairs is expected to improve representation of the travelers' itinerary choice behavior. Furthermore, the prediction accuracy depends on the quality of the input data and the validity of the processes constituting the NCA

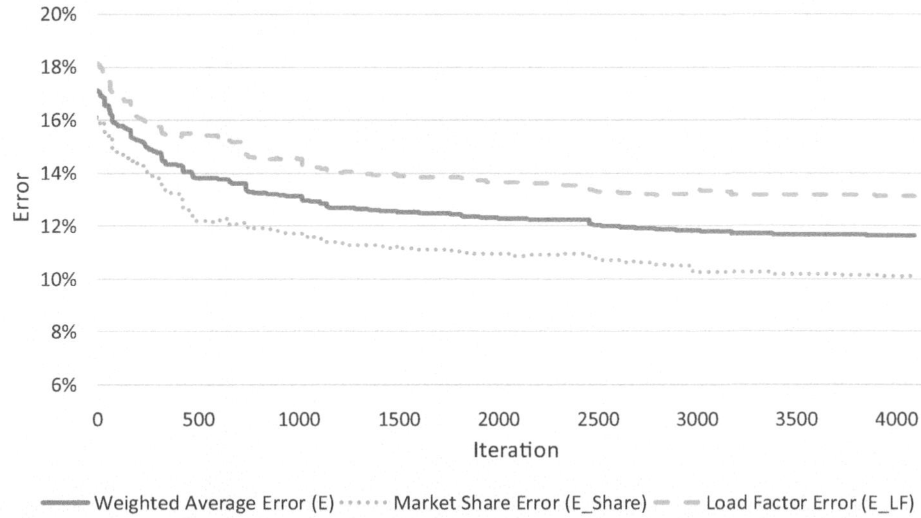

Fig. 5 Convergence pattern of the RL solution algorithm

model, which estimates the validation metrics based on the results of the imbedded ICM. As mentioned earlier, the input data of the NCA model include the OD demand matrix data, the airline schedule data, and the fare data. These data are typically collected by third-party vendors and validated from different sources. However, the accuracy of these historical data is not fully guaranteed, which could affect the quality of the estimation results. Similarly, the NCA model implements several heuristic rules for itinerary building and enumeration. These rules might not be comprehensive across the global airline network, resulting in a mismatch between the itinerary choice sets built in the model and the actual ones considered by the travelers. A further validation of these rules is expected to reduce this mismatch and enhance the overall predictability of the NCA model. Finally, the NCA outputs an estimate of the flights' load factor and market shares based on a given OD demand matrix and the airlines' weekly schedule. However, the OD demand matrix data are input as an average for the whole month and might not be precisely available for the exact week for which the flight schedule is selected for the analysis. The discrepancy between the demand and the schedule data could explain part of the recorded error between the model results and the ground truth data. As mentioned above, upon availability, using other forms of data (e.g., disaggregate booking data and flight load factor data) is expected to enhance the model calibration process and improve the model quality.

Error distribution

Figure 6 shows the distribution of the error in load factor at the route level (i.e., $LF_i^{est} - LF_i^{act}$). As shown in the figure, the error distribution is a bell-shaped distribution with about 48% of the observations have an error of 5% or less and about 66% of observations have an error of 10% or less. The figure also shows that error distribution is slightly skewed to the left, which indicates that the load factor metrics are slightly underestimated by the model. Figure 7 shows example of the distribution of the load factor error for ten different airlines. These airlines are selected arbitrarily to represent airlines with different sizes, structure, underlying business models, and serving different regions. These airlines include American Airlines, Spirit Airlines, Air Canada, Avianca, Lufthansa, Ryanair, EgyptAir, China Southern Airlines, Qantas, and All Nippon Airlines. In each graph, the x-axis gives the route's PPDEW (to represent the route size in terms of its demand) and the y-axis gives the error in the route load factor. As shown in the figure, most of the errors are within $\pm 20\%$. In addition, the error tends to decrease as the size of the route increases. One can expect this result as the algorithm gives more weight to minimizing the load factor error of high-demand routes. The results recorded for most of these airlines confirm the observation that the load factor metrics tend to be underestimated by the model, where there are more negative errors compared to the positive ones. This underestimation could be contributed inaccuracy of the OD demand matrix used by the NCA model.

Figure 8 shows the distribution of the error of the itinerary market shares (i.e., $\text{share}_j^{est} - \text{Share}_j^{act}$), where the itinerary is defined at the airline-stop level. The figure shows the distribution for four different cumulative demand levels, where the demand is represented by the PPDEW. For example, Fig. 8a shows the error distribution for itineraries with PPDEW greater than or equal 2 passengers (831,880 itineraries). As shown in Fig. 8a, the error distribution is in the form of a bell-shape where about 50% of the observations have an error of 5% or less and about 65% of observations have an error of 10% or less. Similarly, Fig. 8b–d, respectively, shows the error distribution for itineraries with PPDEW greater than or equal 10 passengers (591,200

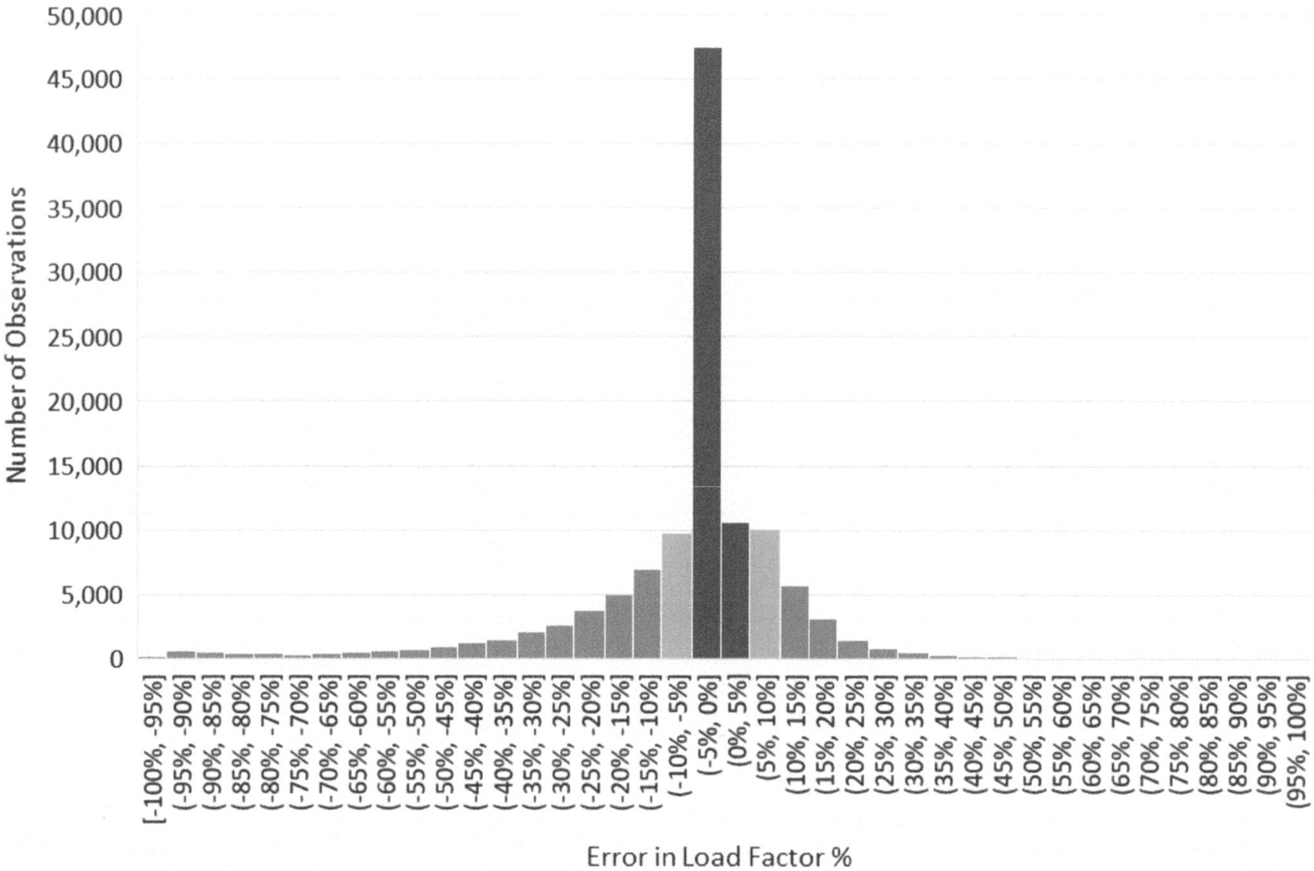

Fig. 6 The distribution of the error in route load factor

itineraries), 50 passengers (332,930 itineraries), and 100 passengers (228,010 itineraries). In all cases, the error distribution follows a bell-shape form, where most itineraries have error less than 5% in market share. As the model gives higher weight to high-demand markets, one can also notice that the variability of the error distribution tends to decrease with the increase of the itinerary's PPDEW.

Figure 9 shows the distribution of market share of itineraries (defined at the airline-stop level) in ten different airport pairs. The airport pairs are represented by the 3-letter IATA airport code identification on the top of each figure. These airport pairs are selected arbitrarily to represents different markets of different sizes, levels of competition, and geographical locations. For each airport pair, the x-axis shows the different itineraries (defined at the airline-stop level) serving in the airport pair and the y-axis gives both the estimated and actual market share of each itinerary. The airlines that are serving the different itineraries are represented by their 2-letter code and the itinerary intermediate stops (if any) are represented by the 3-letter IATA airport code identification. As shown in these figures, the model is generally predicting itinerary shares with adequate accuracy. For example, for the DFW-ORD airport pair, the model estimates a market share of the itineraries of American Airlines (AA) to be 54%, while the actual market share is 59 percent. Also, the model estimates a market share of the itineraries of United Airlines (UA) to be 24%, while the actual market share is 22%. Finally, the model estimates a market share of the itineraries of Spirit Airlines (NK) to be 20%, while the actual market share is 17 percent. For another example, for the KUL-SYD market, the estimated market shares for the two major airlines serving that market (AirAsia (D7) and Malaysia Airlines (MH)) are recorded at 81 percent and 14 percent, compared to actual shares of 83 percent and 14%, respectively.

Model parameters

Table 3 gives the estimated parameter values at convergence for the thirteen ICM models developed in this study. All values are statically tested for significance and tracked in the RL algorithm to make sure that the mapping mechanism is conversing systematically for each parameter. Incorporating the seat capacity constraints in the model estimation process complicates the interpretation of the signage and the value of the ICM parameters, compared to the case in which these

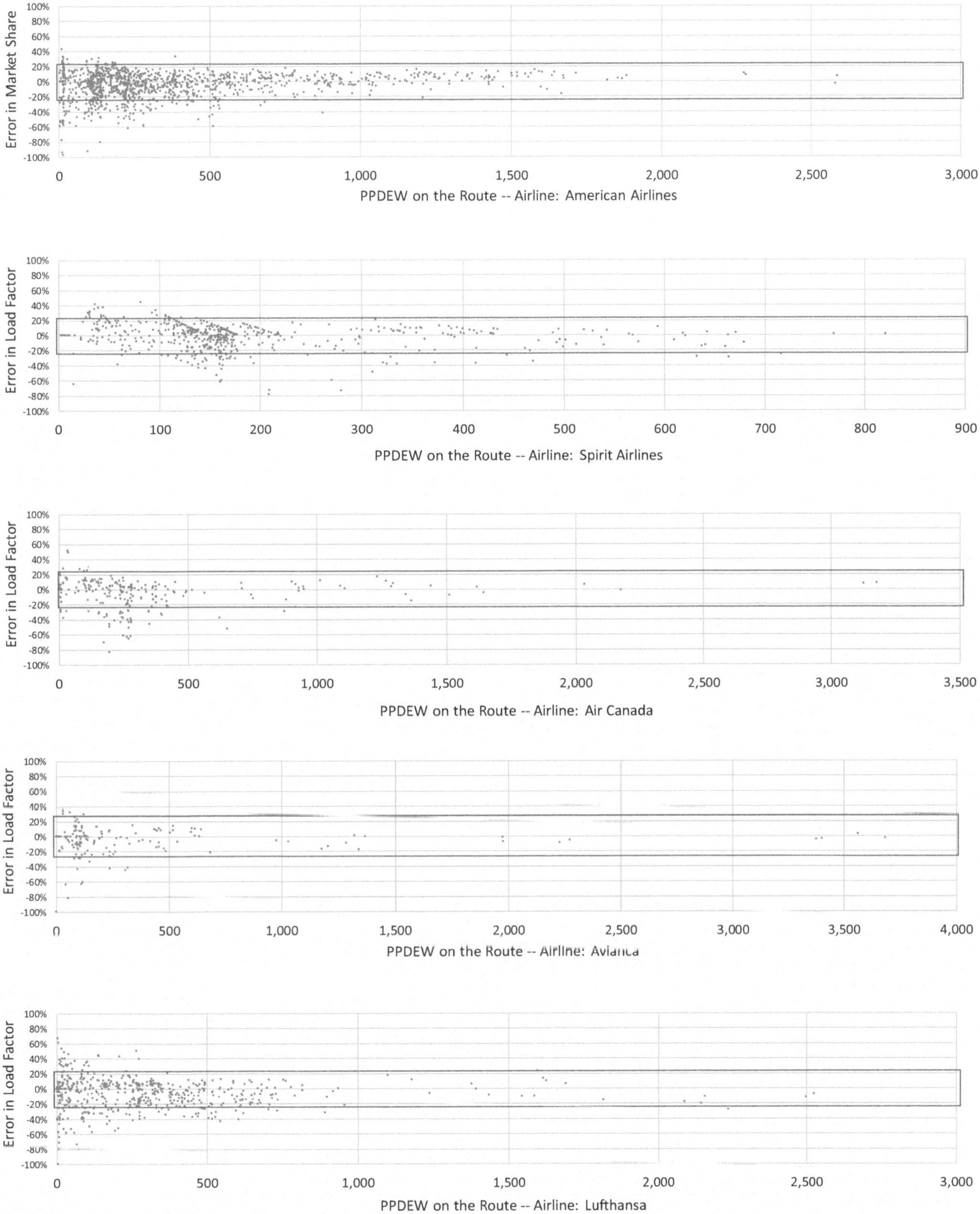

Fig. 7 The distribution of the error in route load factor for ten different airlines worldwide

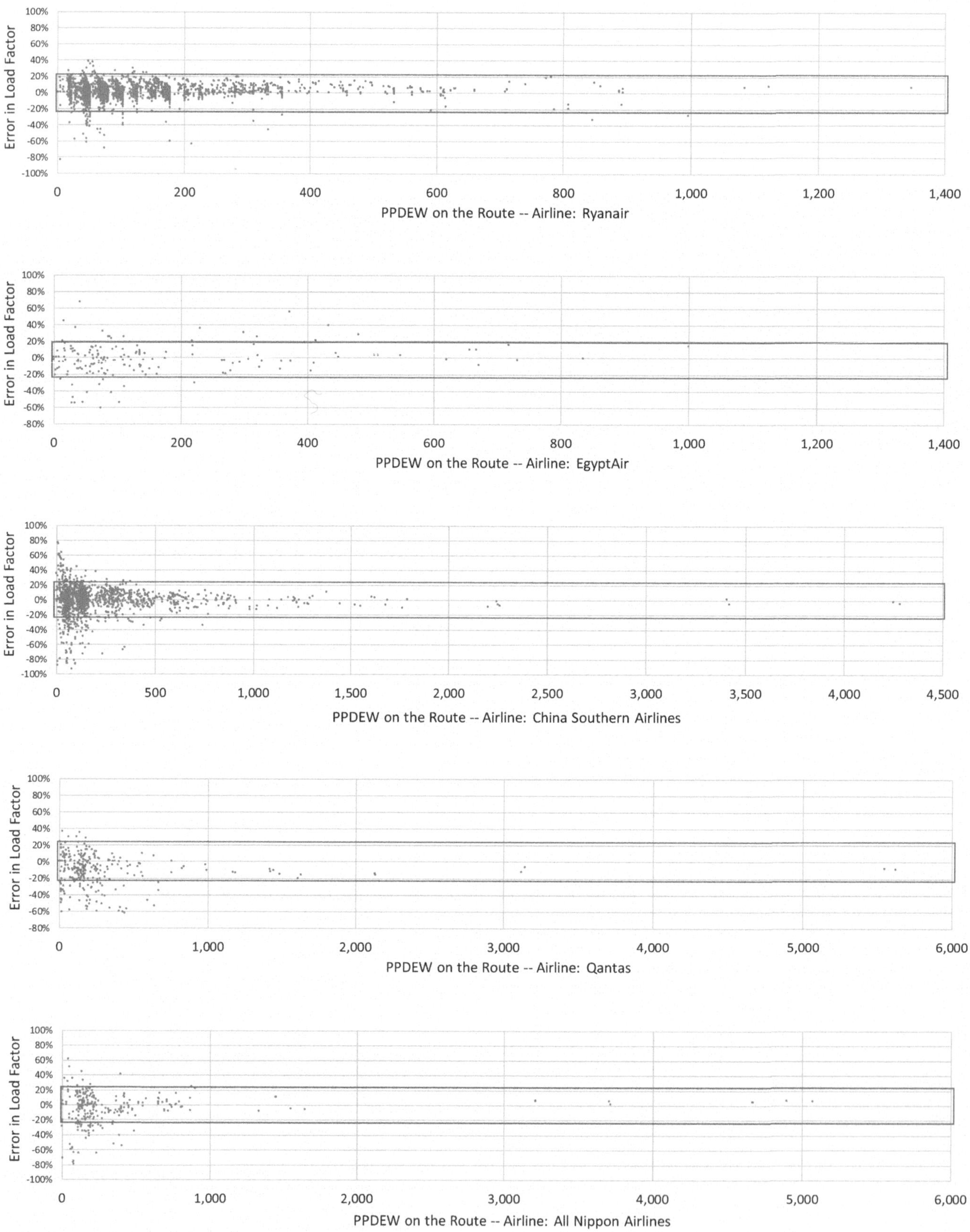

Fig. 7 (continued)

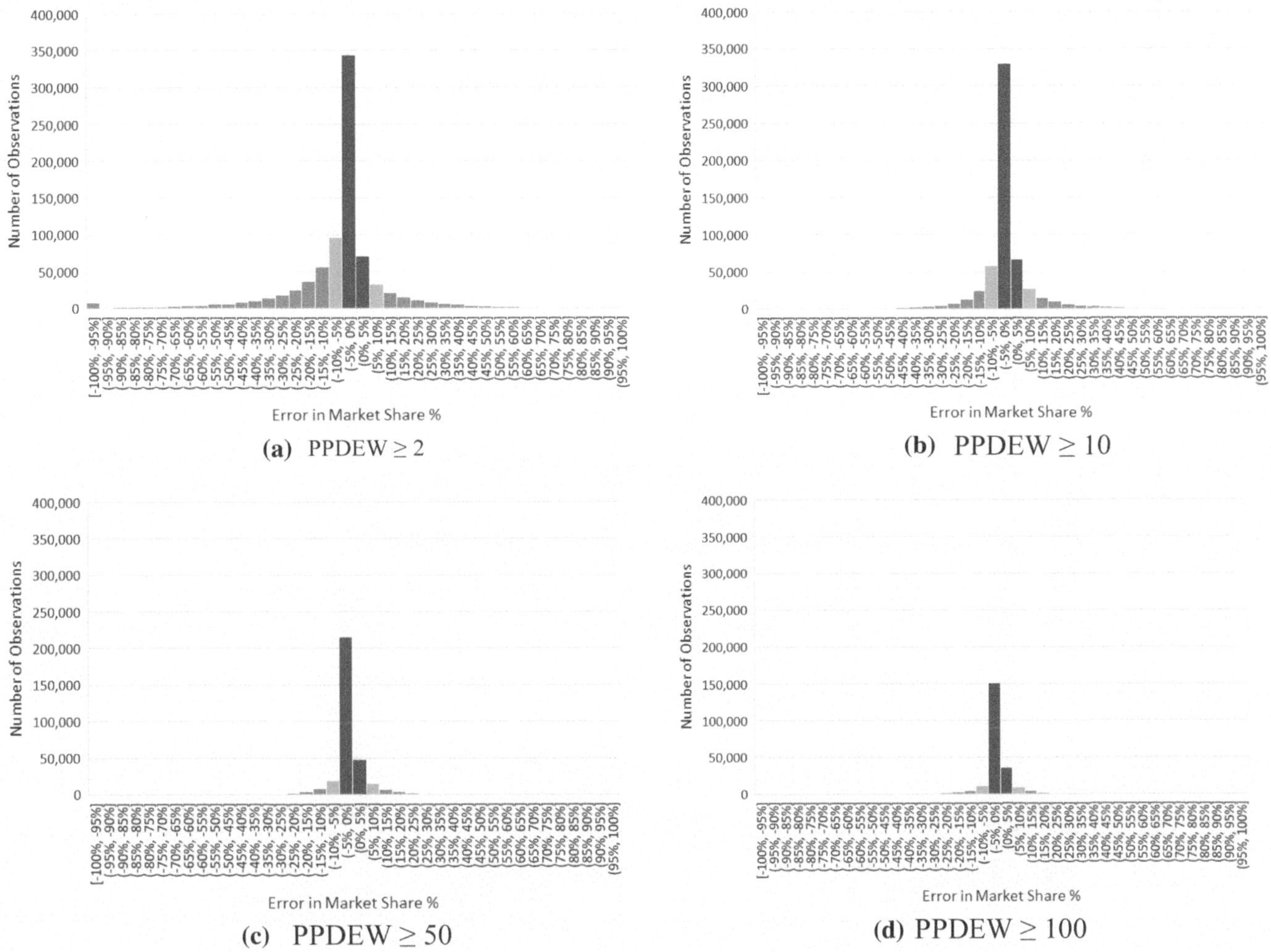

Fig. 8 The distribution of the error in itinerary market share

constraints are not considered as practiced in the traditional utility-maximization-based models (Coldren et al. 2003; Coldren and Koppelman 2005). As shown in the table, the values of the parameters of the Level of Service variables are mostly negative, which imply that itineraries with no stops or minimum number of stops attract more demand. The results also show that the values of the parameters of the Code Share variables are also mostly negative indicating that code-share itineraries attract less demand compared to the non-code-share itineraries. Similarly, the results suggest that longer itineraries and itineraries with longer connections attract less passenger demand. In Model 1 (i.e., airport pairs with non-stop short-haul frequent service), the values of the parameters of the Time of Day variables indicate that morning itineraries attract more demand compare to evening and night itineraries. The values of the parameters of the Fare Ratio variables are mostly negative implying that cheaper itineraries attract more demand. The values of the parameters associated with the Presence variable are mostly positive, which imply that the higher presence of airlines in the airport pair attracts more demand. Finally, the Airline Name variables (which are represented by the airline 2-letter code) have mixed results. Based on the obtained results, the values of the parameters associated with the low-cost airlines are generally positive, while full-service airlines mostly have negative values for their corresponding parameters.

Summary and conclusions

This paper presents a novel methodology to model the air-travel itinerary choice behavior for air travelers that addresses the limitation associated with adopting the utility-maximization theory. The methodology integrates a reinforcement learning algorithm and an airline network competition analysis model. The reinforcement learning algorithm iteratively searches for the optimal values of the itinerary choice model parameters in order to maximize a reward function that is measured as the negative of the difference between the estimated and observed system metrics. The

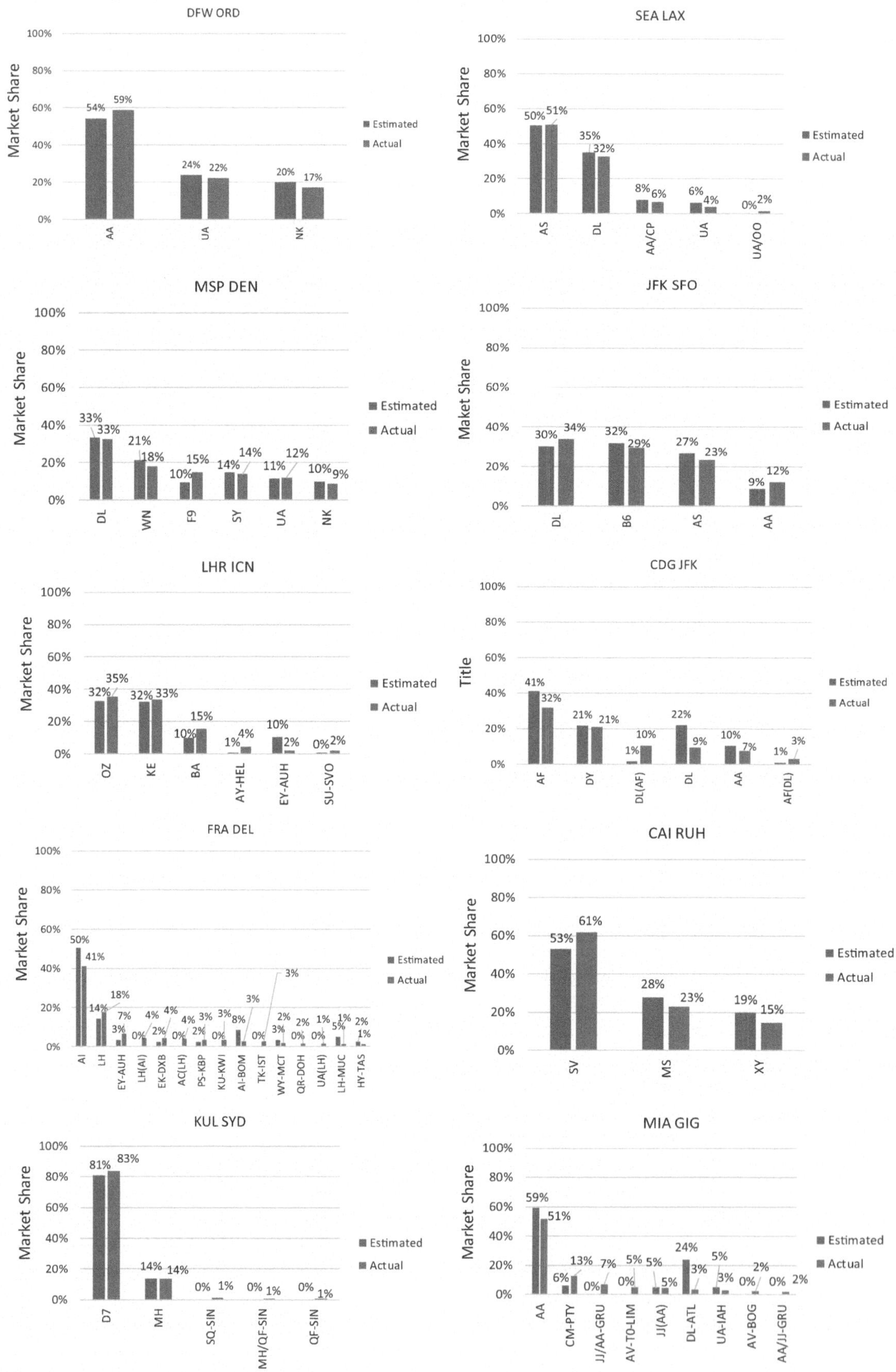

Fig. 9 Comparing the actual and estimated distribution of market share of itineraries in ten different airport pairs

Table 3 The values of the estimated parameters

Variable	Model 1	Model 2	Model 3	Model 4	Model 5	Model 6	Model 7	Model 8	Model 9	Model 10	Model 11	Model 12	Model 13
Level of service													
Less than non-Stop	−2.2	−4.3	−2.6	−3.7	−8.7	−0.3	−1.2	−1.6	−9.4				
Less than one Stop										−0.2	−0.7	−4.9	
Itinerary type													
Code share	−4.8	−4	−2.4	−3.3	−3.5	−7.8	−3.9	−6.5	−9.3	−6.5	−5.5	−5.3	1.8
operated by regional	−0.6	−0.6	0.2	3.6	−6.6	−0.5	3.8	6.5	−0.3	5.1	6.3	4.5	1.8
Connection quality													
Distance ratio	−2	3.2	−5.3	−5.9	−2.6	−9.2	−4	−8.7	−9	−1.4	−2.5	−8.5	−3.8
Connection duration:	−6	−5.6	−3.2	−7.3	−1.1	−2.7	−0.6	0.8	2.8	−2.5	−0.5	−1.3	−1.1
Time of day													
<5:00 AM	−5.1												
5:00–6:00	−0.4												
6:00–7:00	−0.5												
7:00–8:00	0												
8:00–9:00	2.5												
9:00–10:00	2.7												
10:00–11:00	2.6												
11:00–12:00	2.3												
12:00–13:00	2.9												
13:00–14:00	3.1												
14:00–15:00	0.5												
15:00–16:00	3.0												
16:00–17:00	0.4												
17:00–18:00	−9.9												
18:00–19:00	−9.7												
19:00– 20:00	−9.9												
20:00–21:00	−9.8												
21:00–22:00	−5.1												
22:00–24:00	−5.0												
Airline presence and brand													
Fare ratio	−1.1	−0.1	−0.5	−1.8	−3.1	−4.1	−0.4	−4.4	−3.2	−0.9	−4.2	−1.4	−2.3
Airline presence	−0.7	−0.8	1	2.4	1.5	2	3.5	7.3	9.8				
Airline name													
UA	−0.2												
AA	−0.8												
DL	−0.1												

Table 3 (continued)

Variable	Model 1	Model 2	Model 3	Model 4	Model 5	Model 6	Model 7	Model 8	Model 9	Model 10	Model 11	Model 12	Model 13
WN	0.3												
F9	1.4												
NK	0.3												
BA	−2												
LH	−0.9												
AF	−0.2												
KL	−0.2												
TK	−0.2												
FR	0.8												

airline network competition analysis model is a passenger-itinerary assignment model that captures the demand–supply interactions at the network level to estimate several system metrics that can be compared against ground truth data. An ICM system is calibrated using the developed framework considering the global airline network which includes more than 500,000 airport pairs. Validating the model against ground truth data showed that developed model system adequately captures the travelers' itinerary choice behavior and replicate the competition pattern among airlines. While this paper focuses primarily on presenting the methodology for estimating itinerary choice models while considering constrained demand, it is believed that this research allows for several research extensions. For example, research is underway to test different model structures related to mechanisms for the clustering the airport pairs, model specification (explanatory variables), and settings of the error term. In addition, additional research is underway to test the performance of different probability mapping mechanisms used in the reinforcement learning algorithm. Also, additional research is needed to incorporate disaggregate bookings and fare data at the itinerary level in the calibration of the model. These data will capture important phenomenon such as the price elasticity. Furthermore, additional research is needed to consider information on booking patterns and trends in the different airport pairs (e.g., based on advance purchase, length of stay, day of week of departure/return). Including these data is expected to enhance model accuracy. Finally, effort is needed to verify the quality of the used input data through fusing data from multiple sources.

References

Abdelghany, A., and K. Abdelghany. 2007. Evaluating airlines ticket distribution strategies: A simulation-based approach. *International Journal of Revenue Management* 1 (3): 231–246.

Abdelghany, A., and K. Abdelghany. 2008. A micro-simulation approach for Airline Competition Analysis and Demand Modelling. *International Journal of Revenue Management* 2 (3): 287–306.

Abdelghany, A., and K. Abdelghany. 2016. *Modeling applications in the airline industry*. London: Routledge.

Abdelghany, A., and K. Abdelghany. 2018. *Airline network planning and scheduling*. New York: John Wiley & Sons.

Abrahams, M. 1983. A service quality model of air travel demand: An empirical study. *Transportation Research Part A* 17 (5): 385–393.

Algers, S., & Beser, M. 1997. A model for air passengers choice of flight and booking class a combined stated preference and reveled preference approach. In ATRG Conference Proceedings, Vancouver.

Anderson, J.E., and M. Kraus. 1981. Quality of service and the demand for air travel. *The Review of Economics and Statistics* 92: 533–540.

Ben-Akiva, M. E. & Lerman, S. R. (1985). Discrete choice analysis: theory and application to travel demand (Vol. 9). MIT Press.

Busoniu, L., Babuska, R., De Schutter, B., & Ernst, D. (2010). Reinforcement learning and dynamic programming using function approximators (Vol. 39). CRC press.

Carrier, E. (2008). Modeling the choice of an airline itinerary and fare product using booking and seat availability data (Doctoral dissertation, Massachusetts Institute of Technology, Department of Civil and Environmental Engineering).

Coldren, G.M., F.S. Koppelman, K. Kasturirangan, and A. Mukherjee. 2003. Modeling aggregate air-travel itinerary shares: Logit model development at a major US airline. *Journal of Air Transport Management* 9 (6): 361–369.

Coldren, G.M., and F.S. Koppelman. 2005. Modeling the competition among air-travel itinerary shares: GEV model development. *Transportation Research Part A* 39 (4): 345–365.

Corsi, T., Dresner, M., & Windle, R. (1997). Air passenger forecasts: Principles and practices. In Journal of the Transportation Research Forum (Vol. 36, No. 2).

Ghobrial, A., and S.Y. Soliman. 1992. An assessment of some factors influencing the competitive strategies of airlines in domestic markets. *International Journal of Transport Economics* 24: 247–258.

Delahaye, T., R. Acuna-Agost, N. Bondoux, A.Q. Nguyen, and M. Boudia. 2017. Data-driven models for itinerary preferences of air travelers and application for dynamic pricing optimization. *Journal of Revenue and Pricing Management* 16 (6): 621–639.

Diio by Cirium (2020), Diio Mi https://www.diio.net/products/diio-mi/index.html, Retrieved March 1st, 2020.

Hess, S., T. Ryley, L. Davison, and T. Adler. 2013. Improving the quality of demand forecasts through cross nested logit: A stated choice case study of airport, airline and access mode choice. *Transportmetrica A* 9 (4): 358–384.

Ippolito, R.A. 1981. Estimating airline demand with quality of service variables. *Journal of Transport Economics and Policy* 13: 7–15.

Nason, S. D. (1981). The airline preference problem: an application of disaggregate logit. In AGIFORS PROCEEDINGS.

Nako, S.M. 1992. Frequent flyer programs and business travellers: An empirical investigation. *Logistics and Transportation Review* 28 (4): 395.

Nikseresht, A., and K. Ziarati. 2017. A demand estimation algorithm for inventory management systems using censored data. *Engineering, Technology & Applied Science Research* 7 (6): 2215–2221.

Proussaloglou, K., and F. Koppelman. 1995. Air carrier demand. *Transportation* 22 (4): 371–388.

Proussaloglou, K., and F.S. Koppelman. 1999. The choice of air carrier, flight, and fare class. *Journal of Air Transport Management* 5 (4): 193–201.

Sabre (2020). Sabre Intelligence 6.3 https://www.sabreairlinesolutions.com/home/software_solutions/product/intelligence_exchange/, Retrieved March 1st, 2020.

Sutton, R.S., and A.G. Barto. 2018. *Reinforcement learning: An introduction*. Cambridge: MIT Press.

Suzuki, Y., J.E. Tyworth, and R.A. Novack. 2001. Airline market share and customer service quality: A reference-dependent model. *Transportation Research Part A* 35 (9): 773–788.

Szepesvári, C. 2010. Algorithms for reinforcement learning. *Synthesis Lectures on Artificial Intelligence and Machine Learning* 4 (1): 1–103.

Vulcano, G., G. Van Ryzin, and R. Ratliff. 2012. Estimating primary demand for substitutable products from sales transaction data. *Operations Research* 60 (2): 313–334.

Warburg, V., C. Bhat, and T. Adler. 2006. Modeling demographic and unobserved heterogeneity in air passengers' sensitivity to service attributes in itinerary choice. *Transportation Research Record* 1951 (1): 7–16.

Wiering, M., and M. Van Otterlo. 2012. Reinforcement learning. *Adaptation, Learning, and Optimization* 12: 3.

Yoo, K.E., and N. Ashford. 1996. Carrier choices of air passengers in pacific rim: Using comparative analysis and complementary interpretation of revealed preference and stated preference data. *Transportation Research Record* 1562 (1): 1–7.

Zeni, R. H. (2001). Improved forecast accuracy in airline revenue management by unconstraining demand estimates from censored data. Universal-Publishers.

Ahmed Abdelghany is the Associate Dean for Research and Professor of Operations Management at the David B. O'Maley College of Business at Embry-Riddle Aeronautical University. Prior to his academic appointment, he worked as a senior operations research analyst at United Airlines in Chicago, IL. He authored two books titled modeling applications in the airline industry (2010) and airline network planning and scheduling (2019)

Khaled Abdelghany is a Professor at the Department of Civil and Environmental Engineering at the Lyle school of engineering, Southern Methodist University. He has extensive research experience in airline strategic planning and operations management, multimodal transportation network modeling, real-time traffic network management systems, and connected vehicle applications.

Ching-Wen Huang is a masters student in aviation finance and graduate research assistant at the David B. O'Maley College of Business at Embry-Riddle Aeronautical University. She has complete her internship at Airbus and holds a certification in Aircraft Trading & Leasing from ISTAT University.

Journal of Revenue and Pricing Management (2021) 20:248–255
https://doi.org/10.1057/s41272-021-00310-5

RESEARCH ARTICLE

Decoupling the individual effects of multiple marketing channels with state space models

Melvin Woodley[1]

Received: 18 June 2020 / Accepted: 26 September 2020 / Published online: 7 April 2021

Abstract
An important problem in marketing is understanding the impact of various marketing efforts on sales and revenue. If specific measures have not been taken before hand to distinguish responders across the various marketing initiatives, it becomes increasingly difficult to assess the effectiveness of investment across individual marketing channels. We present State Space model to estimate the individual channel effects using aggregate sales or response data across all channels. The proposed framework allows for varying carry over effects across marketing channels. Also, the proposed framework allows for differing rates of decay across marketing channels. We demonstrate its use when data on sales due to individual marketing channels is not available and only aggregated sales data are available. The proposed State Space modeling approach offers the advantage of: (1) allowing for varying rates of decay across marketing channels, and (2) allowing for a natural way of modeling time series dynamics. The approach also opens the way for more comprehensive marketing-mix optimization by allowing varying rates of decay.

Keywords Marketing-mix · Koyck · State space · Marketing–sales relationship

Introduction

A fundamental problem faced by many companies is understanding the effectiveness of multiple marketing initiatives simultaneously executed. Companies that invest millions of dollars in marketing each year have to be able to assess the relative impact of their investment across multiple channels. This problem becomes increasingly important when budget allocations to the various marketing channels need to be reviewed. Although companies invest in multiple marketing efforts such as print, media and email campaigns, the sales received as a result of those efforts are not generally distinguishable. Companies are usually faced with the inability to filter sales and revenue data according to the respective contributions of various marketing initiatives.[1] This is because it has become common place for companies to have various marketing efforts simultaneously active. Yet marketers and executives are faced with the decision of which marketing initiatives to continue and which ones they should reduce funding or abandon altogether. This study is concerned with modeling and estimating the impact of individual marketing initiatives where only cumulative sales or revenue data is available.

The problem of decoupling and estimating the effects of various marketing campaigns has a long history in the marketing research literature. It has been addressed by many researchers with various techniques. Our proposed approach is aimed at addressing some of the short comings of these approaches and to also shed light on the often not recognized advantages of using state space models in such a setting.

While time series models have formed part of the core of traditional aggregate advertising response models, many researchers do not fully exploit the advantages that state space models present. In practical applications of traditional time series models, it is difficult to reconcile univariate sales and revenue forecasts with investments made in marketing channels. For these reasons, it is important to bridge the gap between underlying time series dynamics and associated

✉ Melvin Woodley
melvin.woodley@gmail.com

[1] Sabre Research, Southlake, TX, USA

[1] Some companies use promotion codes to track specific marketing campaigns. However, many responders to marketing campaigns may not use promotion codes even when they are available. Designed experiments are also used to estimate marketing effects. We consider the case where Designed Experiments are not a plausible option for companies.

Chapter 4 was originally published as Woodley, M. Journal of Revenue and Pricing Management (2021) 20:248–255. https://doi.org/10.1057/s41272-021-00310-5.

forecasts with the estimates from aggregate market response models. The most common approach to dealing with seasonal variation and underlying time series dynamics is to employ indicator variables to estimate seasonal variation. This approach implies a static seasonal impact and ignores the reality that seasonal changes are often evolving over time. In addition to dealing with this dynamic variation in seasonality, many marketing response models do not attempt to model underlying trends. While there are some studies that incorporate first order autoregressive error in modeling the error structure, this is usually done after indicator variables have been applied as a remedial measure to model any additional auto-correlation that was remained. Adhering to such protocol does not place enough emphasis on the role that underlying time series dynamics plays in driving sales and revenue and in some cases can cause the effects of marketing efforts to be overestimated. While using indicator variables is easy to incorporate in many otherwise sophisticated modeling strategies, it gives limited insight into how the effects of marketing efforts compares with changes in sales and revenue due to organic economic changes.

A common method used to assess the impact of marketing is transfer modeling. In transfer function modeling, a baseline time series model is chosen using data from the pre-intervention period. This model is usually in the form of an autoregressive integrated moving average (ARIMA) model. Then a functional form of the intervention is specified and incorporated with the ARIMA model. Finally, parameters of the fully specified model along with intervention specification are estimated with the entire dataset. Various functional forms for the intervention are tested. A study that employ intervention to assess the marketing impact on sales is 'Bivariate Time Series Analysis of the Relationship Between Advertising and Sales' by Dominique Hanssens.

Literature review

Marketing-mix models have become very pervasive among marketing agencies, where statistical models are developed to assess the impact of marketing investments on sales. Although these models are conceptually clear in their intent, because of the competitive landscape of the marketing agency arena and the use of proprietary data, there is very little knowledge sharing and transparency regarding methods and models used. Naik et al. (2007) points out that it is common practice to employ ordinary least squares estimation and hence yield biased estimates in estimating such models.

Within the marketing research literature, there has been much work done to develop marketing-mix models for example Edell et al. (1999), Gatignon (1993), Gatignon et al. (1987), Hanssens et al. (2001), Hanssens (1980), Leeflang et al. (2000), Naik (1998). The Integrated Marketing Communications (IMC) model of Naik and Raman (2003) addresses many of the shortcomings of ignoring the interaction or cross media synergy role in many additive marketing models. The IMC model builds on the Koyck distributed lag model which was extended by Montgomery and Silk (1972) to incorporate multiple channels of marketing.

One limitation of some marketing-mix models is that they rest on the assumption of constant carry over effect as in Montgomery and Silk (1972). This assumption is quite restrictive and is unlikely to be appropriate in practice. A more flexible framework would allow for varying carry over effect. In fact, the problem of estimating varied carry over effects was discussed at length in Naik and Raman (2003). They address the issue of constant carry over effect by introducing additional parameters and restrictions on sales due to specific marketing channels. They note that the challenge of estimating varying carry over effect is due to the fact that contributions of the individual marketing channels are unobserved. Total sales is the only data that is usually available. Sales from individual marketing channels cannot be collected since there is no way of collecting this data outside of scanner data or designed experiments Naik and Raman (2003).

Although the IMC model addresses the cross synergy limitation of other modeling frameworks, it does not allow for a convenient way of incorporating varied carry over effects for individual marketing channels. To date, very few marketing-mix frameworks allow for varied carry over effect. Recently, Kappe et al. (2014) proposed a framework that allows for multiple carry over effects and demonstrate its use in understanding game attendance from ticket sales. However, the modeling framework does not allow for modeling the underlying time series dynamics of the data. Any framework that estimates the effect of marketing should be built on the idea of separating the incremental sales due to marketing investment from the usual time series and seasonal dynamics of the sales. Most studies attempt to resolve this issue with the use of dummy variables to account for seasonal variation. However, dummy variable use assumes a static time series structure, which is often impractical for future planning and forecasting. Assuming a static baseline dummy variable model makes the restrictive assumption that future sales will be the same as current baseline sales which may not be a reasonable one in practice, especially in the context of a growing or evolving market. A competitive landscape would also merit the use of a more dynamic modeling framework. Smith et al. (2006) provide supporting evidence when they explored how the effects of marketing can be very different depending on economic conditions. The state of the economy ultimately affects the underlying time series structure of sales and thus makes the case for a more dynamic approach to marketing-mix modeling. While many marketing-mix models seek to measure the incremental effect of marketing on sales, very few frameworks attempt to model the underlying time series structure.

A limitation of some marketing-mix modeling frameworks is that the models are developed and applied on real world data without first validating the model performance in a simulation study. While the emphasis on practical applications with real data is well placed, it is of equal importance that models be validated via simulation in advance. This would provide a means of ensuring that the estimates from the application of the model to real world data can be trusted. A model that does not estimate the true effects in a simulation study cannot be trusted in practical applications.

This study adds to the existing literature in several ways. Firstly, it adds to the limited use of state space models in marketing studies. Although state space models are very flexible tools for modeling, there have been very few studies that examine their use in a marketing context as noted in Dekimpe and Hanssens (2000). Additionally, our study demonstrates the use of the superposition principle in combining state space models in an additive manner to obtain and larger model.

The second way that this study differentiates itself from other studies is that it proposes a novel framework for allowing for differentiated carry over effects for multiple channels of marketing. Constant carry over effect is not addressed in many multimedia marketing communications models. Thirdly, we add to the limited literature that investigate multiple marketing vehicles.

Fourthly, this study demonstrates an important use of Kalman Filtering in State Space models as a tool for estimating unobserved effects within marketing data. Although this is commonly done in the field of time series modeling, the technique has very useful practical applications in the marketing field where underlying effects are constantly being estimated.

The fifth way in which this study is unique is that it demonstrates the use of an underlying time series model to capture the underlying dynamic time series structure of the data. This provides a reasonable way of reconciling naive forecast performance with the estimates from marketing-mix models. This fills a gap in the literature where most marketing-mix models do not explicitly model the underlying time series dynamics.

Additionally, this study focuses on the use of simulation to validate the framework used to estimate the effects of marketing on sales.

Model development

Let f_t denote the sales at time t that a company receives as a result of their marketing efforts. Let us initially assume for simplicity that the company has one main channel of marketing. A marketing channel is defined as any distinctive form of marketing that the company engages in. Let μ_t denote the units of investments made in marketing for this one marketing channel at time period t. Therefore, μ_t represents total marketing.

A convenient functional form for the sales at time t due to marketing is:

$$f_t = \alpha + \beta\mu_t + \lambda f_{t-1} + \varepsilon_t \tag{1}$$

We note that this is the classic Koyck model Wooldridge (2003). Quite a number of studies in the marketing literature have incorporated this Koyck model in the estimation of marketing effects. For example, Naik and Raman (2003), Naik et al. (2007) and Bucklin and Gupta (1999) to name a few. This model is also known as the geometric lag model. While many studies employ specification (1), some authors have pointed out the shortcomings of using (1) instead of the unrestricted Koyck model where the error terms follows an MA (1) model with coefficient equal to the negative coefficient of f_t. For a discussion of the implicit bias in (1) see Frances and van Oest (2004) and page 635 of Wooldridge (2003). The unrestricted version of the Koyck model is:

$$f_t = \alpha + \beta\mu_t + \lambda f_{t-1} - \lambda\varepsilon_{t-1} + \varepsilon_t \tag{2}$$

$$\varepsilon_t \sim N(0, \sigma^2) \tag{3}$$

For the reasons aforementioned we let f_t assume the functional form above. That is we assume the unrestricted version of the Koyck model.

The Koyck model was extended by Montgomery and Silk (1972) to incorporate multiple marketing channels. In the case of two marketing channels the model becomes:

$$\begin{aligned} f_t = \alpha &+ \beta_1\mu_{1,t} + \beta_2\mu_{2,t} + \lambda f_{t-1} \\ &- \lambda\varepsilon_{t-1} + \varepsilon_t \end{aligned} \tag{4}$$

The equation can be extended to incorporate interaction effects $\mu_{1,t}\mu_{2,t}$ as illustrated in Eq. 5.

$$\begin{aligned} f_t = \alpha &+ \beta_1\mu_{1,t} + \beta_2\mu_{2,t} + \gamma_1(\mu_{1,t}\mu_{2,t}) + \lambda f_{t-1} \\ &- \lambda\varepsilon_{t-1} + \varepsilon_t \end{aligned} \tag{5}$$

The above formulation allows for multiple marketing channels to be considered. This is particularly convenient where a company engages in marketing across multiple channels and they are interested in obtaining a measure of the relative effectiveness of each marketing channel. One limitation of the above extension is that all variables (channels) have the same rate of decay λ. However, in practice it is unlikely that λ would be the same across all channels. Our goal in what follows is to demonstrate in part how one can use the state space model formulation to accommodate various rates of decay and estimate them.

Consider the case where a company engages in k different marketing initiatives, where $k > 1$. For each marketing channel C_i, let $\mu_{i,t}$ denote the units invested at time t by the company in marketing channel i. Further, let $f_{i,t}$ denote the respective contribution to sales at time t due to spending on marketing channel i at time t.

We therefore, have the following system of equations:

$$\begin{aligned} f_{1,t} &= \alpha_{1,t} + \beta_1 \mu_{1,t} + \lambda_1 f_{1,t-1} - \lambda_1 \varepsilon_{1,t-1} + \varepsilon_{1,t} \\ f_{2,t} &= \alpha_{2,t} + \beta_2 \mu_{2,t} + \lambda_2 f_{2,t-1} - \lambda_2 \varepsilon_{2,t-1} + \varepsilon_{2,t} \\ f_{3,t} &= \alpha_{3,t} + \beta_3 \mu_{3,t} + \lambda_3 f_{3,t-1} - \lambda_3 \varepsilon_{3,t-1} + \varepsilon_{3,t} \\ &\vdots \\ f_{k,t} &= \alpha_{k,t} + \beta_k \mu_{k,t} + \lambda_k f_{k,t-1} - \lambda_k \varepsilon_{k,t-1} + \varepsilon_{k,t} \end{aligned} \tag{6}$$

The contribution of all k marketing channels at time t to sales is the sum of all the individual effects:

$$M_t = \sum_{i=1}^{k} f_{i,t} \tag{7}$$

Note that $f_{i,t}$ represents the sales contribution at time t due to marketing initiative i. However, $f_{i,t}$ does not represent the total sales but instead represents the incremental contribution of the individual marketing channel to the baseline sales. Although the individual $f_{i,t}$ are not observed and hence not known, the total sales is known and is sufficient to estimate all the parameters in (5) above.

The estimation problem has been discussed in Naik and Raman (2003) where estimation heuristics were adapted to overcome this estimation problem. In the next section we present a State Space model formulation that addresses this estimation problem.

State space model

The State Space model is a very flexible tool for modeling dynamic relationships that has many applications in Statistics and Engineering. The state space modeling frame work provides a very elegant way of handling a very broad range of statistical models. Despite its appeal and flexibility state space models are not often used in practice since state space modeling requires a more customized programming and are not as easily applied as other statistical modeling approaches. A State Space model consists of two equations of the following form:

$$y_t = Z\alpha_t + d_t + z_t \tag{8}$$

$$\alpha_t = T\alpha_{t-1} + c_t + \eta_t \tag{9}$$

In the above equations, y_t denotes unit sales observed. d_t and c_t are input variables. Z, T are matrices. And the error terms are such that $z_t \sim N(0, R)$ and $\eta \sim N(0, Q)$.

There are many references that describe state space modeling, for example, Harvey (1989) and Shumway and Stoffer (2000) to name a few. In general α_t is not observable. However, the Kalman Filter provides optimal estimates of α_t. All the parameters of the model are estimated by combining maximum likelihood estimation with the Kalman Filtering.

State space modeling provides a means for representing many different classes of statistical models by casting these models in state space form. For example, Vector Autoregressive models, ARIMA models, time varying regression models, stochastic volatility models and mixed models are different classes of models that can be cast in state space form and estimated. A particularly attractive feature of State Space models is the use of the Kalman Filtering recursive equations to optimally estimate the unobserved state vector α_t. Typically the state vector consists of unobserved or unmeasurable latent components. The Kalman filtering recursion estimates the values of α_t optimally.

The following are some benefits of using state space models and Kalman filtering:

- Ability to assume the functional form of other classes of models.
- Ability to optimally estimate unobserved components.
- Ability to estimate standard model parameters.
- Ability to incorporate time varying parameters.
- Allows for recursive estimation of α_t
- Ability to handle missing values.

For all of these reasons, the State Space model is a very flexible modeling framework.

Consider the unrestricted Koyck model in Eq. 2. Note that the model has very similar structure to a MA(1,1) model with regression inputs. Hence, we note that the model can be easily cast into state space form of Eqs. (6) and (7) using the following specification:

$$y_t = Z\alpha_t \tag{10}$$

$$\alpha_t = T\alpha_{t-1} + c_t + \eta_t \tag{11}$$

$$Z = \begin{bmatrix} 1 & 0 & 0 \end{bmatrix} \tag{12}$$

$$T = \begin{bmatrix} \lambda & 1 & -1 \\ 0 & 1 & 0 \\ 0 & 0 & 0 \end{bmatrix} \tag{13}$$

$$d_t = 0, \quad x_t = \mu_{t,1}, \quad R = 0 \tag{14}$$

$$\alpha_t = \begin{bmatrix} f_t \\ \alpha_t \\ \lambda\varepsilon_t \end{bmatrix} \tag{15}$$

$$cov(\eta) = Q = \begin{bmatrix} \sigma^2 & 0 & \lambda\sigma^2 \\ 0 & 0 & 0 \\ \lambda\sigma^2 & 0 & \lambda^2\sigma^2 \end{bmatrix} \tag{16}$$

Given the above state space representation for a single marketing channel, we can use a very important principle from time series called the Superposition Principle to extend this model. The Superposition principle is mentioned in many text on state space models for example, Hyndman et al. (2008); West and Harrison (1997). Taken from West and Harrison, the Superposition Principle states that any linear combination of independent state space models is itself also a state space model. We state the following theorem adopted from Mike and West:

Superposition Principle *Consider k time series $y_{i,t}$ generated by state space models with state space representation:*

$$M_i = (Z_{i,t}, T_{i,t}, Q_{i,t}, R_{i,t})$$

The state vector $\alpha_{i,t}$ is of dimension n_i and the observation and state errors are $z_{i,t}$ and $\eta_{i,t}$ respectively. The state vectors are distinct and for all $i \neq j$, the series $z_{i,t}$ and $\eta_{i,t}$ are not equal to the series $z_{i,t}$ and $\eta_{i,t}$.

$$Y_{i,t} = \sum_{i=1}^{k} y_{i,t}$$

follows the n dimensional state space model with system matrices (Z_t, T_t, Q_t, R_t), where $n = n_1 + n_2 + n_3 + \ldots + n_k$ with the following matrices:

$$\alpha_t = \begin{bmatrix} \alpha_{1,t} \\ \alpha_{2,t} \\ \alpha_{3,t} \\ \vdots \\ \alpha_{k,t} \end{bmatrix}, \quad Z = \begin{bmatrix} Z_{1,t} \ldots Z_{k,t} \end{bmatrix}$$

$$T_t = bdiag(T_1, T_2, \ldots, T_k)$$
$$Q_t = bdiag(Q_1, Q_2, \ldots, Q_k)$$
$$z_t = z_{1,t} + \ldots + z_{k,t}$$

We now extend this model by considering the case of k marketing channels. Making use of the superposition principle, we note that for the k variable case we can write the following:

$$Z = \begin{bmatrix} \alpha_{1,t} \ldots \alpha_{k,t} \end{bmatrix} \tag{17}$$

$$T = bdiag(T_1, \ldots, T_k) \tag{18}$$

$$\alpha_t = \begin{bmatrix} f_{1,t} \\ \alpha_{1,t} \\ \lambda_1\varepsilon_{1,t} \\ \vdots \\ f_{k,t} \\ \alpha_{k,t} \\ \lambda_k\varepsilon_{k,t} \end{bmatrix} \quad x_t = \begin{bmatrix} \mu_{1,t} \\ \vdots \\ \mu_{k,t} \end{bmatrix} \tag{19}$$

$$\beta = \begin{bmatrix} \beta_1 & 0 & \ldots & 0 \\ 0 & 0 & & \vdots \\ 0 & 0 & & \\ 0 & \beta_2 & 0 & \\ \vdots & 0 & \ddots & \\ & & & \beta_k \\ & & & 0 \\ 0 & \ldots & & 0 \end{bmatrix} \tag{20}$$

$$d_t = 0, \quad R = 0, \quad c_t = \beta x_t \tag{21}$$

$$Q = bdiag(Q_1, \ldots, Q_k) \tag{22}$$

We note that for k equal to 2 when we expand the above equations we have the following:

$$\begin{aligned} y_t &= \begin{bmatrix} 1 & 0 & 0 & 1 & 0 & 0 \end{bmatrix} \alpha_t \\ &= \lambda_1 f_{1,t-1} + \alpha_{1,t} - \lambda_1\varepsilon_{1,t-1} + \beta_1\mu_{1t} + \varepsilon_{1,t} \\ &\quad + \lambda_2 f_{2,t-1} + \alpha_{2,t} - \lambda_2\varepsilon_{2,t-1} + \beta_2\mu_{2t} + \varepsilon_{2,t} \\ &= f_{1,t} + f_{2,t} \\ &= M_t \end{aligned} \tag{23}$$

The state space representation in Eqs. (10) and (11) has some particularly advantageous properties. The different marketing vehicles $\mu_{i,t}$ are allowed to have different geometric rates of decay. That is the λ_i's do not have to be equal.

Although the individual $f_{i,t}$ are not observable, the state space modeling framework is designed to handle this particular situation and the Kalman Filter recursions provides optimal estimates of the state variable α_t. Estimating the $f_{i,t}$ could be thought of as similar to estimating the unobserved components of a time series. These topics are discussed in Harvey for example.

Baseline time series dynamics

In practical applications of marketing models, there are underlying dynamics that influence the behavior of sales that are not necessarily attributable to marketing efforts or any related investment. As an example, seasonal variations should not be attributed to the effect of marketing investments. Also, long term changes in sales due to economic growth or economic contractions are changes that one should not attribute to marketing efforts. These such effects need to be accounted for in order to avoid biased estimates of the impact of marketing investments. Essentially, the underlying time series dynamics of the data should be modeled simultaneously with any marketing impact. Modeling the underlying dynamics also means that any known market interventions that affected sales should be accounted for.

Because marketing efforts can be thought of as a intervention that occurs over a period of time, it is recommended that a portion of the data (usually the observations prior to the commencement of marketing) be used to estimate a time series model that is representative of the underlying time series dynamics of the data. This is in line with common statistical practice of estimating a baseline model when investigating the impact of an intervention or when considering transfer function modeling of an exogenous variable.

Consider the case where a firm significantly increases its marketing investment relative to the previous year's levels. However, sales for the year in which marketing was increased can be forecasted within a small margin of error without accounting for the significant increase in marketing (only using historical sales data without exogenous input factors). Such occurrences brings into question the relative effectiveness of the marketing investment since the increased marketing budget had little influence on disrupting the naive sales forecast. This is a typical benchmark that some executives use in validating the truthfulness of deductions made from marketing-mix models. How does one reconcile the estimates of impact of marketing and the general trajectory and dynamics of the time series.

To this end, we further extend the above model to incorporate the time series dynamics of the data. Let $y_t = \zeta_t + M_t$, where ζ_t represent the underlying stochastic process and M_t represent the change in y_t due to marketing dynamics. We model M_t using (9) and (10) and we model ζ_t using 1) an estimated ARIMA model or 2) the basic structural model of Harvey (1989). The basic structural model takes the following form:

$$\zeta_t = \mu_t + \gamma_t + \epsilon_t \tag{24}$$

$$\begin{aligned} \gamma_{t+1} &= -\sum_{i=1}^{s-1} \gamma_{t+1-j} + \omega_t \\ \mu_{t+1} &= \mu_t + \nu_t + \xi_t \\ \nu_{t+1} &= \nu_t + \eta_t \end{aligned} \tag{25}$$

As noted in West and Harrison (1997) and Hyndman et al. (2008), if a state space model can be broken into components C_1 and C_2, each having a state space representation, then the sum $y_t = C_1 + C_2$ also has state space representation by combining the individual components C_1 and C_2. Although here we assume a basic structural model, the state space modeling framework is flexible enough to accommodate many different formulations.

Estimation

The Kalman Filtering recursions are a set of equations used to used to estimate in a recursive manner the elements of the state vector α_t. The Kalman Filter is known to be an optimal estimate of these unobserved states. The Kalman filtering recursions are given by the following equations:

$$\begin{aligned} v_t &= y_t - Z_t a_t - d_t \\ F_t &= Z_t P_t Z_t' + R_t \\ \alpha_{t|t} &= \alpha_t + P_t Z_t' F_t^{-1} v_t \\ P_{t|t} &= P_t - P_t Z_t' F_t^{-1} Z_t P_t \\ \alpha_{t+1} &= T_t \alpha_{t|t} + c_t \\ P_{t+1} &= T_t P_{t|t} T_t' + Q_t \end{aligned}$$

For a given state space representation, the conditional distribution of y_t is normal. Hence, the likelihood function is given by:

$$\begin{aligned} \log L = &-\frac{NT}{2}\log 2\pi - \frac{1}{2}\sum_{t=1}^{T} \log |F_t| \\ &- \frac{1}{2}\sum_{t=1}^{T} v_t' F_t^{-1} v_t \\ v_t &= y_t - Z_t \alpha_{t|t-1} + d_t \\ F_t &= Z_t P_{t|t-1} Z_t' + R_t. \end{aligned}$$

The likelihood is then maximized with respect to the model parameters.

Application to simulated data

In this section we demonstrate the ability of the state space model formulation presented in estimating the individual effects of marketing channels. We simulate at dataset $y_t = f_{1,t} + f_{2,t}$ and we then estimate the model parameters by only using the aggregated data. We simulate the data with the following parameters, for $t = 1, \ldots, 500$.

$$\begin{aligned} f_{1,1} &= 0 \\ f_{1,t} &= 10 + 0.3 f_{1t-1} + 0.7 x_{1t} - 0.3 \epsilon_{1t-1} + \epsilon_{1t} \\ f_{2,t} &= 20 + 0.5 f_{1t-1} + 0.9 x_{1t} - 0.6 \epsilon_{2t-1} + \epsilon_{2t} \\ \epsilon_{1t} &\sim N(0, 2^2) \\ \epsilon_{2t} &\sim N(0, 4^2) \\ y_t &= f_{1t} + f_{2t} \end{aligned} \tag{26}$$

Table 1 Results from Simulation 1

Model Parameters		
Parameter	Actual Value	Estimated Value
β_1	0.7	0.67
α_1	0.3	0.29
β_2	0.9	0.89
α_2	0.6	0.57

Table 2 Results from Simulation 2

Model Parameters		
Parameter	Actual Value	Estimated Value
β_1	0.7	0.72
α_1	0.3	0.51
β_2	0.9	0.90
α_2	0.6	0.59

We then proceed to utilize the framework laid out above to estimate the parameters of the model. Table 1 shows the results of the estimation.

We further test the state space frame work by simulating the following. We simulate a seasonal autoregressive model, namely an ARIMA(0,1,1)x(0,1,1)[12] model to represent underlying time series dynamics. We then add to this process the combined effects of two marketing vehicles as done before. We proceed likewise and estimate the parameters governing the underlying marketing effects using only the aggregate data. The results are presented in Table 2.

The only variables used in the estimation of the pare meters are y_t, $x_{1,t}$, and $x_{2,t}$. We do not use $f_{1,t}$ and $f_{2,t}$ as inputs in the estimation process. The table below shows the actual values and the estimated values using the state space model outlined above.

Optimum budget allocation

Having estimated the effects of the different marketing channels, the next fundamental question is how does one determine the optimal allocation of marketing spending given a budget. The state space formulation detailed above allows us to estimate the model parameters. Moreover, as we shall show in this section, the formulation is rather convenient in that it allows us to formulate the optimum budget allocation problem as a standard linear programming problem as we will now demonstrate. This advantage comes from the fact that the state space model formulation allows us to consider the effects of each marketing channel separately in the measurement equation of the state space model.

Let x_j^i denote the decision variable at time j for marketing channel i. That is,x_j^i represents the amount chosen to invest in channel i at time j. Let β_i represent the estimate of the initial impact of an investment in marketing channel i as estimated by the model detailed above. Also, let α_i be the associated rate of decay for this particular channel i . Let the number of marketing channels by $k > 0$. Let the decision horizon be from time $t = 1$ to time $t = n$, some time in the future. Let x_i^T be the total effect of the decision variable x_i over time. At, each stage t, each of the k decision variables x^i has to be chosen so as to maximize total sales or revenue across the all time horizons. At each stage the effect of the decision variable propagates through the end of the decision horizon. To be precise, theoretically we have assumed that the effect of the decision variables are propagated through infinity. Since the decision horizon is n and there are k decision variables at each stage, then there are a total of $n \times k$ decision variables across the entire time horizon. For each channel i we have the following:

$$x_t^1\left(\beta_1 + \lambda_1\beta_1 + \lambda_1^2\beta_1 + \lambda_1^3\beta_1 + \dots\right) = x_t^1\left(x_t^{1,T}\right)$$
$$x_t^2\left(\beta_2 + \lambda_2\beta_2 + \lambda_2^2\beta_2 + \lambda_2^3\beta_2 + \dots\right) = x_t^2\left(x_t^{2,T}\right)$$
$$x_t^3\left(\beta_3 + \lambda_3\beta_3 + \lambda_3^2\beta_3 + \lambda_3^3\beta_3 + \dots\right) = x_t^3\left(x_t^{3,T}\right)$$
$$\vdots$$
$$x_t^k\left(\beta_k + \lambda_k\beta_k + \lambda_k^2\beta_k + \lambda_k^3\beta_k + \dots\right) = x_t^k\left(x_t^{k,T}\right)$$

The optimization problem can therefore be formulated as:

$$\underset{x}{\text{maximize}} \quad \sum_{j=1}^{t}\sum_{i=1}^{k} x_i^j \frac{\beta_i}{1-\lambda_i}$$
$$\text{subject to} \quad \sum_{j=1}^{t}\sum_{i=1}^{k} x_i^j \leq C$$

Note also that this optimization problem can be formulated with the following contraints instead:

$$\underset{x}{\text{maximize}} \quad \sum_{i=1}^{n}\sum_{i=1}^{k} x_i^j \frac{\beta_i}{1-\lambda_i}$$
$$\text{subject to} \quad \sum_{i=1}^{n} x_i^j \leq C_j$$
$$\sum_{i=1}^{k} C_j \leq C$$

This formulation allows each time period to be allotted a specific amount across all marketing channels.

Conclusion

We have presented a state space model that can be used to estimate the individual effects of marketing channels and hence each channel's effectiveness. We have demonstrated its ability to estimate individual marketing effects when the only data available is aggregate sales data. Additionally, we have shown that the method is robust even when the underlying time series dynamics are not specified.

References

Bucklin, Randolph E., and Sunil Gupta. 1999. Commercial Use of UPC Scanner Data: Industry and Academic Perspectives. *Marketing Science* 18 (3): 247–73.

Dekimpe, Marnik G., and Dominique M. Hanssens. 2000. Time-Series Models in Marketing: Past, Present and Future. *International Journal of Research in Marketing* 17 (2–2): 183–93.

Edell, Julie E. and Kevin L. Keller. 1999. Analyzing Media Interactions: The Effects of Coordinated TV-Print Advertising Campaigns. *Marketing Science Institute*, Working Paper No. 99-120.

Frances, P.H., and R. van Oest. 2004. On the Econometrics of the Koyck Model. *Technical Report*, Economic Institute, Erasmus University Rotterdam.

Gatignon, Hubert. 1993. Marketing Mix Models. In Eliashberg, J. and G.L. Lilien (eds.) *Handbook of OR and MS*, Vol. 5. New York: North-Holland, 697–732.

Gatignon, Hubert, and Dominique M. Hanssens. 1987. Modeling Marketing Interactions with Application to Salesforce Effectiveness. *Journal of Marketing Research* 24 (August): 247–57.

Hanssens, Dominique M., Leonard J. Parsons, and Randall L. Schultz. 2001. *Market Response Models: Econometric and Time Series Analysis*. New York: Kluwer.

Hanssens, Dominique M. 1980. Bivariate Time-Series Analysis of the Relationship Between Advertising and Sales. *Applied Economics* 12: 329.

Harvey, Andrew C. 1989. *Forecasting, Structural Time Series Models, and the Kalman Filter*. Cambridge: Cambridge University Press.

Hyndman, R.J., A.B. Koehler, J.K. Ord, and R.D. Snyder. 2008. *Forecasting With Exponential Smoothing: The State Space*. Berlin: Springer.

Kappe, E., A. Stadler Blank, and W.S. DeSarbo. 2014. A General Multiple Distributed Lag Framework for Estimating the Dynamic Effects of Promotions. *Management Science* 60: 1489–1510.

Leeflang, Peter S.H.., Dick R. Wittink, Michel Wedel, and Philippe A. Naert. 2000. *Building Models for Marketing Decisions*. New York: Kluwer Academic Publishers.

Montgomery, David B., and Alvin U. Silk. 1972. Estimating Dynamic Effects of Marketing Communications Expenditures. *Management Science* 18 (10): 485–510.

Naik, Prasad A., Murali K. Mantrala, and Alan G. Sawyer. 1998. Planning Media Schedules in the Presence of Dynamic Advertising Quality. *Marketing Science* 17 (3): 214–35.

Naik, Prasad A., D.E. Schultz, and S. Srinivasan. 2007. Perils of Using OLS to Estimate Multimedia Communications Effects. *Journal of Advertising Research* 13: 257–269.

Naik, Prasad A., and Kalyan Raman. 2003. Understanding the Impact of Synergy in Multimedia Communications. *Journal of Marketing Research* 40 (November): 375–88.

Shumway, Robert H., and David S. Stoffer. 2000. *Time Series Analysis and Its Applications*. Berlin: Springer.

Smith, Aaron, Prasad A. Naik, and Chih-Ling. Tsai. 2006. Markov-Switching Model Selection Using Kullback–Leibler Divergence. *Journal of Econometrics* 134 (October): 553–577.

West, M., and J.F. Harrison. 1997. *Bayesian Forecasting and Dynamic Models*, 2nd ed. Springer Verlag Series in Statistics. Berlin: Springer.

Wooldridge, Jeffrey M. 2003. *Introductory Econometrics: A Modern Approach*. Stamford: Thomson South-Western.

Melvin Woodley is a Principal Data Scientist at Sabre Research focusing on the application of machine learning to improve travel products and enhance travel technology infrastructure. He recieved his Masters degree in Mathematics from the University of North Texas in 2010. Prior to joining Sabre he worked at Hilton Worldwide and then at Ansira where he focused on building forecasting models and applying machine learning to solve marketing problems.

Journal of Revenue and Pricing Management (2021) 20:256–275
https://doi.org/10.1057/s41272-021-00311-4

RESEARCH ARTICLE

Competitive revenue management models with loyal and fully flexible customers

Ravi Kumar[1] · Wei Wang[1] · Ahmed Simrin[2] · Sivarama Krishnan Arunachalam[2] · Bhaskara Rao Guntreddy[2] · Darius Walczak[1]

Received: 15 June 2020 / Accepted: 30 July 2020 / Published online: 24 March 2021

Abstract
Developing practical models for capturing competitive effects in revenue management and pricing systems has been a significant challenge for airlines and other industries. The prevalent mechanisms of accounting for competitive effects rely on changing the price structure and making manual adjustments to respond to dynamically evolving competitive scenarios. Furthermore, micro-economic models have also not become popular in practice primarily because of the simplistic mechanisms proposed for modeling consumer behavior in a competitive setting. In particular, many of these models assume that the customers always seek the lowest price in the market, that is they are fully flexible. In practice, customers may display some degree of affinity or loyalty to an airline and may pay a premium for their preferred choice. On the other hand, almost all early revenue management models did not explicitly consider competitive effects and assumed that an airline's demand only depends on their prices i.e., demand is fully dedicated to an airline (loyal). This paper develops a model to capture more realistic competitive dynamics by including both these types of customer behavior. We also develop a Bayesian machine learning based demand forecasting methodology for such models with explicit competitive considerations and show the benefit of this approach over traditional models on a real airline data set.

Keywords Bayesian inference · Dynamic pricing · Competitive modelling · Continuous pricing

Introduction

The bulk of revenue management models implemented in practice do not explicitly account for competitors in the market. Instead, the models rely on capturing competitive effects implicitly, which can be reasonable in an environment where markets and the competitive strategies deployed by the market participants are stable. This is not to say that airlines do not realize that there is competition, rather, they choose to use models that are simpler and require less informational sources. We will refer to those models as 'monopolistic' models. The tendency towards monopolistic models is also observed in the published academic research in the area of revenue management (RM) and dynamic pricing (DP). The reasons for this state of affairs were multi-fold. In our opinion, apart from the relative sufficiency of simpler models in a stable competitive environment, the main reason was the lack of reliable and up-to-the minute price information about competitor pricing. While web-scraping has existed for at least two decades, it was not until relatively recently that we saw emergence of third parties providing good quality information about how competitors are pricing in the multitude of markets. There also might have been a legal angle to what can be collected from the web and utilized. Furthermore, the main feature of the existing micro-economic theory was, and is, that demand drops off to zero once an airline prices even a little bit above the competitor, clearly a characteristic that is only rarely observed in practice. The game-theoretic models also tend to be static, which does not fit well with how the airlines book customers. Namely, the existence of the booking horizon necessitates multi-period modelling and thus significantly increases the level of complexity. And even if the level of complexity could be handled with computational advances, the equilibrium solutions proposed by the

✉ Ravi Kumar
rkumar04@pros.com

1 PROS Inc, Suite 600, 3200 Kirby Dr, Houston, TX 77098, USA

2 Etihad Airways, Khalifa city A, PO BOX 35566, Abu Dhabi, UAE

Chapter 5 was originally published as Kumar, R., Wang, W., Simrin, A., Arunachalam, S. K., Guntreddy, B. R. & Walczak, D. Journal of Revenue and Pricing Management (2021) 20:256–275. https://doi.org/10.1057/s41272-021-00311-4.

theory appear drastic and not immediately implementable in the real-life.

Faced with sometimes missing or questionable competitive information and models that make drastic assumptions airlines resorted, as is typical, to business rules. One can guess that the most prevalent one was to match the competitive offering: both in fare class products and often in price. The reliance on business rules to handle competitive situations grew into certain inertia in the industry: while every airline and vendor were extolling the virtues of competitive modeling very few actually implemented any models that offered a level of sophistication above typical business rules.

The aim of this work is to propose an intuitive demand model that utilizes information about competitors' actions. Its main feature is that we extend traditional microeconomics models by allowing for the following customer behavior: buying from the airline in question even if its price is higher than competitors' prices. And we mean 'higher' in a strong sense, i.e. potentially significantly higher. In other words, we assume that there is a component of demand that is price-sensitive but loyal to the airline. In contrast to that we will refer to customers who are seeking the lowest price in the market as the fully flexible customers. There could be a number of reasons why customers are loyal to an airline, and membership in its frequent flyer program definitely stands out, especially for customers who live near the airline's hub and for whom the frequent flyer miles can be very useful in getting to many destinations world-wide. Other reasons may include level of service, safety record, on-time arrivals, and reliability.

Our model is formulated for single demand forecasting entity such as an O&D market or a single-leg flight. Therefore, its main concept extends easily, if not the effort needed to solve it, to general airline network setting. We cast the concept of loyal customers in two versions: one where an existing class structure is present and airlines close or open classes as they see fit, and perhaps the more relevant one nowadays where airlines price dynamically in the absence of fare classes, i.e. where we assume class-free context and the RM system produces a single price in response to customer request.

There are many other important aspects that may impact an airline's demand, such as spill between flights for a given departure date or across itineraries for a given origin-destination pair. Many of these aspects can be handled by current generation of demand forecasters through various mechanisms. Spill between flights can be addressed, for instance, by properly grouping flights into departure time windows and modeling customer preferences for flights within a departure time window using choice models or considering departure time of a flight as a covariate in the demand forecasting model. A similar approach can be taken to model customer choice across itineraries for a given origin-destination pair. The focus of our work in this paper is mainly on bringing competitor price awareness to airline demand forecasting and we do not discuss some of these other aspects further.

The article is organized as follows, by section:

- Motivation and Literature Review
- Review of relevant demand models that include competitive information
- Class-based forecasting with market information
- Class-free model with loyal customers
- Numerical analysis and forecast accuracy study.

Motivation

Historically, the competitive information has not been used explicitly in RM. Abstracting out, for the sake of argument, from the lack of reliable competitive information, we posit that this was because available models were inadequate. Certainly, RM on its long voyage from models assuming independent-class behavior to more realistic demand models has been more focused on modeling price-sensitivity of customers first, with the inclusion of competitive behavior more of an afterthought until that first hurdle is overcome (as we all can guess it is still work in progress at many airlines). The very real and immediate needs of accounting for customer choice between competitors were handled by the business rules based pricing strategies.

In general sense, the business rules, or ways of accounting for competitive effects were created and imposed from one level up from the tactical-in-nature RM: the airline pricing departments. These would utilize various business intelligence analytics that provided insights about competitors' operations in a market, fare products that were offered (level of service etc.), and fares filed with Airline Tariff Publishing Company (ATPCO, airlines typically distribute fares to Global Distribution Systems worldwide by filing fares to ATPCO databases) to determine how to respond in terms of flight frequencies, product offers, and class fares. The role of RM was then to control on a tactical level which price points or even products to close as a function of estimated future demand for the airline. Clearly, the controls exercised by the RM system were much more frequent than the more strategic in nature decisions by the pricing department.

This is not to say that RM has historically nothing to say about responding to competitive actions, but these were often an application of a business rule that was implemented as part of airlines strategy. For example, on market AAA-BBB, always match competitor X. So, once some information about competitor X's pricing was available, the airline might adjust the open classes to match what that competitor was doing.

In relatively stable markets, with competitive effects being managed by rule-based price-matching strategies, models that did not explicitly use competitor price information still made sense especially if reliable competitive price information is not available. However, with the increasing complexity of airline offerings, easy availability of good quality competitor price information and increased focus on automating pricing strategies, including competitor price awareness in RM and pricing models has become important. Moreover, with important advances made in the modeling the price-sensitive customers, the next logical step is to further refine demand models with the assumption that some, if not all, customers are sensitive to competitors' prices too. It is a step that is a necessary one to take too: we know from Cooper et al. (2015) that for airlines competing in a market, forecasting and optimizing as monopolists may not bring them to any sort of equilibrium that they will be satisfied with. That is, there is no guarantee that the "Market Response Hypothesis" that monopolistic models will capture competitive effects if the models are estimated with historical data collected under competition holds.

Developing equilibrium RM and pricing policies considering the game-theoretic impact under competitor price-aware demand models is another important and open challenge for both industry and academia. In this work we do not go all the way to that level of sophistication. However, as an intermediate step, we discuss the implication of utilizing competitor price-aware demand model on availability and pricing policies under the assumption that the competitors do not have the capability to immediately respond to the price that the airline is charging to the booking client. We do realize that this is not a typical setting when modeling competition but at the same time we believe that it is very realistic: competing airlines do not react immediately to changing prices and the time lag that is necessarily involved means that an exogenous customer who is making a request will not be going back and forth between competitors while making a booking. We show that with this assumption, the competitor price aware availability and dynamic pricing policies under the proposed demand model show certain intuitive properties which make this model desirable from a practical perspective.

Therefore, taking that next step where we try to include available competitor information is necessary if our objective is to improve revenues.

Review of demand models that include competitive information

In the literature that is available on the subject we can observe a certain dichotomy. On the one hand, there is the assumption in the monopolistic setting that all or some customers are price-sensitive but without the concept of switching to a competing airline; on the other hand, the game theoretic literature assumes that all customers switch if there is a competitor that has a price that is even a proverbial epsilon smaller. So, either all price-sensitive customers always buy from the airline (monopolistic setting) or all switch (typical game-theoretic setting).

Motivated by the airline practice where we know that there exist product-oriented as well as price-oriented customers (Boyd and Kallesen 2004), we next introduce another demand component: customers that are sensitive to the lowest price in the market. So, the model (in its class-based setting) will have customers who are loyal and want only to buy a particular fare class product from an airline, those who are loyal but only will buy the cheapest product the airline has to offer, and those who always seek the cheapest airfare in the market.

Thanks to this new demand component, we will see that the customer behavior does not have to be the "all or nothing" as is often seen in game-theoretic models, which allows for a more realistic modeling where there will be at least some demand even if the airline's own price is higher than competitors'. In other words, we allow for a drop in demand (discontinuity) but to some possibly nonzero value. This behavior will be emphasized, and the airline may choose to price higher than competitors, if the airline's estimated future marginal displacement cost ('the bid price') becomes higher.

Several discrete customer choice models can include the choice of buying from competitors, either implicitly as part of the 'no purchase' choice, or as an explicit option in itself (Fiig et al. 2019). However, Multinomial logit model and its various flavors, the most popular in this category, may still not be able to model the discontinuity in behavior.

Our model, particularly the one based on an existing class structure can be thought of as an extension of the 'lowest available class' model (Talluri and Van Ryzin 2004a) expanded to include the product-oriented as well as the fully flexible customers. To estimate these intuitive and realistic models we chose the Bayesian inference methodology that makes forecasting robust and relatively easily implementable.

There is a limited number of publications available as far as equilibrium pricing in the setting that is relevant to RM where demand arrives over a finite booking horizon and buys from finite inventories of competitors (often assumed symmetric); this necessitates subgame perfect equilibrium analysis and makes useful results harder to come by. In this category, Dudey (1992) offers results for the case where customer willingness to pay ('WTP', or customer reservation price) is known to the competitors and the market size (number of customers) is deterministic. Under these assumptions the structure of equilibrium is such that the competitor with less capacity sells out first, followed by the competitor with more capacity (who then becomes a monopolist for a possibly short amount of time). Given this structure it is advantageous to start with less capacity and the model may lead to negative prices for some demand and capacity scenarios. Similar results were later obtained in Martínez-de-Albéniz and Talluri (2011) allowing a random market size. More recently, Singh (2019) confirmed the same equilibrium structure when the customer WTP is random (and unknown to both competitors) and the market size is also random. Under the assumption of an exponentially distributed WTP with no discontinuity when a competitor undercuts, similar conclusions were reached (Isler and Imhof 2008).

The setting with random WTP and random market size with customers arriving sequentially is closest to what we need to realistically model competitive phenomena in RM, and it was natural to expand it to consider loyal demand in hopes of obtaining a more realistic equilibrium structure. Such a relatively straightforward extension was investigated (Singh and Walczak 2019), where it was shown that pure strategy equilibrium may fail to exist in general when a non-negligible proportion of demand is loyal. We therefore believe that investigating competitive demand models where competitors do not necessarily follow any type of equilibrium strategies is important and relevant to the industry. At the core of any such investigation is determining how sensitive the airline's demand is to competitors' prices and this is what we set out to achieve in this work.

Modeling contexts

There are two main modeling contexts when demand is price-sensitive (both extend naturally to fare families). The class-based one where we can accommodate not only an existing fare class structure but also the customers who are product-oriented (Boyd and Kallesen 2004) constituting the so-called yieldable or traditional demand component, and those who are price-oriented, cf. *ibid.*, captured in the priceable demand component and, effectively, modeled by a discrete demand curve with price points corresponding to class fares.

The two demand components within the class-based context are assumed loyal to the airline and only consider its products, and so are not sufficient to model the behavior where customers seek the lowest price in the market. To account for those fully flexible customers we introduce a third demand component called market priceable. Our class-based model will be the hybrid of the three components, and we will refer to it, accordingly, as Market Hybrid. To summarize we have three demand components in Market Hybrid:

1. Yieldable, product-oriented and loyal
2. Priceable, price-oriented and loyal
3. Market Priceable, price-oriented and never loyal (fully flexible).

The other modeling context considered in this work assumes no existing fare class structure and we call, appropriately, class-free. Here, the price can change continuously within an interval (a possibly unbounded one). Demand is exclusively price-oriented and buys the lowest fare offered for a product such as a fare family. Proceeding in a manner analogous to the class-based case we will assume that there is a proportion of customers, who we call loyal, that will buy from the airline even if competitors in the market offer lower price. In what follows we first analyze the class-based context and then follow with the class-free model and report on the forecast study results.

We want to emphasize that competitive market information such as lowest price available in the market is crucial in either context. In the numerical study section we present an analysis, based on real airline data, to determine the relevant market reference price.

Class-based Bayesian learning demand model

In the section we sketch out how Bayesian learning improves forecasting of the Market Hybrid demand model. We base our analysis on the Bayesian Dynamic Linear Model with covariates ('Bayesian DLM') that we introduce below. It is a well-established and well-researched framework that allows (among other things) for sequential forecast updates as more data becomes available. There are connections between it and other models; for example, the updates can be shown equivalent to filtering equations in the Kalman filter approach that is based on minimizing variance. West and Harrison (1997) provide a good discussion on the subject.

Class-based forecasting with market information (market hybrid)

Forecast dimensions determine the entities for which forecasts are produced. These may differ depending on whether an airline is a network carrier or a leg-carrier, with either origin & destination (OD) or a flight leg being the main dimension, respectively. Other dimensions typically are: days prior to departure or a data collection period (DCP), day of week (DOW), departure date, departure time or departure time window, and possibly others, e.g. point of sale (POS) or point of commencement (POC). For ease of exposition, in what follows we will assume that an OD, or a market has been fixed, along with other relevant dimensions, so that we do not have to refer explicitly to those.

Assuming that dimensions of a given forecasting entity have been fixed, we will now describe, at a very intuitive level, what data we need in order to model all three demand components within the Bayesian DLM. Unlike the class-free model discussed later, here the classes serves as tiers, and the model is agnostic of what particular class fares are. This is useful in modeling when customers compare product price tiers within the airline itself and in the market.

We start with the yieldable, or the product-oriented demand. To unconstrain and create observations for each class we need to know percentage of time that this class was open in a DCP (Open%). The price-oriented, or priceable, demand observed for each class depends on class availability for all classes at or below its class fare value. In other words, we need to know the percentage of time within DCP that the class was the lowest available in the DCP (LA%). We can produce the forecast either as a cumulative number, i.e. what demand to expect when the class is the lowest open for 100% of the DCP or its increments. In the latter case the interpretation is that the (incremental) priceable forecast is the expected number of customers willing to pay the class fare but not the fare of the class immediately higher in hierarchy.

For the new third component of demand, the so-called market priceable demand, what matters is whether the class was the lowest available in the entire market. This is in contrast to the yieldable and the priceable components which model two different flavors of customer behavior, but who still remain loyal and will not switch to competitors. Thus, here we need to know the percentage of time in the DCP where the class was the lowest available in the market (MLA%). An incremental market-priceable forecast should be understood as those customers who are willing to pay the class fare only if it is the lowest available in the market (but not the fare of the class above it in hierarchy, should that become the lowest available in the market). Depending on a market (be it leg, or OD) different demand components may become prominent.

Bayesian dynamic linear model

In Bayesian approach to forecasting we start with a prior belief about model parameters μ and as observational data x (assumed distributed according to $f(x|\mu)$) becomes available we use it to update our belief $\pi(\mu)$ in a rational fashion that is captured by the Bayes Rule:

$$\pi(\mu|x) = \frac{\pi(\mu)f(x|\mu)}{\int \pi(\mu)f(x|\mu)d\mu}.$$

Once our beliefs have been updated with the most recent data, to forecast we integrate over the uncertainty in μ using its current distribution $\pi(\mu)$ and obtain the predictive distribution $m(x)$:

$$m(x) = \int f(x|\mu)\pi(\mu)d\mu.$$

The predictive distribution captures both the data variation as well as the uncertainty in the parameters. Effectively, it computes a weighted average over all parameter values using the updated posterior belief as the weighing distribution.

While departing from the classical "plug and play", what we obtain here is not just a point forecast but a proper probability distribution of future values that provides a measure of randomness (data) and uncertainty (in parameters) and allows us to derive many useful statistics about forecasted quantities including confidence bounds.

The Bayesian learning framework provides a very intuitive learning process. As well, the process is storage and runtime efficient since we only need to store the most recent system state and apply incremental new data to recalculate the updates; in other words, it learns as it goes along. It also provides for a much-needed stability and robustness and as long as the priors are reasonable (at least vaguely resembling reality) it precludes bad forecasts.

More specifically, we assume that all we know about the random quantities (variables) we set out to forecast is captured by a vector of parameters ¯. Our uncertainty about the vector is encapsulated in a probability distribution on the ¯ and it is this distribution that gets updated as we make new observations. Knowing updated ¯ (the posterior) and its associated variances, we can obtain the predictive distribution of a future value of our dependent variables at all future times.

The particular flavor of Bayesian Learning framework that we focus on here is the Bayesian Dynamic Linear Model (DLM) that builds on the simple Bayes Rule by allowing dependence on covariates, that is, utilizing other random variables that may contain information about the quantities we want to forecast. The 'Dynamic' part of the model refers to the fact that we allow the distributions of parameters to be non-stationary, i.e. we allow them to change with time. One can refer to West and Harrison (1997), Chapter 4, Section 1 for the following distinguishing features of the Bayesian DLM:

1. Parametric models with meaningful dynamic parameters
2. A probabilistic representation of information about parameters
3. A sequential model definition utilizing conditional independencies
4. Robust conditionally independent model components
5. Forecasts derived as probability distributions
6. A facility for incorporating expert information
7. Model quality control.

The Bayesian DLM is a large and comprehensive framework and we do not attempt to present here its full underlying mathematical theory. Instead, we focus on its covariate-modeling aspect that we utilize in estimating the three demand components. The high-level introduction that follows is taken from Chapter 16 in West and Harrison (1997) that will guide our explanations to come.

Let $\mathbf{X_t}$ be a vector of r observations at time t of a series that follows the DLM, and let D_t be the vector of data available at time t. The model is described by two linear relationships, the observational equation 1 and the system equation 2, along with initial information $(\boldsymbol{\mu}_0|D_0)$ that is independent from the two error sequences.

$$\mathbf{X}_t = \mathbf{F_t'}\boldsymbol{\mu}_t + \nu_t,\ \nu_t \sim N[\mathbf{0}, \mathbf{V}_t], \tag{1}$$

$$\boldsymbol{\mu}_t = \mathbf{G_t'}\boldsymbol{\mu}_{t-1} + \omega_t,\ \omega_t \sim N[\mathbf{0}, \mathbf{W}_t], \tag{2}$$

$$(\boldsymbol{\mu}_0|D_0) \sim N[\boldsymbol{\theta}_0, \mathbf{C}_0]. \tag{3}$$

The observational and system noise sequences, respectively, ν_t and ω_t are assumed to be independent and mutually independent and $\boldsymbol{\mu}_t$ will be referred to as the parameter vector. The first equation thus describes how our observations depend on the state, and the second equation describes the evolution of the system. We note that both equations are linear since the dynamic regression matrices $\mathbf{F}_t$ and the state evolution matrices $\mathbf{G}_t$ correspond to linear operators. Thus the vector of observations depends linearly on covariates (regressors). For our setting, the dependent variable $\mathbf{X}$ that we observe will be bookings for each forecasting entity and ultimately we will make them depend on seasonality as well as availability covariates.

The quadruple $\{\mathbf{F}, \mathbf{G}, \mathbf{V}, \mathbf{W}\}_t$, specified for each t, characterizes the general normal dynamic linear model (DLM). It relates the X_t to the parameter vector $\boldsymbol{\mu}_t$ through the following distributions:

$$(X_t|\mu_t) \sim N[\mathbf{F_t'}\boldsymbol{\mu}_t, \mathbf{V}_t] \tag{4}$$

$$(\boldsymbol{\mu}_t|\boldsymbol{\mu}_{t-1}) \sim N[\mathbf{G_t'}\boldsymbol{\mu}_{t-1}, \mathbf{W}_t]. \tag{5}$$

The generality and flexibility of the Bayesian DLM builds on the assumption of normally distributed random variables involved in the system. Modelling bookings requests as arriving according to a Poisson distribution is common in modern RM system, especially on the network since it is well suited for handling forecasting entities that see small number of observations, such as bookings for some (origin, destination, itinerary, fare class) combinations. Therefore, additional work is needed to make the booking data look normal and we achieve it through the well-known Box-Cox transformation. The power of the DLM makes this additional effort a good tradeoff.

Update equations

Once the initial prior information is given at $t = 0$, for any subsequent t the information set is $D_t = \{X_t, D_{t-1}\}$. Following Theorem 4.1 of West and Harrison (1997) (albeit with a slightly changed notation) we can describe how the learning about system parameters happens through updates (for the ease of exposition univariate DLM is considered here, cf. *ibid.* Theorem 16.1 for the multivariate version):

- Posterior at $t-1$: $(\boldsymbol{\mu}_{t-1}|D_{t-1}) \sim N[\boldsymbol{\theta}_{t-1}, \mathbf{C}_{t-1}]$, for some mean $\boldsymbol{\theta}_{t-1}$ and variance matrix $\mathbf{C}_{t-1}$.
- Prior at $t-1$: $(\boldsymbol{\mu}_t|D_{t-1}) \sim N[\mathbf{a}_t, \mathbf{R}_t]$, where $\mathbf{a}_t = \mathbf{G}_t\boldsymbol{\theta}_{t-1}$ and $\mathbf{R}_t = \mathbf{G}_t\mathbf{C}_{t-1}\mathbf{G}'_t + \mathbf{W}_t$
- One Step Forecast: $(X_t|D_{t-1}) \sim N[f_t, Q_t]$ where $f_t = \mathbf{F}'_t\mathbf{a}_t$ and $Q_t = \mathbf{F}'_t\mathbf{R}_t\mathbf{F}_t + V_t$.
- Posterior at t : $(\boldsymbol{\mu}_t|D_t) \sim N[\boldsymbol{\theta}_t, \mathbf{C}_t]$, with $\boldsymbol{m}_t = \mathbf{a}_t + \mathbf{A}_t e_t$ and variance matrix $\mathbf{R}_t - \mathbf{A}_t Q_t \mathbf{A}'_t$, where $\mathbf{A}_t = \mathbf{R}_t\mathbf{F}_r Q_t^{-1}$ and $e_t = X_t - f_t$.

This very general and flexible framework underlies the time-series aspect of the Bayesian learning that we present both for the class-based and class-free settings. The dependent variable $\mathbf{X}_t$ in the former case will be the bookings observed and in the latter it will be unconstrained (detruncated or uncensored) parameter observations constructed from the observed bookings and prices offered. As we explain in the next section, in the class-based setting we will introduce the availability covariates to control for the effects of class closures directly and thus will not need a self-standing unconstraining algorithm.

In the simplest case with no covariates, the matrices $\mathbf{F}$ and $\mathbf{G}$ will just be the identity matrices of appropriate dimensions, and the covariance matrix $\mathbf{V}_t$ will be of the form $\sigma^2 I$, where I is an identity matrix. If the time dynamics and hierarchical dependence are not modelled, the system covariance matrix $\mathbf{W}_t$ can also be assumed to be constant and of a diagonal form. In the presence of covariates all these will become more complex.

A number of further assumptions can be made about the observational and system covariances in an increasing order of complexity. For us the most important aspect is to account for system parameters possibly changing in time. This is a very realistic requirement and we achieve that by introducing the drift into the model where the system variance grows linearly with time. In other words we assume that the parameter vector is disturbed with noise of zero mean and with variance multiplied by a factor of k where k is the number of steps ahead in the future. This model feature ensures that the uncertainty in the parameters does not converge to zero as the number of observations increases but instead asymptotically results in an exponential smoothing-type relationship.

Last but not least, it is often desired to apply conjugate theory to assure that posterior updates are of the same form as prior with only the parameter vector (the state now) changing. For the discrete case this is exactly what we will use by assuming that all distributions involved are normal. When we consider the class-free model we will face an added difficulty of not having a closed form posterior updates and will have to resort to the Variational Bayesian inference. We also want to point out that in a general case with a discrete class structure the normal assumption is not limiting: in practice, applying Box-Cox transformation to booking data that most of the time is consistent with being Poisson distributed enables us to keep things within the normal domain.

Availability covariates

To summarize, the three availability-type covariates that capture customer behavior with respect to product, price, as well as competitive prices are:

- Open%, the fraction of time a class was open in a DCP
- LA%, the fraction of time a class was lowest available in a DCP (ignoring competition)
- MLA%, the fraction of time a class was the lowest available in the market in a DCP.

To see the role the three availability scenarios play, it helps to look at the following availability scenarios for any class:

1. Open but not the lowest open, in which case all the bookings observed are yieldable
2. Open and the lowest open, where both yieldable and priceable bookings are seen
3. Open, lowest open, and the lowest open in the market (among the competitors in the market) to observe bookings in all three demand components.

We will formalize the notation now. Let t be the departure date and let the vector of observed bookings for departure date t be $X_t = (x_{t,1}, x_{t,2}, \ldots, x_{t,n})$, where $x_{t,i}$ are the bookings for t and DCP i, $i = 1, 2, \ldots, n$.

For each of the three covariates at the level of departure time t we observe three covariate vectors:

$$\begin{aligned} Open_t &= (Open_{t,1}, Open_{t,2}, \ldots, Open_{t,n}) \\ LA_t &= (LA_{t,1}, LA_{t,2}, \ldots, LA_{t,n}) \\ MLA_t &= (MLA_{t,1}, MLA_{t,2}, \ldots, MLA_{t,n}), \end{aligned}$$

as well as the realized bookings vector X_t. We can now set the main ingredients of our model as follows.

$$\mathbf{X}^* = (\mathbf{X}, \mathbf{C}) = \begin{bmatrix} DCP_1 \begin{cases} x_{t,1} \\ Open_{t,1} \\ LA_{t,1} \\ MLA_{t,1} \end{cases} \\ \ldots \\ DCP_n \begin{cases} x_{t,n} \\ Open_{t,n} \\ LA_{t,n} \\ MLA_{t,n} \end{cases} \end{bmatrix} = \begin{bmatrix} \begin{bmatrix} x_{t,1} \\ \ldots \\ x_{t,n} \end{bmatrix} \\ \begin{bmatrix} Open_{t,1} \\ \ldots \\ Open_{t,n} \end{bmatrix} \\ \begin{bmatrix} LA_{t,1} \\ \ldots \\ LA_{t,n} \end{bmatrix} \\ \begin{bmatrix} MLA_{t,1} \\ \ldots \\ MLA_{t,n} \end{bmatrix} \end{bmatrix}$$

$$= \begin{bmatrix} \mathbf{X_t} \\ \mathbf{Open_t} \\ \mathbf{LA_t} \\ \mathbf{MLA_t}. \end{bmatrix}$$

The prior information is contained in the system parameter vector θ_0 where we group its components first starting with prior information about the data distribution (θ_0^X)and then the three availability covariates,

$$\theta_0 = \begin{bmatrix} \theta_0^X \\ \theta_0^{Open} \\ \theta_0^{LA} \\ \theta_0^{MLA} \end{bmatrix} = \begin{bmatrix} \begin{bmatrix} \theta_{0,1}^X \\ \ldots \\ \theta_{0,n}^X \end{bmatrix} \\ \begin{bmatrix} \theta_{0,1}^{Open} \\ \ldots \\ \theta_{0,n}^{Open} \end{bmatrix} \\ \begin{bmatrix} \theta_{0,1}^{LA} \\ \ldots \\ \theta_{0,n}^{LA} \end{bmatrix} \\ \begin{bmatrix} \theta_{0,1}^{MLA} \\ \ldots \\ \theta_{0,n}^{MLA} \end{bmatrix} \end{bmatrix}$$

On a high level, the covariance matrix can be broken down into blocks of covariances between each group of covariates, which we further consolidate into blocks involving $\mathbf{X}$ and then all the covariates together ('$Covs$') as shown below:

$$\Psi = \begin{bmatrix} \psi_X & \psi_{X,Open} & \psi_{X,LA} & \psi_{X,MLA} \\ \psi_{Open,X} & \psi_{Open} & \psi_{Open,LA} & \psi_{Open,MLA} \\ \psi_{LA,X} & \psi_{LA,Open} & \psi_{LA,LA} & \psi_{LA,MLA} \\ \psi_{MLA,X} & \psi_{MLA,Open} & \psi_{MLA,LA} & \psi_{MLA,MLA} \end{bmatrix}$$

$$= \begin{bmatrix} \psi_X & \psi_{X,Covs} \\ \psi_{Covs,X} & \psi_{Covs,Covs} \end{bmatrix}.$$

In our overview here, the focus is only on the availability covariates but there is no restriction as to what types of covariates can be used in general. For example, another type of covariates that is very useful in RM forecasting are the seasonality covariates, often in the form of finite Fourier series with their frequencies corresponding to annual, semi-annual number, as well as smaller cycles.

For the sake of clarity of exposition, we skip the more complex technical details of the update equations for our model with covariates and with a hierarchical structure assumed for DCPs. However, once the new incremental observations have been used for the updates, the expected bookings given covariate values can be expressed succinctly on a high-level in the following equation.

$$E[X_t|Covs_t = c] = \theta_t^X + \psi_{X,Covs} \times (\psi_{Covs,Covs})^{-1}(c - \theta_t^{Covs}).$$

The availability covariates each range from 0 to 100% and for each forecasting entity we will have the record of bookings (by class) for a given DCP along with fractions of the DCP during which the class was open, lowest open, and lowest open in the market. The model assumes that the rates of arrivals are uniform within each DCP, for instance if class Q were open for 50% of a given DCP then we would expect to see half of the forecasted bookings as compared to when the class was open for 100% of time (availability of other classes being unchanged).

Majority of existing RM models consider demand as exogenous but the actual data records only include bookings, that is, we know what was booked but not necessarily all the requests that had arrived. In other words, we only see that exogenous demand that is recorded through the lens of the RM controls that were applied. So, in order to arrive at good estimates of what the demand 'out there' will be one has to undo the effect of controls that were in place when bookings were happening. This process is often called unconstraining, uncensoring or detruncation of the observed (booked) demand. Many practical techniques have been developed for it, cf. Zeni (2001), including an application of the famous EM algorithm, cf. Talluri and Van Ryzin (2004b).

In the class-structure setting considered in this section we circumvent the process of explicitly unconstraining by estimating the correlation structure of bookings with respect to the three availability covariates. This process allows us to predict demand at various combinations of the availability covariates including the case when the class was 100 % open, lowest available or lowest available in the market which is akin to calculating the full ('unconstrained') demand. We describe these calculations in more detail next. Also note, that the availability covariates are the results of a control policy and the range of covariate values seen in the history will depend on it. Thus the accuracy of the estimates of correlation relationship between bookings and the three availability covariates depends to some extent on the variability in these covariates under the implemented control policy.

Once the model has learnt about the correlations between availability covariates and the bookings we are ready to feed the forecasts into optimization. To this end we need to know unconstrained demand from all components which we obtain by conditioning the joint distribution of the booking variable and covariates ($X, Open\%, LA\%, MLA\%$) on particular values of the three availability covariates, $E[X|C = c]$, and then by performing simple calculations.

For the unconstrained yieldable forecasts, we set $Open\% = 100\%$, and the other two to zero: $LA\% = 0$, $MLA\% = 0$. Having retrieved that forecast we subtract it from what we obtained by keeping the $Open\%$ still at 100% and setting the lowest available to 100%, ($LA\% = 100\%$), and the MLA% to zero.

Finally, in order to arrive at the market priceable forecast we set all three availability covariates to 100% ($Open\% = 100\%, LA\% = 100\%, MLA\% = 100\%$) and subtract from it the previously obtained yieldable and priceable forecasts. A sample forecast structure for a single entity is in Table 1.

Table 1 Market hybrid forecast example

Forecasts	Fare	Yieldable	Priceable	Market priceable
Y Class	\$1000	7	5	1
B Class	\$600	10	7	2
M Class	\$450	21	13	21
Q Class	\$250	20	5	12

The knowledge of the three components of demand provides us a more precise picture of the magnitude of each of those building blocks of the behavior and with proper optimization model what to expect and how to control it. To give a broader context, most economics literature assumes in their modeling that demand is exclusively what we termed market priceable. That is the demand always buys from the competitor offering the lowest price and with various rules on how to split the demand if prices (effectively) match. In the simple example of demand forecast in our framework that typically modelled demand would be just the last column.

Certainly, any combination of components is possible which speaks to the flexibility of the model. The Bayesian forecaster can be configured to reflect that. For example, if it is known that all customers are price-oriented, then the yieldable component could be removed, and we will not have to collect the $Open\%$ information in the system.

Market hybrid examples

To further elucidate the basic concepts of the demand model it is useful to look at how these perform in practice. We will

Table 2 Forecast and expected revenue example

Forecasts	Fare	Yieldable	Priceable	Market priceable	Nest open	Expected revenue
Y Class	$1000	7	5	1	Y	$12,000
B Class	$600	10	7	2	Y, B	$20,200
M Class	$450	21	13	21	Y,B, M	$44,500
Q Class	$250	20	5	12	Y,B,M,Q	$43,950

consider several very simple examples below. The examples are static in nature, that is, we do not account (yet) for decisions made later, we involve only one forecast entity, and we do not include capacity constraints (Wang et al. 2019).

Market positions and optimal actions

We continue analyzing the model and its features in a simplified world. The data used in these examples are actual forecasts of the yieldable and priceable demand components for particular OD market, point of sale ('POS'), and departure date. To protect anonymity, we have adjusted the market priceable component by making it as small as possible but still important enough to show how different optimal decisions can arise depending on how it changes. We think it emphasizes the power of the model.

We begin by assuming that the lowest price in the market is $500, and that it is known and not immediately changing in response to airline's actions. As we discussed in the Introduction, this is a reasonable and practical assumption that is key to the analysis in this work. With class fares as given, we won't capture the market priceable component if we only open class Y, or class Y and B. We will capture it though if we open classes Y, B, and M together, or open all classes.

Static and deterministic market scenario

In the next equally simple extension of the market hybrid forecast example, we look at potential optimal decisions given the forecasts. We amended Table 1 by adding two columns: one that explicitly spells out classes that are open for each nest and the other with expected revenue corresponding to each open class nest. For completeness, below we show a sample calculation of the expected revenue based on the new Table 2, note that all demands listed are incremental.

1. Y lowest open: E[Revenue] $= (7+5)*\$1,000 = 12,000$
2. B lowest open: E[Revenue]$= 7*\$1000 + (10+5+7)*\$600 = \$20,200$
3. M lowest open: E[Revenue]$= 7*\$1000 + 10*\$600 + (21+5+7+13+1+2+21)*\$450 = \$44,500$
4. Q lowest open: E[Revenue]$= 7*\$1000+10*\$600+21*\$450+(20+5+7+13+5+1+2+21+12)*\$250 = \$43,950$.

Table 3 Expected demands under dominant market position

Open Nest/ Expected demand	Y	B	M	Q
Y	2.91	0	0	0
Y, B	2.58	4.75	0	0
Y, B, M	2.58	3.77	3.24	0
Y, B, M, Q	2.58	3.77	2.26	1.28

We can read off from Table 2 what the optimal action of a revenue maximizer would be if we assumed the same $500 market price. In this particular sample scenario the expected revenue-optimal decision under this market scenario is to open classes Y, B, and M. The optimal decision will change depending on the lowest available price in the market.

Dominant market position

In some markets there are well-established, dominant carriers, due to brand recognition, quality of service, scheduling, and capacity offered. In such markets we expect the market priceable component of demand to be relatively small compared to the airline's priceable and yieldable components.

Now assume that there is an aggressive Low-Cost Carrier in the market and that the market fare is at $360. Under this competitive price scenario, we have calculated the expected revenue from each feasible action (open nest of classes).

We can see that in this scenario the market priceable demand is captured only if we open all classes down to Q. Opening Q, though, will cause all the priceable demand to book at Q class fare which will contribute to dilution. Again, for completeness, we have listed expected bookings by class resulting from each feasible action in Table 3.

The optimal decision is to open Y, B and M, closing Q, as can be easily seen from Table 4. In this case, the extra demand that could be generated by matching in the market does not overcome the fare dilution of our priceable demand. The last point cannot be overemphasized: with enough demand in the loyal components matching or undercutting the competitor can be sub-optimal.

Table 4 Actual forecast and revenue example

Forecasts	Fare	Yieldable	Priceable	Mkt priceable	Nest open	Expected revenue
Y	603	2.58	0.33	0.02	Y	1751.90
B	529	3.77	0.65	0.07	Y, B	4067.32
M	464	2.26	0.00	0.00	Y, B, M	5050.96
Q	351	0.00	0.18	0.04	Y, B, M, Q	5045.83

Table 5 Actual forecast and revenue example, medium market position

	Fare	Yieldable	Priceable	Mkt Priceable	Nest open	Expected revenue
Y	603	2.58	0.33	0.03	Y	1,751.90
B	529	3.77	0.65	0.10	Y, B	4067.32
M	464	2.26	0.00	0.00	Y, B, M	5050.96
Q	351	0.00	0.18	0.05	Y, B, M, Q	5069.42

Medium market position

In this example we increase, but only slightly, the amount of the expected market priceable demand. As the Table 5 illustrates the additional demand that can be generated by matching the market does overcome the fare dilution from the (loyal) priceable demand. This fact will only become exacerbated if we include the marginal capacity value (the 'bid price') in the consideration, i.e. if we drop the simplifying zero bid price assumption.

Class-free demand model with competitor information

In this section we present an approach for modeling customer behavior and demand modeling for an airline under competitive consideration in a class-free setting. Similarly to the class-based model, here we also explicitly model the fact that some of the customers are loyal to the airline, while the remaining customers are fully flexible, and will choose the lowest priced product available when they arrive (i.e. they constitute a market-priceable demand component in the terminology introduced in the previous section).

Furthermore, we assume that every customer is price-sensitive within a differentiated product (seat with privileges and restrictions) offered by the airline for the itinerary under consideration e.g., fare-families such as Basic, Flex, Premier etc. In particular, we assume that the probability of purchase for a differentiated product is a decreasing function of the price charged and depends on the willingness-to-pay of arriving customers.

We allow for the fact that the loyal and flexible customers' willingness-to-pay distributions may be governed by different parameters. In particular, we assume that the willingness-to-pay distribution of the loyal customers is exponentially distributed with parameter β^l while the willingness-to-pay

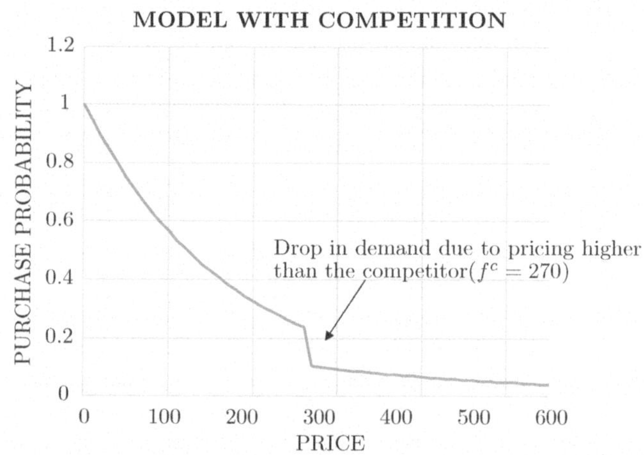

Fig. 1 Purchase probability for the competitive model with loyal and market priceable customers

distribution for the market-priceable customers is exponentially distributed with parameter β^m. We also assume that the fraction of loyal customers is δ i.e., an arriving customer is loyal to the airline with probability δ. Since any customer may be loyal or flexible, the probability of purchase by a given customers is not continuous in the fare charged by the airline, f, given that the competing fare is f^c. Figure 1 shows this behavior where demand drops off when the airline starts charging a fare higher than the competitor ($f > f^c = 270$).

The resulting probability that an arriving customer will purchase a seat is

$$\pi(f, f^c) = \begin{cases} \delta e^{-\beta^l f}, & \text{if } f > f^c \\ \delta e^{-\beta^l f} + (1-\delta)e^{-\beta^m f}, & \text{if } f \leq f^c \end{cases}, \quad (6)$$

where we have assumed that if the airline matches the competition, the market priceable customers will prefer to buy from the airline rather than the competition.

To get to the full demand model form, we need to define an arrival process for the customers. In that respect we assume that the customers arrive according to a Poisson process with rate λ. Combining the arrival process assumption with the purchase probability expression given above, if the airline offers a fare f and the competing fare is f^c, the bookings for the airline occur according to a Poisson process with the following instantaneous rate:

$$d(f, f^c) = \begin{cases} \lambda\delta e^{-\beta^l f}, & \text{if } f > f^c \\ \lambda\left(\delta e^{-\beta^l f} + (1-\delta)e^{-\beta^m f}\right), & \text{if } f \le f^c. \end{cases} \tag{7}$$

Therefore, the demand process is completely specified by the parameters λ, β^l, β^m and δ. To more fully relate the above model (7) to the class-based model described in the previous section, we note that discretizing it would result in a discrete model with no yieldable demand component, and with the priceable component being a discretization of the $\lambda\delta e^{-\beta^l f}$ while the market priceable component would be a discretized version of the $\lambda(1-\delta)e^{-\beta^m f}$ curve.

The demand model in (7) could be made more realistic by multiplying the $(1-\delta)e^{-\beta^m f}$ term in (7) by a known constant, $\kappa \in (0, 1]$, when the airline exactly matches the competitor ($f = f_c$). We interpret this parameter κ as the probability that a market priceable customer will buy from the airline given that the airline matches the competitor's price. Moreover, the criteria for match can also be made more robust to small differences in airline and competitor's prices by assuming the purchase probability equal to $\delta e^{-\beta^l f} + \kappa(1-\delta)e^{-\beta^m f}$ whenever the airline's price is within $\pm\eta\%$ (e.g., $\pm 5\%$) of the competitor's price instead of the strict match condition of $f = f_c$.

Bayesian inference for the class-free demand model

We next describe a Bayesian approach for forecasting for the parameters of the class-free demand model. In the model described in (7), we include the seasonal dependence of the volume parameter (λ), the price-sensitivity parameters (β_l, β_m) and the competitive parameter (δ) using the Fourier basis on week of year (WOY); similarly the dependence on booking days prior (DCP) is also modeled using the Fourier basis. To further refine the demand model, we generate forecasts for unique combinations of Origin-Destination (OD), Path, Compartment, Day of Week (DOW), Point of Sale (POS) and Fare-Family. Including the dependence of price-sensitivity and competitive parameters on booking days prior and week of year is extremely important to ensure that these parameters capture the right relationship between demand, airline's price and the competitor's price. For example, we typically see that the customers' willingness to pay increases as we approach closer to departure date and therefore we may see bookings increase even when the prices increase closer to departure so not modeling the dependence on booking days prior will lead to biased estimates of price-sensitivity parameters. Given this structure, the model in (7) is a Poisson Generalized Additive Model (GAM) (Hastie and Tibshirani 1990). Point estimation methods for estimating the parameters of the Poisson GAM models have been proposed by many authors (see for example Wood 2000; Hastie and Tibshirani 1990) and advanced software packages for efficiently estimating these models exist in R and other languages e.g., mgcv (Wood 2017).

However, developing fast and scalable methods for Bayesian inference for these models, though desirable due to reasons mentioned in Sect. 4, remains a challenging problem since the Poisson assumption along with the non-linear dependence on price related terms makes the posterior updates of the parameter distributions computationally difficult. Markov Chain Monte Carlo (MCMC) methods for sampling from the posterior distribution can be extremely slow for the dynamic forecast updates required for the airline application. To overcome this challenge, we leverage the recent advances made in Bayesian Variational Inference (Blei et al. 2017) to efficiently estimate approximate posterior distributions of the demand model parameters. The main idea behind this approach is that instead of computing exact posterior distribution, we approximate the true posterior by a distribution from a tractable class of approximating distributions e.g., factored Normal distributions and find the closest approximation to the posterior by minimizing the Kullback-Leibler (KL) divergence. We use the Bayesian variational inference approach with mean-field approximation (Bishop 2006; Wainwright and Jordan 2008) for computing the posterior updates for the model parameters. In particular, we introduce an independent Normal prior distribution on the parameters, $(\lambda, \beta^l, \beta^m, \delta)$, of the model described in (7). Let $\boldsymbol{\nu} = (\mu_\lambda, \sigma_\lambda, \mu_{\beta^l}, \sigma_{\beta^l}, \mu_{\beta^m}, \sigma_{\beta^m}, \mu_\delta, \sigma_\delta)$ denote the parameters of this Normal prior. Then the prior distribution, $q(\cdot)$, is

$$q(\lambda, \beta^l, \beta^m, \delta; \boldsymbol{\nu}) = \mathcal{N}(\mu_\lambda, \sigma_\lambda) \cdot \mathcal{N}(\mu_{\beta^l}, \sigma^2_{\beta^l}) \cdot \mathcal{N}(\mu_{\beta^m}, \sigma^2_{\beta^m}) \cdot \mathcal{N}(\mu_\delta, \sigma^2_\delta), \tag{8}$$

where $\mathcal{N}(\mu, \sigma^2)$ is the probability density function for a normally distributed random variable with mean μ and variance σ^2. The distribution of observed bookings is

$$X|\lambda, \beta^l, \beta^m, \delta, f, f_c \sim Poisson(d(f, f_c)), \tag{9}$$

where $d(f, f_c)$ given in (7) is the mean demand rate. Note that under these assumptions the exact posterior distribution of the parameters given observed bookings X, $p(\lambda, \beta^l, \beta^m, \delta|X)$, doesn't follow the form of the prior, $q(\cdot)$.

Table 6 Market characteristics for the airline data set

Name	Type	Competing airlines	Compartment	history
Market A	Long-Haul	4	Economy	21 months
Market B	Long-Haul	4	Economy	22 months

Instead, as described earlier, we use the variational inference approach to find the closest approximation to the exact posterior $p(\lambda, \beta^l, \beta^m, \delta|X)$ in the class of distributions specified by the assumed prior form of $q(\cdot)$ in (8) by minimizing the KL divergence:

$$KL(q(\lambda, \beta^l, \beta^m, \delta; \mathbf{v})||p(\lambda, \beta^l, \beta^m, \delta|X)) = E_q\left[\log \frac{q(\lambda, \beta^l, \beta^m, \delta; \mathbf{v})}{p(\lambda, \beta^l, \beta^m, \delta|X)}\right]. \quad (10)$$

Note that the procedure of estimating the posterior distribution of model parameters given the observed bookings described above is essentially a mechanism for unconstraining (sometimes also referred to as uncensoring or detruncation) for the fully price-sensitive demand model in (7) defined over a continuum range of prices. In other words, for parametric models like the one in (7), posterior updates of the model parameters via the Variational Bayes approach allow us to estimate quantities like mean demand rate at any given combination of airline and competitor's price. Furthermore, we utilize the online variational inference approach (Wang et al. 2011) for updating the parameters of the posterior distribution in an online fashion as new booking data arrives (typically nightly).

In contrast to the class-based model developed in Sect. 4 in this approach we do not utilize availability covariates $Open$, LA, MLA that were deployed there. Instead, only the airline's fare, reference competitor's fare and bookings information is used, and the generation of observations for parameters β^l, β^m, λ and δ is handled using the Variational Bayes framework. The time series forecasting aspect of the demand model parameters still relies on the Bayesian DLM framework presented in the previous section, including handling of seasonality covariates and time nonstationarity (we omit the technical details here).

Numerical study with airline data

To illustrate the class-free demand model methodology, we applied this approach to an airline data set containing bookings, fares paid for the bookings, and daily snapshots of offered fares for competing airlines as well as for the airline itself; we considered two different markets. An overview of the characteristics of these markets is provided in Table 6. The data set consists of slightly less than two years of historical data.

In conducting this numerical study, we have two objectives in mind. Firstly, we want to demonstrate qualitatively that the methodology generates reasonable forecasts in terms of capturing historical seasonal patterns, sensitivity of consumers to competition and airline prices etc. Secondly, using quantitative metrics we want to show the value of incorporating additional information of competitive fares in terms of improvement of forecast accuracy over methods which do not consider this additional information.

In the study we also make the simplifying assumption that the willingness-to-pay for the loyal customers and the market-priceable customers is the same i.e., $\beta^l = \beta^m = \beta_c$. Under this assumption the bookings occur according to a Poisson process with the following instantaneous rate:

$$d(f, f^c) = \begin{cases} \lambda_c \delta e^{-\beta_c f}, & \text{if } f > f^c \\ \lambda_c e^{-\beta_c f}, & \text{if } f \le f^c \end{cases} . \quad (11)$$

As a benchmark, we also applied our methodology (with necessary adjustments) to the scenario where the airline doesn't include competitive price information in generating its forecasts. In that case, the airline assumes that the bookings occur according to a Poisson process with the following instantaneous rate that is not affected by the competitor fare and the δ:

$$d_m(f) = \lambda_m e^{-\beta_m f}. \quad (12)$$

We refer to the above demand model that doesn't explicitly explicitly consider competitor's price for demand forecasting as the monopolistic model and differentiate the parameters of this model by using a subscript m. Note that the monopolistic demand model described in (12) is also a Poisson GAM model so we can apply the Bayesian Variational Inference methodology already described for forecasting the parameters of this models as well.

Preliminary data analysis and reference competitor price

In the competitive demand model described in (11), we have used a reference competitor's price f_c. Identifying such a single reference competitive information variable has the advantage that a relatively clean competitive price signal which is highly correlated with the airline's demand can be fed into the core demand forecasting model in a relatively straightforward manner. This approach ensures that the core forecasting model can be updated efficiently using the most relevant competitive information while keeping the model interpretable and reducing the complexity of managing the forecasts.

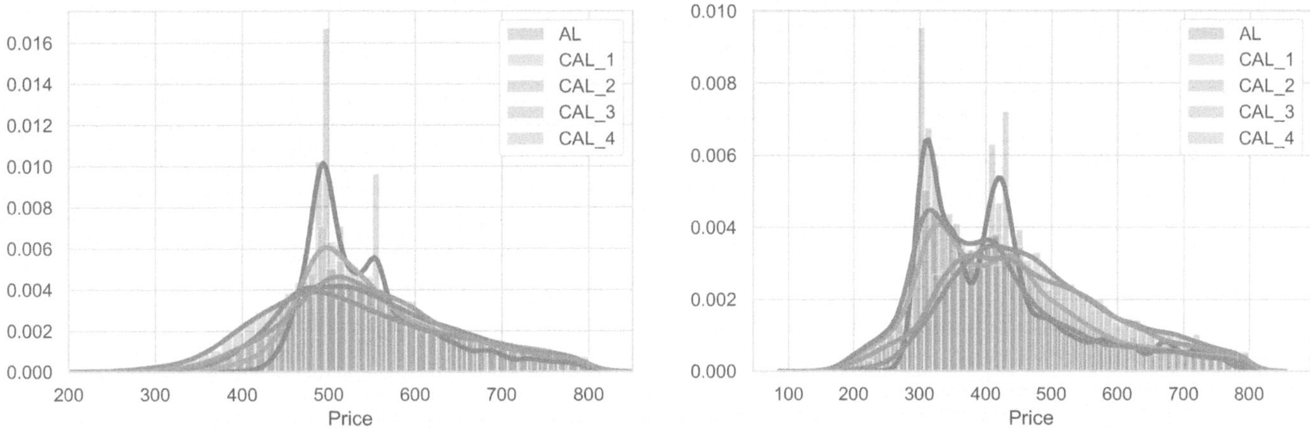

(a) Market A (b) Market B

Fig. 2 Historical fares for the Airline ('AL') and its competitors ('CAL_i')for the two markets

Table 7 Average historical fares for the Airline and its competitors

Name	AL	CAL_1	CAL_2	CAL_3	CAL_4
Market A	545	556	551	533	549
Market B	422	428	410	482	467

Table 8 Coefficient of price terms and associated statistics from the multiple linear-regression analysis of Airline's bookings versus prices of the Airline, its competitors and other features for Market A

Market A	Coefficient	Std. error	t-statistic	p-value
Price AL	− 0.0021	0.00010	− 5.8850	< 0.0001
Price CAL_1	− 0.0002	0.00014	− 2.9140	0.0020
Price CAL_2	0.0009	0.00010	3.0250	< 0.0001
Price CAL_3	0.0020	0.00012	10.9130	< 0.0001
Price CAL_4	− 0.0001	0.00019	− 2.3530	0.0190

Therefore, we recommend conducting a preliminary data analysis to identify an appropriate reference competitor price. In practical setting, the reference competitor price could be identified by the airline based on its knowledge of the dominant airline in the market or perhaps a specific competing airline that happens to have a significant impact on its demand. In case the airline considers multiple competitors impacting its demand, the reference competitor's price could also represent some statistic based on the prices of the set of competing airlines e.g., minimum of the competing airline set or mean of the competing airline set. More data-driven approaches to create the reference competitive price signal can also be used for large-scale implementations.

In this study we conducted a preliminary data analysis to narrow down the set of competitors that have the most impact on the airline's bookings. Figure 2 shows a histogram of historical one-way fares of the airline along with the competitors where the fares have been scaled and are represented in US dollars. The airline's itinerary as well as the competing itineraries were chosen to be of the same type with the same number of connections. We can see that there is considerable variability in the historical fare data for the airline and its competitors, with the average fares shown in Table 7.

We also perform analysis using multiple linear regression for constructing the reference competitor price signal to be used in the forecast model. For this purpose, we fit a simple linear model with airline's bookings as the response variable and include airline's price, competing airlines prices, Fourier basis terms for capturing departure date seasonality, and dependence on booking days prior as predictors in the linear model.

Table 8 shows the multiple regression coefficient estimates for price related terms for Market A. As expected the negative coefficient for airline price indicates that an increase in airline's price leads to decrease in its bookings while the positive coefficient for the CAL_2 and CAL_3 indicates that an increase in the price of these competitors leads to an increase in the airline's bookings. Moreover, the magnitude of coefficients and their respective p-values show that airline's own prices and the prices of CAL_2 and CAL_3 have significant impact on airlines bookings. Therefore, we tested two candidate statistics for the reference competitive price f^c for market A: $f^c = \min\{f^{CAL_2}, f^{CAL_3}\}$ and $f^c = mean\{f^{CAL_2}, f^{CAL_3}\}$.

Table 9 shows the multiple regression coefficient estimates for price related terms for Market B. In this market as well, we see that the negative coefficient is associated with the airline's price while now the relatively large positive coefficients are associated with CAL_2 and CAL_4. The magnitude of coefficients and their respective p-values show that airlines own prices and the prices of CAL_2 and CAL_4 have significant impact on airlines bookings. Therefore, for

Table 9 Coefficients of price terms and associated statistics from the multiple linear-regression analysis of Airline's bookings versus prices of the Airline, its competitors and other features for Market B

Market B	Coefficient	Std. error	t-statistic	p-value
Price AL	− 0.0018	0.00012	− 4.7930	< 0.0001
Price CAL_1	− 0.0001	0.00011	− 0.1480	0.8820
Price CAL_2	0.0013	0.00023	4.7990	< 0.0001
Price CAL_3	− 0.0001	0.00011	− 0.5530	0.9700
Price CAL_4	0.0025	0.00021	10.5690	< 0.0001

market B we tested two candidate statistics for the reference competitive price f^c: $f^c = \min\{f^{CAL_2}, f^{CAL_4}\}$ and $f^c = mean\{f^{CAL_2}, f^{CAL_4}\}$.

Qualitative analysis

As noted, we applied a Bayesian GAM model to the two markets and generated forecasts of the volume (λ_c), price-sensitivity (β_c) and loyal fraction (δ_c) parameters of the model. Figures 3, 4 and 5 show the historical uncensored observations and 52 week future forecasts for the three parameters, respectively, for Market A. We have also shown the historical bookings and historical predictions obtained by calculating the constrained booking values using historical price and competitor price statistic in Fig. 3. Note that there is a shift upward in demand, starting in week 36 in the historical departures. The forecast profiles for the volume parameter and the price-sensitivity parameters have adjusted appropriately to capture this trend. Moreover, the forecasts for the volume and the price-sensitivity parameter are capturing general seasonal patterns in the data, aside from individual spikes in bookings caused by Holidays or Special Events which were not modeled in this analysis. The loyal fraction parameter shown in Fig. 5 has a relatively flat shape and is fluctuating around a mean value of 0.42 in the forecast period.

In Figures 6, 7 and 8 we show historical estimates (uncensored observations fed into the Bayesian DLM) and 52 week future forecasts for the three parameters, respectively, for Market B. Again, note the two distinct upward demand shifts, starting in week 16 and week 40 in the historical departures. The forecast profiles for the volume parameter and the price-sensitivity parameters have adjusted appropriately to capture this trend. Moreover, the forecasts for volume and the price-sensitivity parameter are capturing general seasonal patterns. The loyal fraction parameter shown in Fig. 8 is fluctuating around a mean value of 0.38 in this market.

Forecast accuracy analysis

In this section we conduct a numerical study using the real airline data set for the two markets described in Sect. 5.2. The main aim of the study is to quantify the improvement in forecast accuracy brought about by the use of additional competitive price information, (11), when compared to a simpler

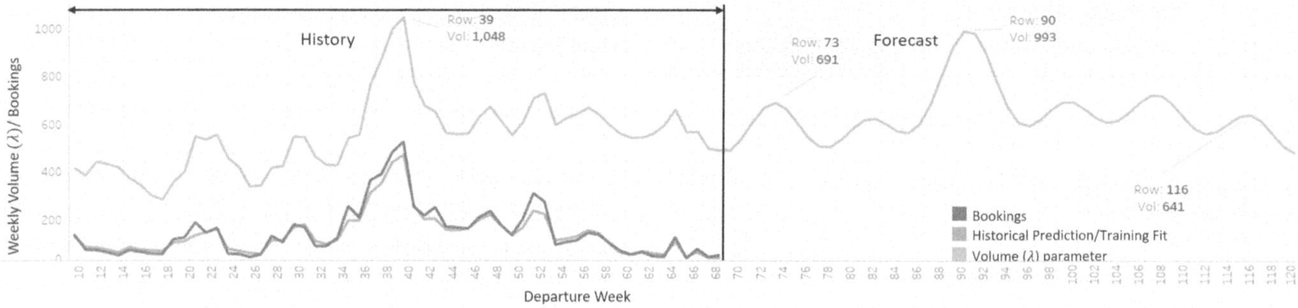

Fig. 3 Weekly bookings and volume (λ_c) parameter for market A

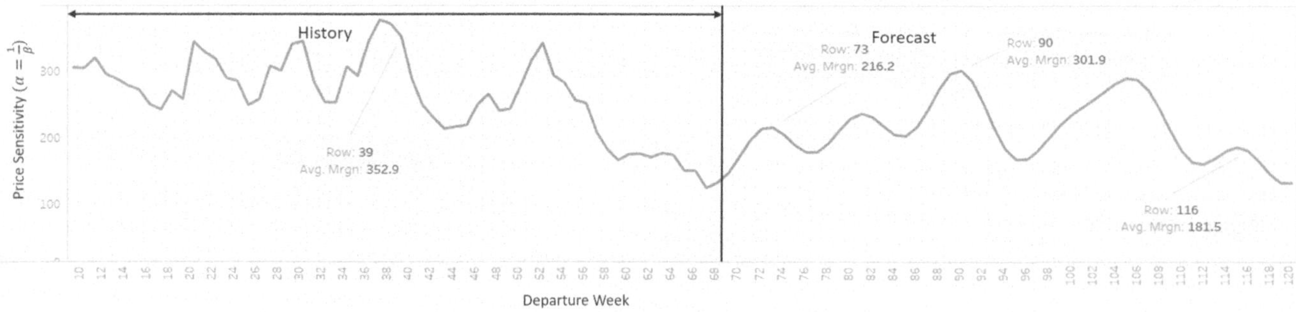

Fig. 4 Weekly-averaged price sensitivity parameter ($\alpha_c = \frac{1}{\beta_c}$) for Market A

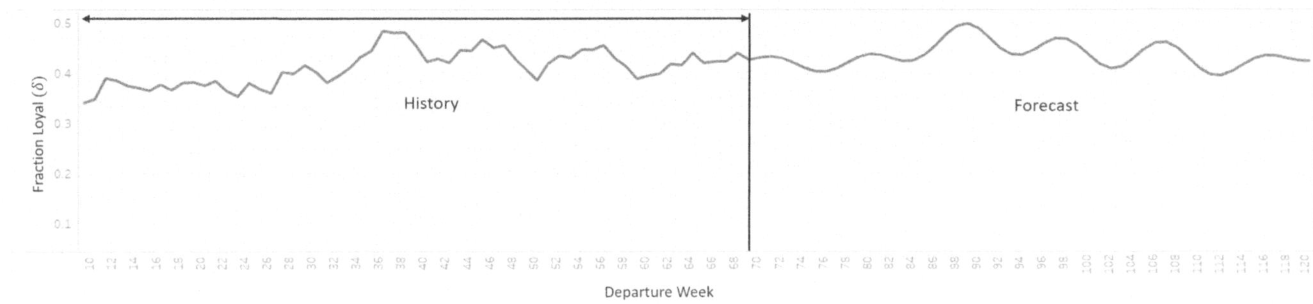

Fig. 5 Weekly averaged loyal fraction parameter (δ) for Market A

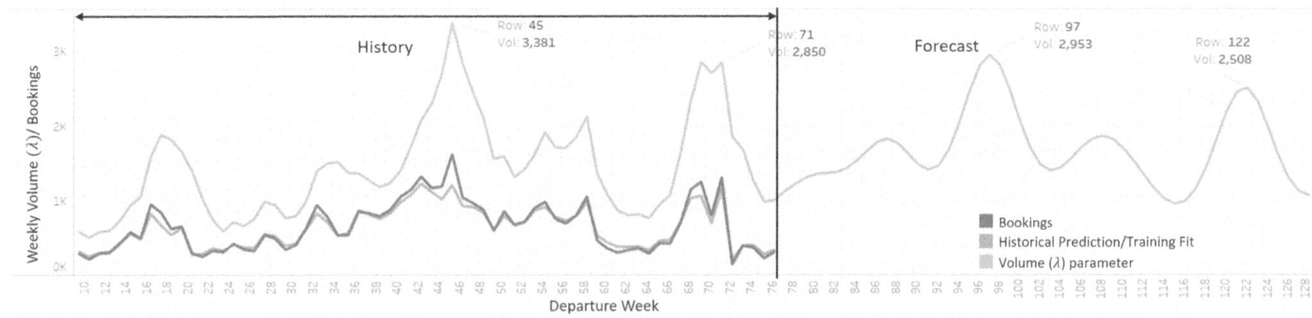

Fig. 6 Weekly bookings and volume (λ_c) parameter for market B

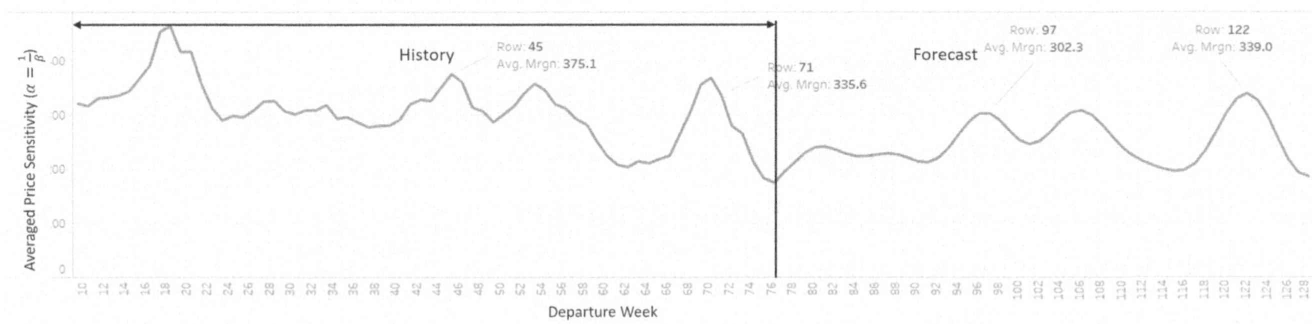

Fig. 7 Weekly averaged price sensitivity parameter ($\alpha_c = \frac{1}{\beta_c}$) for Market B

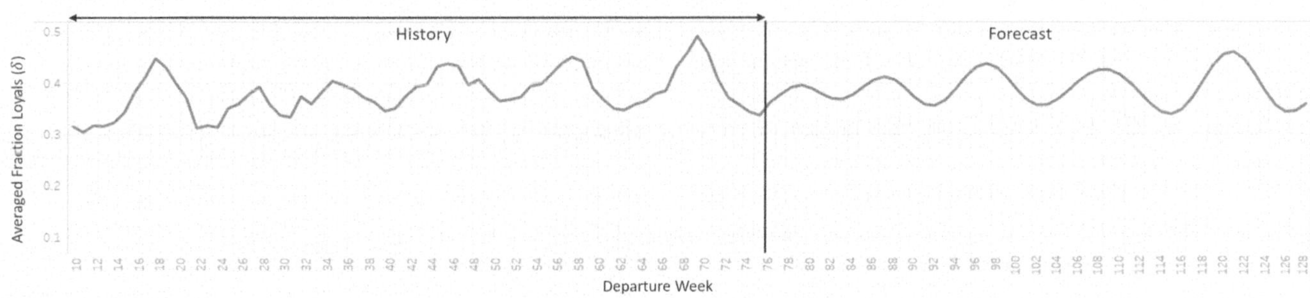

Fig. 8 Weekly averaged loyal fraction parameter (δ) for Market B

monopolistic model which doesn't explicitly consider competitor's prices (12).

For this study we first divide the data set into training and test sets for each market such that the first 70 weeks of the history are in the training set and the next 16 weeks are in the test set. We then train the competitive and monopolistic Bayesian GAM models on the training set and generate the corresponding forecasts for the test set period. We generate the demand predictions (re-constrained demand) for the competitor price-aware forecaster over the test set period

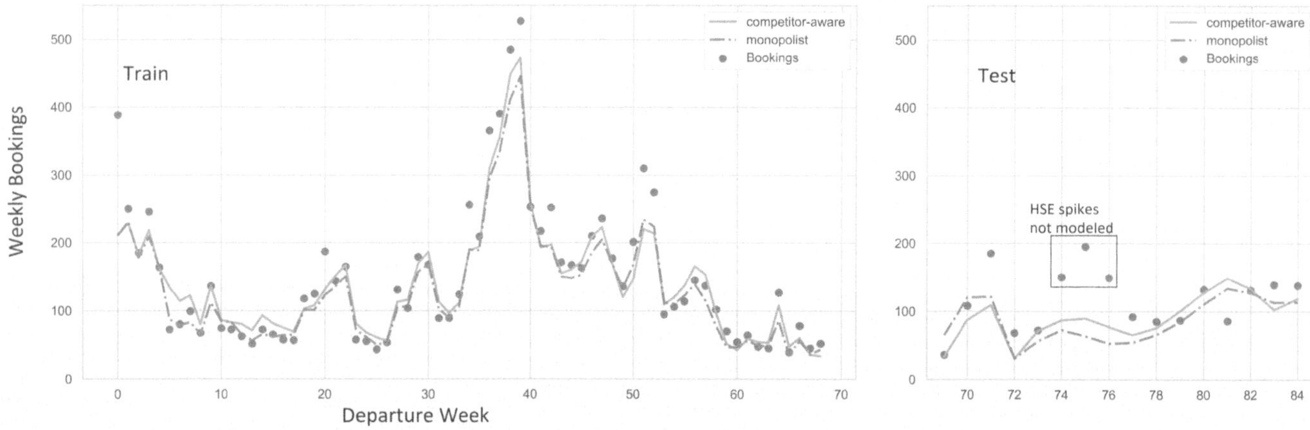

Fig. 9 Weekly historical and predicted bookings in the training and test set for market A

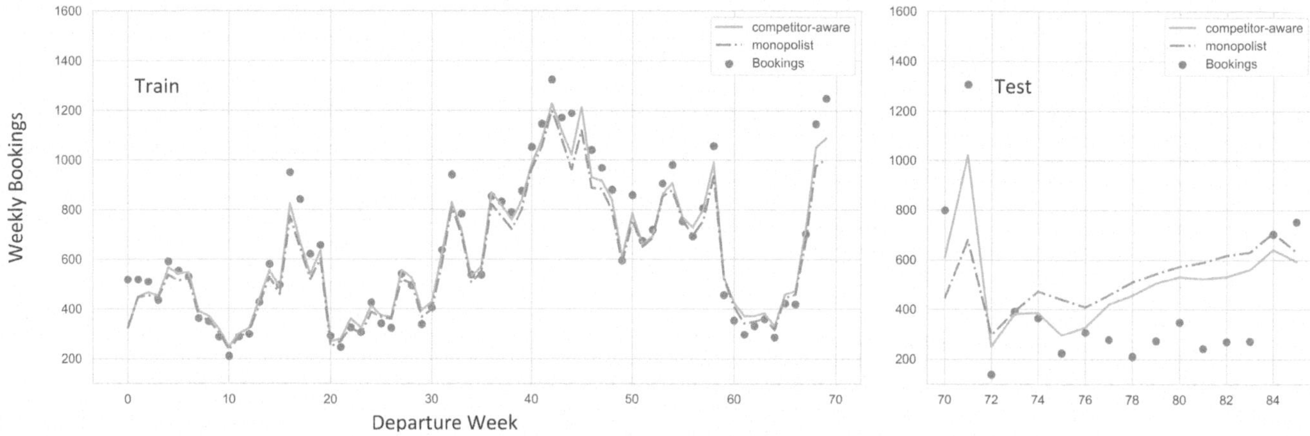

Fig. 10 Weekly historical and predicted bookings in the training and test set for market B

based on the maximum a posteriori probability (MAP) estimate of the model parameters and computing the estimated mean demand under the observed prices of the airline and its competitors using equation (11). For the monopolistic model the demand predictions (re-constrained demand) are similarly computed using the MAP estimate of the parameters but based only on the observed airline's prices in the test set using equation (12). The demand predictions generated at departure date, time of day window, compartment and booking days prior level are then aggregated over days prior for each market and compared to the respective bookings aggregated at the same level using metrics such as Root Mean Squared Error (RMSE), Mean Absolute Deviation (MAD) and Bias.

Figures 9 and 10 show the aggregated (weekly) training and test set predictions for the competitor price-aware and the monopolistic models. The aggregated (weekly) observed bookings are also shown in blue dots. We see that in the test set period, the competitor price-aware model is closer to the observed bookings as compared to the monopolistic model. We also note that given less than one and a half years of history, the test set period bookings observed for Market B are especially challenging to predict since the trend is lower and much flatter than the corresponding historical period between weeks 20 and 36. Still the competitor price-aware model shows better prediction accuracy, partly due to the information gained from the competitive prices. The average reference competitor price in the test set period for Market B is \$416.63 whereas the airline's average price is slightly higher at \$426.76 while in the corresponding historical period between weeks 20 and 36, the average reference competitor price was \$502.11 while the airline's average price was much lower at \$ 451.26, partly explaining why the airline's demand during the test set is much flatter than the corresponding history.

Table 10 shows the test set accuracy metrics comparing re-constrained demand with actual bookings for the two markets. While computing these metrics, a 95% confidence interval was generated via bootstrapping by randomly sampling observations from the test set. We see that the competitor price-aware model generates more accurate forecasts as compared to the monopolistic model. In particular, we

Table 10 Comparison of test set forecast accuracy metrics for the two markets

Metric	Market A			Market B		
	Competitive Min (f^c)	Mean (f^c)	Monopolistic	Competitive Min (f^c)	Mean (f^c)	Monopolistic
RMSE	11.25 ±0.062	11.47 ±0.066	12.06 ±0.060	21.56 ±0.067	20.58 ±0.073	22.12 ±0.072
MAD	7.04 ±0.028	6.79 ±0.026	7.67 ±0.033	15.78 ±0.029	15.39 ±0.027	16.66 ±0.028
BIAS	7.78 ±0.052	7.92 ±0.053	8.38 ±0.055	13.84 ±0.06	13.66 ±0.058	15.19 ±0.060

see that the model with competitive information improves the forecast accuracy by 5–10%. However, there is no conclusive evidence from these tests to show that the competitor reference price generated by using $\min\{f^c\}$ is better or worse than the one generated using $\text{mean}\{f^c\}$.

We also stress that the forecast accuracy study shown here is based on realized competitor's prices during the test set. In practice, for using a competitor price-aware forecasting methodology in the revenue management system to generate bid prices using dynamic programming, it is necessary to predict the competitor's price over the decision horizon. In this case the accuracy of the forecast and, consequently, the quality of the bid prices will also depend on the accuracy of the predicted competitor's prices. We do not explore this important topic further in this paper and instead point the interested reader to the work by Fiig et al. (2019) which shows that the competitor's prices can be predicted reliably. However, the quantitative benefit of the competitor price-aware forecaster demonstrated through the numerical study here along-with the qualitative analysis presented in the previous section shows that our methodology is able to estimate important parameters related to price-sensitivity and competitive parameter δ, the fraction of loyal customers, reliably and robustly. In addition to generating bid prices, these estimates of price-sensitivity and competitive parameters along-with the real-time competitive price information can also be used to construct dynamic competitor price-aware pricing policies. We discuss this important application next.

Optimal pricing policy

The revenue-maximizing pricing policies obtained with the competitor-price aware demand model described in (7) exhibit certain desirable and intuitive properties. In this section we discuss how to compute optimal pricing policies and characterize the nature of these policies under the competitor price-aware demand model. In the analysis presented in this section, we do not consider the game-theoretic aspects and assume that the competitor will not immediately respond to the airline's price change. Moreover, since the focus here is on the dynamic competitor-aware pricing, we assume that the displacement costs or bid prices have already been generated by the RMS either using monopolistic demand forecasts or as mentioned earlier, the competitor-price aware demand forecasts with predicted competitor prices.

Given a real-time reference competitor price f^c and the current bid price c, the optimal price under the competitor-price aware model is obtained by maximizing the expected margin contribution:

$$M(f, f^c, c) = d(f, f^c) \cdot (f - c), \tag{13}$$

and the optimal price is the price achieving this maximum

$$f^*(f^c, c) = \arg\max_{f \in \mathcal{F}} M(f, f^c, c), \tag{14}$$

where $\mathcal{F}$ is the set of feasible prices. Since the demand model has an exponential form with a discontinuity when the airline matches competitor's price ($f = f^c$), the expected margin contribution function is a piece-wise concave function (see Example 5.3.1 and Fig. 11) and leads to a very intuitive and easy to compute pricing policy where one of the following actions is optimal:

- **Under-cut** the reference competitor to obtain the entire market demand
$$f_u^* = \arg\max_{f \in \mathcal{F}} \lambda\left(\delta e^{-\beta^l f} + (1-\delta)e^{-\beta^m f}\right) \cdot (f - c),$$
- **Match** the competitor so as to not lose demand to them $f_m^* = f^c$
- **Ignore and price higher** than the competitor, the offered price in this case is the optimal price for the airline's loyal customers
$$f_i^* = \arg\max_{f \in \mathcal{F}} \lambda\left(\delta e^{-\beta^l f}\right) \cdot (f - f^c) = c + \frac{1}{\beta_l}$$
.

The optimal action out of the three possible ones is determined by comparing the expected margin contribution at these three candidate prices and selecting the one that maximizes it i.e.,

$$f^*(f^c, c) = \arg\max\{M(f_u^*, f^c, c), M(f_m^*, f^c, c), M(f_i^*, f^c, c)\}. \tag{15}$$

Table 11 Estimated parameters for the entity in example 5.3.1

Volume(λ)	Price sensitivity for market-priceable (β^m)	Price sensitivity for loyals (β^l)	Fraction loyal (δ)
44.40	0.00303	0.00254	0.59

To further elucidate the intuition about the optimal pricing policy, we consider simple examples next.

Example: optimal pricing policy

Consider a specific entity with estimated values of the demand model parameters shown in Table 11.

In this example, the fraction of loyal customers for the airline is 0.59, so the fraction of market-priceable demand is 0.31 which is high enough so that the airline doesn't completely dominate the market and competitive effects may play a significant role in shaping their pricing policy. Also, note that $\beta^m > \beta^l$ implies that the market-priceable customers are more price sensitive than the loyal customers, in other words the willingness-to-pay of market-priceable customers is lower than that of the loyal customers.

Now assume that the current reference competitor's fare in the market is \$380. We analyze the optimal pricing policy of the airline under three bid price scenarios: low (bid price = 0), medium (bid price = 100) and high (bid price = 250). Figure 11 shows the expected margin contribution functions as a function of airline's price for each of these three bid price scenarios.

Table 12 shows the undercut, match and ignore prices, the corresponding expected margin contribution function value and the optimal strategy for each of the three bid price scenarios. Note that if the candidate optimal undercut price i.e, $f_u^* = \arg\max_{f\in\mathcal{F}} \lambda\left(\delta e^{-\beta^l f} + (1-\delta)e^{-\beta^m f}\right)\cdot(f-c)$ is higher than the competitor's price, we indicate the undercut price and the corresponding margin contribution by NA in Table 12.

We see from this example that in situations where the airline's bid price is very low, the optimal policy is to undercut the competitor fare and offer a price which is low enough to not only attract the more price-sensitive market-priceable customers but also to take into consideration their lower willingness-to-pay. In the medium bid price scenario, the optimal policy suggests matching the competitor's fare so that we don't lose the market-priceable customers to the competing airline. While in situations where the airline's bid price is high enough, the airline should price higher than the competitor and set optimal price based on the willingness-to-pay of the loyal customers.

Example: real airline data

We next show some examples based on the forecasts generated for the real airline data as part of the numerical study described earlier. Note that in this example we use the simplified model form in (11).

The parameters in Table 13 for the model come from actual forecasts generated in the section on forecast accuracy analysis for the particular market, point of sale (POS) and departure date. The optimal policy for each case is also shown in this table. In the first example (Market A), the optimal policy is to ignore the competitor and price higher. This is because the fraction of airline's loyal customers is very large ($\delta = 0.93$), and even if the airline prices higher than the competitor, it is not going to lose too much demand. Therefore, in this case it is optimal to price based on the higher willingness-to-pay of the loyal customers. While in the second instance (Market B) the fraction of loyal customers are not as high ($\delta = 0.42$). If the airline prices higher than the competitor, the reduction in revenue from losing market-priceable demand is quite significant. Therefore, in this case, the optimal policy is to match the competitor's price. Note that in these example we have considered the bid price to be 0. As shown in Example 5.3.1, if the bid price is high enough the pricing policy for Market B could change from match to ignore and price higher than the competitor.

These examples show that deciding when to match, undercut or ignore the competitor depends crucially on the fraction of loyal customers, price sensitivity of the loyal and market-priceable customers and on the current bid price. Higher

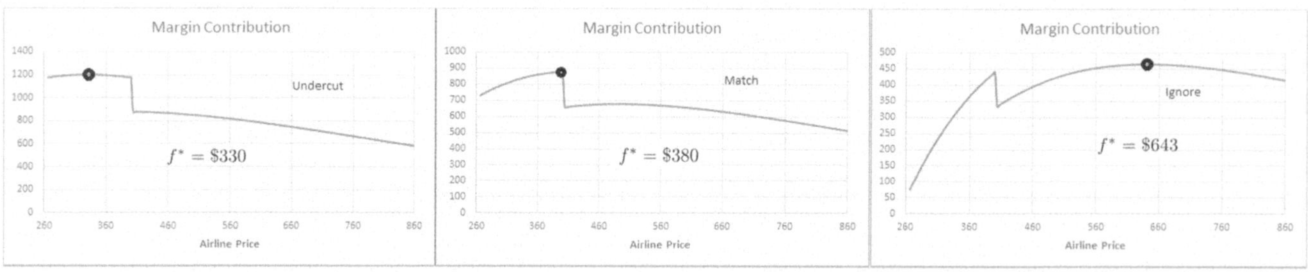

Fig. 11 Expected margin contribution and optimal policy for the example in 5.3.1, left panel image shows the case with bid price = 0, middle panel for bid price = 100 and right panel for bid price = 250

Table 12 Expected margin contribution and optimal prices for example 5.3.1

Bid price	f_u^*	f_m^*	f_i^*	$M(f_u^*, f^c, c)$	$M(f_m^*, f^c, c)$	$M(f_i^*, f^c, c)$	Optimal policy	Optimal price (f^*)
0	\$330	\$380	\$393	\$5336.30	\$5280.10	\$3753.50	Undercut	\$330
100	NA	\$380	\$493	NA	\$3925.73	\$2936.40	Match	\$380
250	NA	\$380	\$643	NA	\$1822.66	\$2004.78	Ignore	\$643

Table 13 Optimal policy for specific examples based on the real airline data presented in the section on forecast accuracy analysis

Market	λ_c	$\alpha_c = \frac{1}{\beta_c}$	δ	f^c	f_i^*	$M(f^c, f^c, 0)$	$M(f_i^*, f^c, 0)$	Optimal action
A	7.35	317.32	0.93	\$150	\$317.32	\$1277.79	\$1280.38	Ignore ($f^* = \$317.32$)
B	13.41	393.44	0.42	\$200	393.44	\$3207.12	\$1355.89	Match ($f^* = \$350$)

fraction of loyal customers (δ) with high willingness-to-pay allow the airline to ignore the competitor and act more "monopolistically". Whereas in markets where the fraction of loyal customers is low and their willingness-to-pay not as high (relative to the market-priceable customers), airlines need to respond to their competitors much more actively in terms of matching or even under-cutting them to ensure that they can attract enough demand. Finally, higher bid price values allow the airline to price based on their loyal customers more and ignore the competitors. In contrast to that, lower bid prices require the airline to manage competitive effects more carefully so as to not lose too much demand and therefore revenue to their competitors.

Conclusion and future research

We have introduced new demand models that make use of available competitive information such as prices charged by competitors in the market for products that can attract some but potentially not all of an airline's customers. We call such customers who buy a product from an airline regardless of how competition is pricing, the loyal customers. We account for that customer behavior explicitly and model its time dynamics within the learning framework of the Bayesian dynamic linear model.

To accommodate the more traditional airline setting we first have shown how to extend existing class-based models to include the loyal customers as well as the fully flexible customers and then provided the much more novel framework for the class-free environment. Within the latter, we have used a continuous exponential willingness-to-pay curve and have applied the Variational Bayes techniques to create observations to update the curve's parameters.

The Bayesian learning principles turn out to be very useful and provide the desired stability and robustness, especially when having to discern whether a lack of a booking was due to there being no request arrival or offered price being too high. We show through a data study based on one airline's real-life data that including market information improves typical measures of forecast accuracy by up to 10%.

The main focus of our research presented here has been the forecasting side of the revenue management, but we also have indicated by means of simple examples how this additional information impacts decision making and control generation. In particular, we have highlighted that both the class-based and the class-free models exhibit features that are desired in practice. Namely, depending on the relative magnitude of the loyal demand, competitors' prices, as well as the marginal displacement cost, an airline may decide to either ignore the competition or match the competition, or undercut them in order to maximize its expected revenue.

In the simple examples presented here we assumed a given (zero) marginal displacement cost (bid price). At the same time the bid price is the main ingredient in airline revenue management where seat inventory and the booking horizon are both finite. While our models can be readily used in bid price generation they require at least a proxy of what competitor's price will be over the booking horizon.

There are a number of possibilities in that regard to explore: from explicitly modeling competitive price path to including simplified competitor response models to very simple (but perhaps effective in practice) static proxies such as average of past competitor prices or the latest competitor price. We believe it is a rich research area and our initial investigation indicates that when a good proxy for competitor's price is not obtainable then it might be better to use a monopolistic forecast model for the purposes of the bid price generation and combine it with the more precise models such as the ones presented here for the real-time request handling when the competitor's price becomes available (Walczak and Kumar 2019).

Finally, to firmly establish the validity of the explicit modeling of loyal customers an exhaustive comparison to the classical models would be very valuable. We include in the latter category models that do not involve discontinuous change in demand when competitor undercuts. An equally important research avenue is comparison to discrete customer choice demand models with continuous prices (Fiig et al. 2019), especially in the broader context of offer optimization.

Acknowledgements We are grateful to Dr. Karl Isler for his insights on Bayesian Variational Inference which helped us develop the methodology for class-free models. We are also grateful to the referees for their valuable comments that helped improve this paper.

References

Bishop, C.M. 2006. *Pattern Recognition and Machine Learning*. New York: Springer.

Blei, D.M., A.D. Kucukelbir, and J. McAuliffe. 2017. Variational inference: A review for statisticians. *Journal of the American Statistical Association* 112 (518): 859–877.

Boyd, E.A., and R. Kallesen. 2004. The science of revenue management when passengers purchase the lowest available fare. *Journal of Revenue and Pricing Management* 3: 171–177.

Cooper, W.L., T. Homem-de-Mello, and A.J. Kleywegt. 2015. Learning and pricing with models that do not explicitly incorporate competition. *Operations Research* 63 (1): 86–103.

Dudey, M. 1992. Dynamic Edgeworth-Bertrand competition. *The Quarterly Journal of Economics* 107 (4): 1461–1477.

Fiig, T., M. Wittman, and C. Trescases. 2019. *Towards a Competitor-Aware RMS, presented at AGIFORS RM Study Group*. Panama: Panama City.

Hastie, T.J., and R.J. Tibshirani. 1990. *Generalized additive models*. Boca Raton: Chapman & Hall/CRC.

Isler, K., and H. Imhof. 2008. A game theoretic model for airline revenue management and competitive pricing. *Journal of Revenue and Pricing Management* 7 (4): 384–396.

Kumar, R., A. Li, and W. Wang. 2018. Learning and optimizing through dynamic pricing. *Journal of Revenue and Pricing Management* 12: 63–77.

Martínez-de-Albéniz, V., and K. Talluri. 2011. Dynamic price competition with fixed capacities. *Management Science* 57 (6): 1078–1093.

Singh, S. 2019. Chapter 3: Towards explicitly incorporating competition under flexible models of demand in dynamic pricing. Doctoral Thesis in Operations Management, Tepper School of Business, Carnegie Mellon University, Pittsburgh, Pennsylvania.

Singh, S., and Walczak, D. 2019. Explicitly incorporating competition and realistic models of customer demand in dynamic pricing, working paper.

Talluri, K., and G. Van Ryzin. 2004. Revenue management under a general discrete choice model of consumer behavior. *Management Science* 50 (1): 15–33.

Talluri, K., and G. Van Ryzin. 2004. *The theory and practice of revenue management*. Dordrecht: Kluwer Academic Publishing.

Wainwright, M., and M. Jordan. 2008. Graphical models, exponential families, and variational inference. *Foundations and Trends in Machine Learning* 1 (1–2): 1–305.

Walczak, D., and R. Kumar. 2019. Degrees of information awareness in revenue management and dynamic pricing, presented at INFORMS RM & Pricing Conference. Stanford, CA: Stanford Graduate School of Business.

Wang, C., J. Paisley, and D.M. Blei. 2011. Online Variational Inference for the Hierarchical Dirichlet Process. *Proceedings of the Fourteenth International Conference on Artificial Intelligence and Statistics, PMLR* 15: 752–760.

Wang, W., S. Arunachalam, R. Kumar, A. Simrin, D. Walczak, and B.R. Guntreddy. 2019. Will they stay or will they go? Competitive RM with loyal and fully flexible customers, presented at AGIFORS RM Study Group. Panama: Panama City.

West, M., and J. Harrison. 1997. *Bayesian Forecasting and Dynamic Models*, 2nd ed. Series in Statistics. New York: Springer.

Wood, S.N. 2000. Modelling and smoothing parameter estimation with multiple quadratic penalties. *Journal of the Royal Statistical Society Series B* 62 (2): 413–428.

Wood, S.. N. 2017. *Generalized additive models: An introduction with R*, 2nd ed. Boca Raton: Chapman & Hall/CRC.

Zeni, R. H. 2001. Improved forecast accuracy in airline revenue management by unconstraining demand estimates from censored data.PhD thesis, Graduate School, Rutgers, State University of New Jersey.

Ravi Kumar is a Lead Scientist at PROS Inc where his work involves research and development of revenue management and dynamic pricing systems. His recent research focuses on reinforcement learning for personalized recommendations and dynamic pricing. He holds a PhD in Operations Research from Cornell University.

Wei Wang is a Senior Scientist at PROS Inc, his work and research include forecasting, overbooking, optimization and dynamic pricing. He received his Ph.D. in Operations Research from the University of Washington.

Ahmed Simrin is a Pricing and Demand Forecasting Manager in the Revenue Management department at Etihad Airways. He is responsible for managing the demand forecast as well as pricing and inventory strategies for a set of markets with the aim of optimizing network and point of sale revenue.

Siva Arunachalam is a Senior Manager of Business Solutions and Innovations in Revenue Management department at Etihad Airways. He is responsible for managing Forecasting, Optimization and Group RM systems and finding Innovative & Incremental Revenue generating solutions.

Bhaskara Rao Guntreddy is the Head of Business Solutions and Innovations in Revenue Management department at Etihad Airways. He is responsible for Operations Research, AI, and value creation with innovative solutions in Dynamic Pricing, Revenue Optimization, and Digital Retail Management.

Darius Walczak is a Principal Research Scientist and a Senior Director at PROS Inc. Since joining PROS in 2001 he has been involved in design and deployment of numerous revenue management (RM) and pricing optimization applications across variety of industries. Darius and his team conduct research in stochastic optimization, data analysis, and forecasting to innovate and improve solutions in the core science platform.

Journal of Revenue and Pricing Management (2021) 20:276–300
https://doi.org/10.1057/s41272-021-00312-3

RESEARCH ARTICLE

Demand estimation from sales transaction data: practical extensions

Norbert Remenyi[1] · **Xiaodong Luo**[2]

Received: 31 May 2020 / Accepted: 10 September 2020 / Published online: 22 March 2021

Abstract

In this paper, we discuss practical limitations of the standard choice-based demand models used in the literature to estimate demand from sales transaction data. We present modifications and extensions of the models and discuss data preprocessing and solution techniques which are useful for practitioners dealing with sales transaction data. Among these, we present an algorithm to split sales transaction data observed under partial availability, we extend a popular Expectation Maximization (EM) algorithm for non-homogeneous product sets, and we develop two iterative optimization algorithms which can handle much of the extensions discussed in the paper.

Keywords Demand estimation · Demand untruncation · Multinomial logit model · EM algorithm · MM algorithm · Frank–Wolfe method · Revenue management

Introduction

Demand estimation using censored sales transaction data has many applications in airline commercial planning process. We refer readers to a survey paper by Sharif Azadeh et al. (2014) for a brief introduction to this topic. In this paper we discuss some practical limitations and extensions of a particular choice-based demand model popular in the literature to estimate demand from sales transaction data. Discrete choice models (e.g., Ben-Akiva and Lerman 1994; Train 2003) have provided a popular approach for estimating demand for different products in a set of substitutable items, especially in transportation and revenue management applications. We look at a common demand model, which appeared in several papers, including Dai et al. (2014), Vulcano et al. (2012), and Abdallah and Vulcano (2016). The motivation of this work came from observing a few shortcoming of these models when applied in practice, on airline revenue management data.

✉ Norbert Remenyi
norbert.remenyi@sabre.com

Xiaodong Luo
xiaodongluo@cuhk.edu.cn

[1] Sabre, 3150 Sabre Dr., Southlake, TX 76092, USA

[2] Shenzhen Research Institute of Big Data, Shenzhen, Guang Dong, China

We will build on and extend the work presented in Vulcano et al. (2012) and Abdallah and Vulcano (2016). They combine a multinomial logit (MNL) choice model with non-homogeneous Poisson arrivals over multiple periods. The MNL model has been used by many practitioners and researchers to represent the underlying choice behavior. Although its property of independence of irrelevant alternatives (IIA) is somewhat restrictive, the model is simple, leading to tractable estimation and assortment optimization (Talluri and van Ryzin 2004). The problem is how to jointly estimate the arrival rates of customers and the preference weights of the products via maximizing the likelihood functions. There are two different likelihood functions. The first one is the incomplete data likelihood function [see (3) and also Eq. (2) in Vulcano et al. 2012] and the second one is the log-likelihood function which is based on the primary demand [see (13) and also Eq. (13) in Vulcano et al. 2012]. The inputs are observed historical sales, availability of the products, and market share information.

Our contribution is to discuss practical limitations of the specific model above, and present some interesting extensions. We will discuss partial availability of products, some relaxation of the IIA assumption, constrained parameter space, non-homogeneous product set, and the interpretation of the no-purchase option and related market share. We hope this discussion can facilitate more research and extension of these models.

Chapter 6 was originally published as Remenyi, N. & Luo, X. Journal of Revenue and Pricing Management (2021) 20:276–300. https://doi.org/10.1057/s41272-021-00312-3.

We will present an algorithm to split sales transaction data observed under partial availability, and an extension of the EM algorithm for the case when we observe a non-homogeneous product set. We develop two iterative optimization algorithms which incorporate partial availability information, non-homogeneous product set, ability to control the availability of outside alternative, and an upper bound on the arrival rates of customers. In the first formulation we use a market share constraint at each time period, and incorporate them into the objective function through the preference weights of the outside alternative. The formulation is solved using the Frank–Wolfe algorithm, leading to a simple coordinate descent algorithm. In the second formulation we use a single, aggregate market share constraint over the time horizon, and assume knowledge of preference weights of the outside alternative. Using this formulation we develop a fast, iterative minorization–maximization algorithm (MM) building on the work in Abdallah and Vulcano (2016).

While the EM algorithm focuses on solving the complete data log-likelihood functions, the two new algorithms (like Abdallah and Vulcano 2016) aim to solve the incomplete likelihood function directly. We remark that both likelihood functions render similar quality solutions, but they involve different intermediate decision variables and require different solution approaches. It is not known that these two methods are mathematically equivalent.

Practical limitations of existing models

In this section we discuss in detail some practical limitations of the choice-based demand model discussed in Vulcano et al. (2012) and in Abdallah and Vulcano (2016). The model combines non-homogeneous Poisson arrivals over multiple periods with a multinomial logit (MNL) choice model. The model assumes a retailer offers a fixed number of n products over a time horizon T, and the products are either available or not available for sale in a time period. From the observed sales data, we estimate the demand, the sales we would have observed if all the products were available for purchase. The total demand of all the products at time t (including outside alternatives and the no-purchase option) is modeled as a Poisson distributed random variable with parameter λ_t. Hence we model the total demand as a non-homogeneous Poisson model, since different time periods are allowed to have different mean demands.

The demand for individual products is modeled using a multinomial distribution, given the total demand of all products. The preferences for different products are assumed to be fixed over time, and the probability that customer chooses product i when $i \in S_t$ is modeled through the simple multinomial logit (MNL) model, that is

$$P_i(S_t, \boldsymbol{v}) = \frac{v_i}{v_0 + \sum_{j \in S_t} v_j}. \tag{1}$$

When $i \notin S_t$, then $P_i(S_t, \boldsymbol{v}) = 0$. Here v_i, $i = 1, \dots, n$ is the positive preference weight for product i, and S_t is the set of products available for sale at time t. The preference weight of outside alternatives and no-purchase option (OA) are embedded into the coefficient v_0. The parameter is set to $v_0 = 1$ in the references above, following a standard approach of normalizing.

The incomplete data likelihood function of the model is defined as

$$L_I(\boldsymbol{v}, \boldsymbol{\lambda}) = \prod_{t=1}^{T} \Bigg[P(m_t \text{ customers buy in period } t | \boldsymbol{v}, \boldsymbol{\lambda}) \frac{m_t!}{z_{1t}! z_{2t}! \cdots z_{nt}!} \prod_{j \in S_t} \left[\frac{P_j(S_t, \boldsymbol{v})}{\sum_{i \in S_t} P_i(S_t, \boldsymbol{v})} \right]^{z_{jt}} \Bigg],$$

where

$$P(m_t \text{ customers buy in period } t | \boldsymbol{v}, \boldsymbol{\lambda}) = \frac{\left[\lambda_t \sum_{i \in S_t} P_i(S_t, \boldsymbol{v})\right]^{m_t} \exp\left(-\lambda_t \sum_{i \in S_t} P_i(S_t, \boldsymbol{v})\right)}{m_t!}.$$

In the above equations z_{it} denotes the number of purchases of product i at time period t, and $m_t = \sum_{i=1}^{n} z_{it}$ denotes the total number of purchases in period t. After some algebra the log-likelihood function can be written as

$$l_I(\boldsymbol{v}, \boldsymbol{\lambda}) = \sum_{t=1}^{T} \left[m_t \log\left(\frac{\lambda_t}{v_0 + \sum_{i \in S_t} v_i} \right) - \lambda_t \frac{\sum_{i \in S_t} v_i}{v_0 + \sum_{i \in S_t} v_i} + \sum_{i \in S_t} z_{it} \log(v_i) \right]. \tag{2}$$

One approach is to directly maximize the log-likelihood function and jointly estimate the preference weights of the products and the arrival rates of customers. The above log-likelihood function, however, is hard to solve in general, therefore the research literature discusses different approaches to estimate the parameters of this model, given sales data and information on what was available for sale. Vulcano et al. (2012) developed an elegant EM algorithm by looking at the problem in terms of primary (first choice) demand, and treating the observed demand as incomplete observations of primary demand. Abdallah and Vulcano (2016) solves the estimation problem by specializing the minorization–maximization (MM) procedure, which is an iterative algorithm for maximizing an objective function by successively maximizing a simpler function that minorizes the true objective function.

It is also interesting to note that the objective function has a continuum of maximizers. For this reason, Vulcano et al. (2012) and Abdallah and Vulcano (2016) imposed additional constraint on the preference weights as a function of the market share

$$s = \frac{\sum_{i=1}^{n} v_i}{v_0 + \sum_{i=1}^{n} v_i}.$$

There are a number of limiting assumptions and possible extensions of the model presented above, when we apply it to real data.

(1) *Products are fully open or closed for sale*
The discussed model assumes that a product is either available or not for sale. Practitioners often work with aggregated data, and unable to capture the sales at a very granular level, every time the assortment changes. It is very practical to extend the model to consume data where we observe partial availability. For example, a product can be 80% open for sale in a time period.
(2) *MNL assumption*
Customers choose from the available products according to an MNL model. One of the properties of the MNL model is the independence of irrelevant alternatives (IIA), which can be unrealistic in real applications. If customers, for instance, always purchase the product with the lowest price, the algorithm in Vulcano et al. (2012) does not converge. If the sell-down is strong, the demand is grossly overestimated.
(3) *Unconstrained parameter space*
In practice we can encounter parameter estimates which are unreasonable in a business scenario. This may be due to the fact that the underlying assumptions of the model (such as MNL) do not exactly fit the true data generating model. Instead of using a more sophisticated model and develop its solution algorithm, we might want to simply constrain the parameters of the simpler model. Abdallah and Vulcano (2016) discusses regularization as an elegant solution to the problem, and they develop algorithms for L_1 regularization (Lasso regression) and L_2 regularization (Ridge regression). It can be also of interest to solve the problem by putting an upper bound on the arrival rate of customers (λ_t). This can be helpful to regularize the model by having an interpretable business fence on these parameters.
(4) *Homogeneous product set*
The model assumes that over different time periods we have the same set of existing products. Some might be unavailable for sale, but there exist an underlying demand for them. In revenue management, we often encounter changing product IDs (unique flight identifiers) due to schedule changes, hence we need to be able to model a non-homogeneous product set over time. Instead of handling the homogeneous parts separately, we would like to use the data over a larger time interval to borrow power to estimate the parameters.
(5) *No-purchase option is always available*
The model above assumes that the no-purchase option is always available, that is fully open. If we include competitor's products into the no-purchase option, this can be an unrealistic assumption. For instance, airline competitors likely control their inventory similarly as the host airline, gradually closing down products as we get closer to departure. This assumption has implications on how the host market share is interpreted in the models.

In this paper we address some of these points, and discuss how we could relax these assumptions, incorporate them into the model, and estimate the parameters. These ideas might foster further extensions of the existing models and rigorous research in the future.

Partial availability

Practitioners can encounter data at an aggregate level, where we can observe partial availability of products. As an example, in a revenue management system we store information on sales and availability at certain pre-departure time points. It is possible that a product becomes closed for sale during the time period, and we observe partial availability. For instance, a booking class on a flight can be open 60% of the time in a time interval, and the bookings we observe are matched to this partial availability. It would be of interest to extend the algorithms to work with this type of data and handle the full spectrum of availability in [0, 1], as opposed to be restricted to either open (1) or closed (0) products. Another possible venue is to modify the data to fit the algorithms already developed in the literature.

Extending the attraction model

A natural way to incorporate partial availability is to extend the MNL purchase probabilities as

$$P^{\star}_{jt}(S_t, \boldsymbol{v}, \boldsymbol{o}) = \frac{v_j \cdot o_{jt}}{v_0 + \sum_{i \in S_t} v_i \cdot o_{it}},$$

where $o_{it} \in [0, 1]$ represents availability of product i at time t. It is obvious that $o_{it} > 0$ is equivalent to $i \in S_t$. This simple formulation modifies the purchase probabilities by linearly adjusting preference weight v_i with open percentage o_{it}. If the open percentage is zero or one, we get back to the original formulation.

We have not derived the EM algorithm of Vulcano et al. (2012) with the extended purchase probability definition, but this could be an interesting venue for research. For demonstration purposes we can resort to directly maximizing the log-likelihood using a solver. The modified likelihood function becomes

$$L_I(\boldsymbol{v}, \lambda) = \prod_{t=1}^{T} \left[P(m_t \text{ customers buy in period } t \mid \boldsymbol{v}, \lambda) \frac{m_t!}{z_{1t}! z_{2t}! \cdots z_{nt}!} \prod_{j \in S_t} \left[\frac{P_j^\star(S_t, \boldsymbol{v}, \boldsymbol{o})}{\sum_{i \in S_t} P_i^\star(S_t, \boldsymbol{v}, \boldsymbol{o})} \right]^{z_{jt}} \right], \tag{3}$$

where

$$P(m_t \text{ customers buy in period } t \mid \boldsymbol{v}, \lambda) = \frac{\left[\lambda_t \sum_{i \in S_t} P_i^\star(S_t, \boldsymbol{v}, \boldsymbol{o}) \right]^{m_t} \exp\left(-\lambda_t \sum_{i \in S_t} P_i^\star(S_t, \boldsymbol{v}, \boldsymbol{o}) \right)}{m_t!}$$

and, after omitting the constant terms, the log-likelihood function modifies to

$$l_I(\boldsymbol{v}, \lambda) = \sum_{t=1}^{T} \left[m_t \log \left(\frac{\lambda_t}{v_0 + \sum_{i \in S_t} v_i \cdot o_{it}} \right) - \lambda_t \frac{\sum_{i \in S_t} v_i \cdot o_{it}}{v_0 \sum_{i \in S_t} v_i \cdot o_{it}} + \sum_{i \in S_t} z_{it} \log(v_i) \right]. \tag{4}$$

Data disaggregation

In the previous section we discussed a simple formulation which could potentially be used to incorporate partial availability information into the purchase probability definition. Another approach to handle partial availability is to disaggregate the data to fully open and closed assortments, and use existing algorithms for the estimation. We can split the observed sales under partial availability by making a simple assumption that the sales are distributed uniformly over time.

To demonstrate this idea on a simple example, let us assume that the observed sales b and open percentages o for three available products are

$$\boldsymbol{b} = \begin{bmatrix} 1 \\ 2 \\ 5 \end{bmatrix}, \quad \boldsymbol{o} = \begin{bmatrix} 1.0 \\ 0.8 \\ 0.5 \end{bmatrix}.$$

We assumed that the elements of $\boldsymbol{o}$ are non-increasing from top to bottom. Then we can represent $\boldsymbol{o}$ as the sum of three fully open and closed assortments with weights α_i as

$$\boldsymbol{o} = \begin{bmatrix} 1.0 \\ 0.8 \\ 0.5 \end{bmatrix} = \alpha_1 \begin{bmatrix} 1 \\ 0 \\ 0 \end{bmatrix} + \alpha_2 \begin{bmatrix} 1 \\ 1 \\ 0 \end{bmatrix} + \alpha_3 \begin{bmatrix} 1 \\ 1 \\ 1 \end{bmatrix} = 0.2 \begin{bmatrix} 1 \\ 0 \\ 0 \end{bmatrix} + 0.3 \begin{bmatrix} 1 \\ 1 \\ 0 \end{bmatrix} + 0.5 \begin{bmatrix} 1 \\ 1 \\ 1 \end{bmatrix}.$$

Note that α_i, $i = 1, \ldots, n$ represent time proportions and $\sum_{i=1}^{n} \alpha_i = 1$. The results indicate that all products were available for sale 50% of the time, two products were available 30% of the time, and one product was available 20% of the time. A graphical representation of this example is shown in Fig. 1.

For the general case, the elements of $\boldsymbol{\alpha}$ can be calculated as the consecutive differences in the open percentages, that is

$$\alpha_i = o_i - o_{i+1}, \quad i = 1, \ldots, n-1,$$
$$\alpha_n = o_n.$$

Following this simple idea we can split the observed sales under partial availability using the following identity

$$\boldsymbol{b} = \begin{bmatrix} \frac{\alpha_1}{\sum_{i=1}^{n} \alpha_i} \\ 0 \\ \vdots \\ 0 \\ 0 \end{bmatrix} \circ \begin{bmatrix} b_1 \\ b_2 \\ \vdots \\ b_{n-1} \\ b_n \end{bmatrix} + \begin{bmatrix} \frac{\alpha_2}{\sum_{i=1}^{n} \alpha_i} \\ \frac{\alpha_2}{\sum_{i=2}^{n} \alpha_i} \\ \vdots \\ 0 \\ 0 \end{bmatrix} \circ \begin{bmatrix} b_1 \\ b_2 \\ \vdots \\ b_{n-1} \\ b_n \end{bmatrix} + \ldots + \begin{bmatrix} \frac{\alpha_{n-1}}{\sum_{i=1}^{n} \alpha_i} \\ \frac{\alpha_{n-1}}{\sum_{i=2}^{n} \alpha_i} \\ \vdots \\ \frac{\alpha_{n-1}}{\sum_{i=n-1}^{n} \alpha_i} \\ 0 \end{bmatrix} \circ \begin{bmatrix} b_1 \\ b_2 \\ \vdots \\ b_{n-1} \\ b_n \end{bmatrix} + \begin{bmatrix} \frac{\alpha_n}{\sum_{i=1}^{n} \alpha_i} \\ \frac{\alpha_n}{\sum_{i=2}^{n} \alpha_i} \\ \vdots \\ \frac{\alpha_n}{\sum_{i=n-1}^{n} \alpha_i} \\ \frac{\alpha_n}{\alpha_n} \end{bmatrix} \circ \begin{bmatrix} b_1 \\ b_2 \\ \vdots \\ b_{n-1} \\ b_n \end{bmatrix},$$

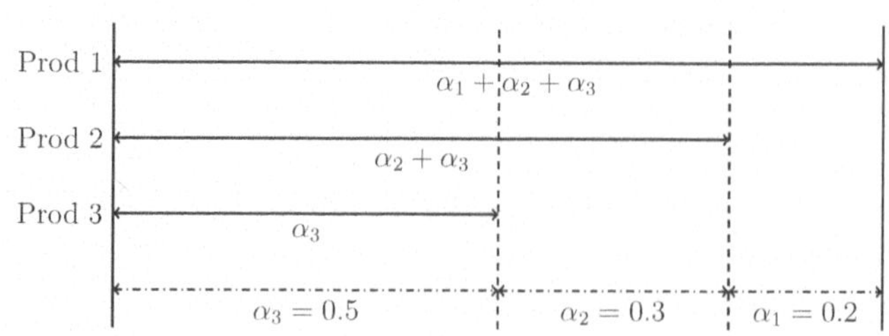

Fig. 1 Split example

where ∘ denotes the elementwise multiplication, or Hadamard product. Note that if $\alpha_i = 0$, we do not need a split to create a new assortment. The sales splitting algorithm under partial availability is described in Algorithm 1.

Algorithm 1 Sales splitting algorithm for partial availability

$\boldsymbol{b}$: vector of observed sales
$\boldsymbol{o}$: vector of observed open percentages
Sort $\boldsymbol{o}$ in decreasing order and apply the order to $\boldsymbol{b}$
Calculate time proportion vector $\boldsymbol{\alpha}$
for $i = 1, \ldots, n$ **do**
    if $i = n$ **then**
        $\alpha_i = o_i$
    else
        $\alpha_i = o_i - o_{i+1}$
    end if
end for
Split sales
for $i = 1, \ldots, n$ **do**
    if $\alpha_i = 0$ **then**
        Continue
    else
        Calculate partial sales vector $\boldsymbol{b}^{(i)}$ and open percentage vector $\boldsymbol{o}^{(i)}$
        for $j = 1, \ldots, n$ **do**
            if $j \leq i$ **then**
                $b_j^{(i)} = \frac{\alpha_i}{o_j} b_j$
                $o_j^{(i)} = 1$
            else
                $b_j^{(i)} = 0$
                $o_j^{(i)} = 0$
            end if
        end for
    end if
end for

Table 1 Example with partial availability

	Period														
	15	14	13	12	11	10	9	8	7	6	5	4	3	2	1
Sales															
1	10	15	11	14	0	0	0	0	0	0	0	0	0	0	0
2	11	6	11	8	20	16	0	0	0	0	0	0	0	0	0
3	5	6	1	11	4	5	14	7	11	0	0	0	0	0	0
4	4	4	4	1	6	4	3	5	9	9	6	9	0	0	0
5	0	2	0	0	1	0	1	3	0	3	3	5	2	3	3
Open percentage															
1	0.7	0.3	1.0	1.0	0.0	0.0	0.0	0.0	0.0	0.0	0.0	0.0	0.0	0.0	0.0
2	0.8	0.5	1.0	1.0	0.3	0.3	0.0	0.0	0.0	0.0	0.0	0.0	0.0	0.0	0.0
3	0.9	0.9	1.0	1.0	0.3	0.5	0.9	0.7	0.7	0.0	0.0	0.0	0.2	0.2	0.0
4	1.0	0.9	1.0	1.0	0.5	1.0	1.0	0.8	0.9	0.6	0.5	0.2	0.3	0.5	0.0
5	1.0	1.0	1.0	1.0	1.0	1.0	1.0	0.9	1.0	1.0	1.0	1.0	1.0	1.0	1.0

Table 2 Estimated demand and parameters using the extended attraction model

Estimates	Period															v_i
	15	14	13	12	11	10	9	8	7	6	5	4	3	2	1	
1	15.02	20.07	12.47	15.70	33.60	22.73	19.60	19.33	24.84	38.14	32.27	84.35	5.74	7.22	35.06	1.000
2	11.23	15.01	9.32	11.74	25.13	17.00	14.66	14.45	18.58	28.53	24.14	63.08	4.29	5.40	26.22	0.748
3	3.91	5.22	3.24	4.09	8.74	5.92	5.10	5.03	6.46	9.93	8.40	21.95	1.49	1.88	9.13	0.260
4	1.97	2.63	1.63	2.06	4.41	2.98	2.57	2.53	3.26	5.00	4.23	11.06	0.75	0.95	4.60	0.131
5	0.40	0.53	0.33	0.41	0.89	0.60	0.52	0.51	0.66	1.01	0.85	2.23	0.15	0.19	0.93	0.026
λ_t	46.48	62.10	38.57	48.57	103.95	70.32	60.65	59.79	76.84	118.01	99.84	260.94	17.76	22.34	108.48	

The algorithm results in partial sales vectors $\boldsymbol{b}^{(i)}$ and fully open and closed assortment vectors $\boldsymbol{o}^{(i)}$ for $i \in \{j : \alpha_j > 0\}$. The splitting logic ensures that

$$\boldsymbol{b} = \sum_{i \in \{j:\alpha_j > 0\}} \boldsymbol{b}^{(i)}.$$

After the splitting logic, we could combine intervals from different time t which have exactly the same product offerings and product availabilities, to improve performance.

Comparison on a simulated example

To demonstrate the estimation from sales data with partial availability, we extended the example from Vulcano et al. (2012) by adding partial availability data to the observed sales. The observed data are presented in Table 1.

We first use the extended attraction model discussed in Extending the attraction model section. Direct maximization of the log-likelihood (4) by a nonlinear solver results in total primary demand of 1194.6. The detailed estimated demands and parameters are presented in Table 2.

Second, we demonstrate the data disaggregation procedure from Sect. Data disaggregation. We first split the observed sales data with partial availability to fully open and closed assortments, apply the EM algorithm of Vulcano et al. (2012), and then aggregate the solution back. To demonstrate this, let us apply Algorithm 1 on the data presented in Table 1. The disaggregated sales are shown in Table 3.

After applying the EM algorithm on the disaggregated data, we end up with the disaggregated solution, as presented in Table 4.

Note that the estimated primary demands do not preserve the split proportions of the sales. For instance, in period 9, $\lambda_{9_1} = 2.55$ and $\lambda_{9_2} = 52.89$, while the sales were split by proportions $\alpha_1 = 0.1$ and $\alpha_2 = 0.9$. Note that the split proportions of sales do not need to match the proportion of demands, because the varying assortments imply varying market shares and spilled demand. However, to reduce the number of parameters to be estimated, it would be possible to modify the EM algorithm and incorporate constraints on λ_{t_i} to preserve the split proportions. Aggregating the results back, we end up with the final results presented in Table 5.

The results are fairly similar to the results in Table 2, where we incorporated partial availability into the attraction model. The total primary demand is 1190.8 as opposed to 1194.6. To benchmark these solutions, a naive approach would be to use the projection method to preprocess the observed sales using the open percentage information, such as

$$\widehat{\text{sales}} = \frac{\text{sales}}{\text{openpct}},$$

and then apply EM algorithm on the preprocessed data. With this approach we estimate the overall primary demand as 1519.9. This shows that incorporating partial availability natively into the attraction model or disaggregating the sales data are much more robust approaches. In case we observe small outliers in partial availability data, we end up with very large preprocessed sales using the projection method. Experiments showed that the disaggregation method dampens the effect of outliers the most.

Strong sell-down

The demand model discussed in Vulcano et al. (2012) and Abdallah and Vulcano (2016) assumes that customers choose from the available products based on the Basic Attraction Model (BAM) with the IIA property. The probability of product selection is governed by (1). This model cannot fit to scenarios with strong sell-down, and the same applies to the Generalized Attraction Model (GAM) discussed in Gallego et al. (2015). Strong sell-down can be a common phenomena in practice, because customers often seek to purchase the cheapest available product. A 100% sell-down example is presented in Table 6.

The EM algorithm developed in Vulcano et al. (2012) does not converge on this example, because the attraction model is unable to fit to the scenario presented in the data. The estimated first choice demands will converge to an unbounded solution.

A natural way to handle sell-down is by introducing an additional parameter l for the lowest available product. The attraction model (1) would be extended as

$$P_i(S_t, \boldsymbol{v}, L_t, l) = \frac{v_i + \mathbb{1}(i \in L_t) \cdot l}{v_0 + \sum_{j \in S_t} \left(v_j + \mathbb{1}(j \in L_t) \cdot l\right)}. \tag{5}$$

Table 3 Disaggregated sales of Table 1

Sales	Period 15				14				13	12	11			10			9	
1				10.00				15.00	11.00	14.00								
2			1.38	9.62			2.40	3.60	11.00	8.00			20.00			16.00		
3		0.56	0.56	3.89		2.67	1.33	2.00	1.00	11.00			4.00		2.00	3.00		14.00
4	0.40	0.40	0.40	2.80		1.78	0.89	1.33	4.00	1.00		2.40	3.60	2.00	0.80	1.20	0.30	2.70
5	0.00	0.00	0.00	0.00	0.20	0.80	0.40	0.60	0.00	0.00	0.50	0.20	0.30	0.00	0.00	0.00	0.10	0.90

	8			7			6		5		4		3			2			1
1																			
2																			
3			7.00			11.00									0.00			0.00	
4		0.63	4.37		2.00	7.00		9.00		6.00		9.00		0.00	0.00		0.00	0.00	
5	0.33	0.33	2.33	0.00	0.00	0.00	1.20	1.80	1.50	1.50	4.00	1.00	1.40	0.20	0.40	1.50	0.90	0.60	3.00

Table 4 Estimated demand and parameters of disaggregated sales data

Estimates	Period 15				14				13	12	11			10			9		v_i
1	0.81	0.91	1.33	10.00	2.41	5.02	2.86	15.00	11.00	14.00	6.03	5.27	15.88	4.05	2.68	11.50	0.81	16.85	1.000
2	0.59	0.67	0.94	9.62	1.76	3.65	1.64	3.60	11.00	8.00	4.39	3.83	13.63	2.95	1.95	10.90	0.59	12.26	0.727
3	0.24	0.25	0.38	3.89	0.71	1.20	0.91	2.00	1.00	11.00	1.77	1.55	2.73	1.19	0.90	2.04	0.24	6.29	0.294
4	0.14	0.18	0.27	2.80	0.36	0.80	0.61	1.33	4.00	1.00	0.91	0.85	2.45	0.71	0.36	0.82	0.11	1.21	0.150
5	0.00	0.00	0.00	0.00	0.06	0.36	0.27	0.60	0.00	0.00	0.15	0.07	0.20	0.00	0.00	0.00	0.04	0.40	0.026
λ_t	2.55	2.87	4.16	37.59	7.57	15.76	8.97	32.19	38.57	48.57	18.94	16.54	49.85	12.73	8.41	36.09	2.55	52.89	

	8			7			6		5		4		3			2			1	v_i
1	4.02	1.94	13.13	0.00	4.05	17.23	14.48	21.90	18.10	15.21	48.27	20.27	16.89	0.41	0.38	18.10	1.82	0.57	36.20	1.000
2	2.93	1.41	9.55	0.00	2.95	12.54	10.53	15.93	13.17	11.06	35.11	14.75	12.29	0.29	0.28	13.17	1.33	0.42	26.33	0.727
3	1.18	0.57	3.15	0.00	1.19	4.95	4.26	6.44	5.32	4.47	14.19	5.96	4.97	0.12	0.00	5.32	0.54	0.00	10.64	0.294
4	0.60	0.22	1.97	0.00	0.71	3.15	2.17	3.20	2.72	2.14	7.24	3.20	2.53	0.00	0.00	2.72	0.00	0.00	5.43	0.150
5	0.10	0.12	1.05	0.00	0.00	0.00	0.37	0.64	0.46	0.53	1.23	0.36	0.43	0.07	0.18	0.46	0.32	0.27	0.92	0.026
λ_t	12.62	6.10	41.19	0.00	12.73	54.09	45.45	68.72	56.81	47.72	151.49	63.63	53.02	1.27	1.20	56.81	5.73	1.80	113.62	

Parameter l is an additional preference weight for the product with the lowest price, or the product with excess attraction in the assortment. In practice we could add additional preference weights by group of products, depending on the structure of the problem. L is a set of the indices for the lowest available product over time, and $\mathbb{1}(i \in L_t)$ is an indicator function, representing whether product i is the lowest available product at time t.

For the example in Table 6, $L = (1, 1, 2, 2, 3, 3)$. $L_3 = 2$ means that at $t = 3$ the lowest available class is the second class from the top.

To present a motivating example with strong sell-down: assuming that the true parameters are $v_0 = 1$, $v_1 = 0.4$, $v_2 = 0.7$, $v_3 = 0.1$ and $l = 10$, the purchase probabilities for various assortments, induced by (5), are shown in Table 7.

These purchase probabilities can describe a realistic scenario, where the lowest available product receives majority of the demand. The products available above, however, still maintain the IIA property. This extended model fits data better with strong sell-down, which cannot be described by the basic MNL model.

Essentially the same idea, with more parameters than just the scalar l and a slightly more general formulation, the spiked-MNL model was considered in Cao et al. (2019). The spiked-MNL model extends the classical MNL model by having a separate attractiveness parameter for the cheapest available fare class on each flight.

The likelihood function for the extended model with parameter l becomes

$$L_l(\lambda, \mathbf{v}, l) = \prod_{t=1}^{T} \Big[P(m_t \text{ customers buy in period } t | \mathbf{v}, \lambda, l)$$

$$\frac{m_t!}{z_{1t}! z_{2t}! \cdots z_{nt}!}$$

$$\prod_{j \in S_t} \left[\frac{P_j(S_t, \mathbf{v}, L_t, l)}{\sum_{i \in S_t} P_i(S_t, \mathbf{v}, L_t, l)} \right]^{z_{jt}} \Bigg],$$

where

$$P(m_t \text{ customers buy in period } t | \mathbf{v}, \lambda, l)$$
$$= \frac{\left[\lambda_t \sum_{i \in S_t} P_i(S_t, \mathbf{v}, L_t, l)\right]^{m_t} \exp\left(-\lambda_t \sum_{i \in S_t} P_i(S_t, \mathbf{v}, L_t, l)\right)}{m_t!}$$

and the log-likelihood function modifies to

Table 5 Estimated demand and parameters using data disaggregation

Estimates	Period															v_i
	15	14	13	12	11	10	9	8	7	6	5	4	3	2	1	
1	13.05	25.29	11.00	14.00	27.19	18.23	17.66	19.09	21.29	36.38	33.31	68.54	17.68	20.50	36.20	1.000
2	11.82	10.64	11.00	8.00	21.85	15.80	12.85	13.89	15.48	26.46	24.22	49.85	12.86	14.91	26.33	0.727
3	4.76	4.82	1.00	11.00	6.05	4.14	6.53	4.90	6.14	10.70	9.79	20.15	5.09	5.86	10.64	0.294
4	3.39	3.10	4.00	1.00	4.21	1.89	1.32	2.79	3.86	5.38	4.85	10.45	2.53	2.72	5.43	0.150
5	0.00	1.29	0.00	0.00	0.43	0.00	0.44	1.27	0.00	1.01	1.00	1.59	0.68	1.05	0.92	0.026
λ_t	47.17	64.50	38.57	48.57	85.33	57.23	55.43	59.92	66.82	114.17	104.53	215.12	55.50	64.34	113.62	

Table 6 Example with 100% sell-down

Sales	Period					
	6	5	4	3	2	1
1	2	6	0	0	0	0
2			13	15	0	0
3					20	22

Table 7 Purchase probabilities with additional preference for lowest available product

Probability	Assortment		
	$L_1 = 1$	$L_2 = 2$	$L_3 = 3$
P_0	8.8%	8.3%	8.2%
P_1	91.2%	3.3%	3.3%
P_2		88.4%	5.7%
P_3			82.8%

$$l_I(\mathbf{v}, \lambda, l) = \sum_{t=1}^{T} \left[m_t \log \left(\frac{\lambda_t}{v_0 + \sum_{i \in S_t} \left(v_i + \mathbb{1}(i \in L_t) \cdot l \right)} \right) - \lambda_t \frac{\sum_{i \in S_t} \left(v_i + \mathbb{1}(i \in L_t) \cdot l \right)}{v_0 + \sum_{i \in S_t} \left(v_i + \mathbb{1}(i \in L_t) \cdot l \right)} + \sum_{i \in S_t} z_{it} \log \left(v_i + \mathbb{1}(i \in L_t) \cdot l \right) \right]. \tag{6}$$

Note that in this paper we are not providing an algorithm for maximizing the log-likelihood (6), but we can rely on available nonlinear solvers, if needed. It would be of practical interest, however, to extend the EM (Vulcano et al. 2012) and MM (Abdallah and Vulcano 2016) algorithms to estimate the parameters of this extended model.

Constrained parameter space

Algorithms based on the model assumptions above can lead to unreasonable estimates in practice. For instance, on airline data we observed estimated demands which were high in a business scenario. This can happen due to data sparsity, outliers, data preprocessing steps, and most likely that the assumptions of the MNL model do not fit the true data generating model. Therefore, in practical applications, we might want to constrain the parameters of the model. Abdallah and Vulcano (2016) discusses regularization as an elegant solution to the problem, and they develop algorithms for L_1 regularization (Lasso regression) and L_2 regularization (Ridge regression). It can be also of interest to apply constraints on the arrival rate of customers (λ_t), having an interpretable business fence on these parameters.

In practice we could solve a constrained maximum likelihood problem, such as

$$\max_{\lambda, \mathbf{v}} \; l_I$$

s.t.

$$\sum_{t=1}^{T} \lambda_t \le L,$$

where the constraint ensures that the overall arrival rate or mean total demand over the time horizon is less than an upper bound L. L could be defined, for instance, as C times the total observed sales, that is

$$L = C \frac{1}{s} \sum_{t=1}^{T} \sum_{i=1}^{n} z_{it}.$$

C is a regularization parameter which has to be set based on practical considerations. Note that λ_t includes the no-purchase alternative, so we scale the total sales with market share s. Alternatively, we could constrain the arrival rates at each time period and solve constrained maximum likelihood problem

$$\max_{\lambda, \mathbf{v}} \; l_I$$

$$s.t.$$

$$\lambda_t \le L_t, \quad t = 1, \ldots, T.$$

Similarly, we could set $L_t = C \frac{1}{s} \sum_{i=1}^{n} z_{it}$, a constant times the observed sales at time period t. It would be of practical interest to extend the EM algorithm of Vulcano et al. (2012) to solve this constrained problem. In Sect. Constrained optimization we will extend the MM algorithm in Abdallah and Vulcano (2016) to this constrained problem and also discuss how to solve the optimization problem using the Frank–Wolfe algorithm.

Non-homogeneous product set

In revenue management, we often encounter changing products due to changes in the system, hence we need to deal with a non-homogeneous product set over time. Instead of dividing the observed data to homogeneous parts, we want to

Table 8 Schedule change example

	Sales	Period														
		1	2	3	4	5	6	7	8	9	10	11	12	13	14	15
1	flt1-prod1	10	15	11	14											
2	flt1-prod2	11	6	11	8	20	16									
3	flt1-prod3	5	6	1	11	4	5	14	7	11						
4	flt1-prod4	4	4	4	1	6	4	3	5	9	9	6	9			
5	flt1-prod5	0	2	0	0	1	0	1	3	0	3	3	5	2	3	3
6	flt2-prod1															
7	flt2-prod2															
8	flt2-prod3															
9	flt2-prod4															
10	flt2-prod5															
11	flt3-prod1	20	30	22	28											
12	flt3-prod2	22	12	22	16	40	32									
13	flt3-prod3	10	12	2	22	8	10	28	14	22						
14	flt3-prod4	8	8	8	2	12	8	6	10	18	18	12	18			
15	flt3-prod5	0	4	0	0	2	0	2	6	0	6	6	10	4	6	6

	Sales	Period														
		16	17	18	19	20	21	22	23	24	25	26	27	28	29	30
1	flt1-prod1															
2	flt1-prod2															
3	flt1-prod3															
4	flt1-prod4															
5	flt1-prod5															
6	flt2-prod1	10	15	11	14											
7	flt2-prod2	11	6	11	8	20	16									
8	flt2-prod3	5	6	1	11	4	5	14	7	11						
9	flt2-prod4	4	4	4	1	6	4	3	5	9	9	6	9			
10	flt2-prod5	0	2	0	0	1	0	1	3	0	3	3	5	2	3	3
11	flt3-prod1	20	30	22	28											
12	flt3-prod2	22	12	22	16	40	32									
13	flt3-prod3	10	12	2	22	8	10	28	14	22						
14	flt3-prod4	8	8	8	2	12	8	6	10	18	18	12	18			
15	flt3-prod5	0	4	0	0	2	0	2	6	0	6	6	10	4	6	6

extend the algorithms to handle non-homogeneous product offerings, allowing us to use the whole data set and borrow power to accurately estimate the parameters.

Let us look at a hypothetical airline sales example in Table 8, which was created from the synthetic example in Vulcano et al. (2012). We did this, so the estimated results are easy to compare.

We have 3 flights, each of them having 5 different products, a total of $n = 15$ products. The products of flight 1 only exist at the first 15 time periods, and after that the label changes to flight 2. This can happen for various reasons, for example schedule change. It is not always possible to preprocess the data and match the products of flight 1 to flight 2. We can see that the product offerings are non-homogeneous over time, or unbalanced. Product 1 of flight 1 is not available for sale for time periods 5–15, which is distinguished from not being part of the product offer set at time periods 16–30. Notice that the observed sales data of flight 1 for time periods 1–15 is the same as the synthetic example created in Vulcano et al. (2012), and the observed sales for flight 3 are just twice of that, like a flight with double the demand and capacity.

To introduce non-homogeneous product set into the notation, let I_t denote the set of existing products of the retailer at time t. Note that this is different from S_t, which is the set of products available for purchase at time t, therefore $S_t \subseteq I_t$. For instance, in Table 8, $I_7 = \{1, 2, 3, 4, 5, 11, 12, 13, 14, 15\}$ and $S_7 = \{3, 4, 5, 13, 14, 15\}$. The set of all products is $I = \bigcup_{t=1}^{T} I_t = \{1, 2, \ldots, 15\}$.

In Sect. Extended EM algorithm we will extend the EM algorithm of Vulcano et al. (2012) for non-homogeneous product offerings. The MM and Frank–Wolfe algorithms discussed in Sect. Constrained optimization will also be able to natively handle this product structure.

Handling market share constraint and no-purchase option

We mentioned that the objective function (2) has an infinite number of maximum likelihood estimates, therefore an additional constraint is applied in Vulcano et al. (2012) on the preference weights as a function of the market share

$$s = \frac{\sum_{i=1}^{n} v_i}{v_0 + \sum_{i=1}^{n} v_i} \tag{7}$$

using the standard scaling of $v_0 = 1$. The market share constraint is linear but the objective function is non-convex; therefore, Abdallah and Vulcano (2016) formulates the problem using $v_i = \exp(\beta_i)$ in the objective function, so the market share constraint becomes

$$s = \frac{\sum_{i=1}^{n} \exp(\beta_i)}{1 + \sum_{i=1}^{n} \exp(\beta_i)}.$$

They establish that the objective function in this space is a concave function, and they show how to deal with the non-convex market share constraint by solving the problem without the constraint and then using a transformation to satisfy the constraint.

The outside alternative could represent the competitor's product, or both the competitor's product and the no-purchase option, which would lead to different interpretations of s and λ (Vulcano et al. 2012). The outside alternative is treated as a separate product that is always available, therefore v_0 is constant and the standard scaling $v_0 = 1$ is used.

To look further, let I denote the set of all products of the retailer, that is $I = \{1, 2, \ldots, n\}$, so it follows that $S_t \subseteq I$. With the market share constraint above, the probability of selecting one of the retailer's product, when everything is available for sale, is s, that is

$$P_I(I, \boldsymbol{v}) = \frac{\sum_{j\in I} v_j}{v_0 + \sum_{j\in I} v_j} = s.$$

When the retailer offers an assortment S_t with a subset of all the products, it follows that

$$\tilde{s}_t = P_{S_t}(S_t, \boldsymbol{v}) = \frac{\sum_{j\in S_t} v_j}{v_0 + \sum_{j\in S_t} v_j} < s$$

given v_0 is fixed. This means that the model induced market share ($\tilde{s}_t$) of the retailer at time t is less than s, and the share of outside alternative is larger than $1 - s$. If we think of the outside alternative as the competitor's product, one could argue that the competitor's product might not always be available for purchase either, so the preference weight v_0 is not constant over time. Also, the estimated market share s is calculated over all possible sets of assortments, not only when all the retailer's products are available for sale.

To extend this model, we can relax the assumption of constant, time-independent outside alternative and introduce v_{0t}, allowing the preference weight to change over time. We also need this extension to be able to incorporate non-homogeneous product sets into the market share constraint, by using I_t, the offer set of retailer at time t (Sect. Non-homogeneous product set). The market share constraint modifies to a set of constraints

$$s = \frac{\sum_{i\in I_t} v_i}{v_{0t} + \sum_{i\in I_t} v_i}, \quad t = 1, \dots, n \tag{8}$$

which ensure that the share of outside alternative is $1 - s$, for all time points t, when all the retailer's products are available for sale. That is

$$P_0(I, \boldsymbol{v}) = \frac{v_{0t}}{v_{0t} + \sum_{i\in I_t} v_i} = \frac{\frac{1-s}{s}\sum_{i\in I_t} v_i}{\frac{1-s}{s}\sum_{i\in I_t} v_i + \sum_{i\in I_t} v_i} = 1 - s.$$

Now let us consider an edge case scenario where the retailer's model induced market share is constant over time regardless of the assortment he offers. This could be achieved with the set of market share constraints

$$s = \frac{\sum_{i\in S_t} v_i}{v_{0t} + \sum_{i\in S_t} v_i}, \quad t = 1, \dots, n \tag{9}$$

leading to

$$\tilde{s}_t = P_{S_t}(S_t, \boldsymbol{v}) = \frac{\sum_{j\in S_t} v_j}{v_{0t} + \sum_{j\in S_t} v_j} = \frac{\sum_{j\in S_t} v_j}{\frac{1-s}{s}\sum_{j\in S_t} v_j + \sum_{j\in S_t} v_j} = s.$$

The retailers's share remains s and the share of the outside alternative remains $1 - s$, at each time t, regardless of what products are available for purchase. This edge case might not make sense for all applications, but it is interesting to consider as an alternative to the assumption that the outside alternative is always fully available. In the airline retail context we could think of the outside alternative as the competitor's products whose availability can be as limited as the host airline's products.

To introduce a continuum of cases between (8) and (9), let us introduce parameter $\alpha \in [0, 1]$ controlling the availability of outside alternative. $\alpha = 0$ represents the case where the outside alternative is always available for sale (8), and naturally, $\alpha = 1$ represents the case where the outside alternative limits availability the same fashion as the retailer. This is more of a hypothetical case, given the outside alternative could include the no-purchase option, which is always an available choice. Combining (8) and (9), the set of market share constraints become

$$s = (1-\alpha)\frac{\sum_{i\in I_t} v_i}{v_{0t} + \sum_{i\in I_t} v_i} + \alpha\frac{\sum_{i\in S_t} v_i}{v_{0t} + \sum_{i\in S_t} v_i}, \quad t = 1, \dots, n. \tag{10}$$

Using $\alpha = 0$, $I_t = I = \{1, \dots, n\}$ and $v_{0t} = 1$, the constraints simplify to (7).

The models require the knowledge of market share s, which in practice can be difficult to acquire, and the estimate itself can be inaccurate. Abdallah and Vulcano (2016) included a study on the sensitivity of model estimates to the input market share. We mentioned that, in practice, market share s is most likely estimated from data observed over various sets of assortments, not only when all the retailer's products are available for sale. Therefore, it could also make sense to use a market share constraint aggregated over time horizon T, that is

$$s = \frac{\sum_{t=1}^{T}\sum_{i\in I_t} v_i}{\sum_{t=1}^{T}\left(v_{0t} + \sum_{i\in I_t} v_i\right)}. \tag{11}$$

This constraint ensures that the market share over the time horizon T is s, taking into account the changing offer set I_t over time. This is less restrictive and could be a more reasonable assumption than using (7), which forces the market share at each time point t with a set of constraints. Adding α to control availability of the outside alternative, we would use

$$s = \frac{\sum_{t=1}^{T}\left[(1-\alpha)\sum_{i\in I_t} v_i + \alpha\sum_{i\in S_t} v_i\right]}{\sum_{t=1}^{T}\left[v_{0t} + (1-\alpha)\sum_{i\in I_t} v_i + \alpha\sum_{i\in S_t} v_i\right]}. \tag{12}$$

Finally, we want to point out that instead of solving the market share constrained optimization problem

$$\max_{\lambda,v} \sum_{t=1}^{T} \left[m_t \log\left(\frac{\lambda_t}{v_0 + \sum_{i\in S_t} v_i} \right) - \lambda_t \frac{\sum_{i\in S_t} v_i}{v_0 + \sum_{i\in S_t} v_i} + \sum_{i\in S_t} z_{it} \log(v_i) \right]$$

s.t.

$$s = \frac{\sum_{i=1}^{n} v_i}{v_0 + \sum_{i=1}^{n} v_i}$$

$$v_0 = 1$$

we can directly incorporate the market share constraint into the objective function through $v_0 = r\sum_{i=1}^{n} v_i$, with $r = (1-s)/s$, and solve

$$\max_{\lambda,v} \sum_{t=1}^{T} \left[m_t \log\left(\frac{\lambda_t}{r\sum_{i=1}^{n} v_i + \sum_{i\in S_t} v_i} \right) - \lambda_t \frac{\sum_{i\in S_t} v_i}{r\sum_{i=1}^{n} v_i + \sum_{i\in S_t} v_i} + \sum_{i\in S_t} z_{it} \log(v_i) \right]$$

s.t.

$$\sum_{i=1}^{n} v_i = 1$$

or by setting $v_1 = 1$. The scaling constraint on $\boldsymbol{v}$ is required to avoid multiple solutions.

Incorporating non-homogeneous product set and control of availability of outside alternative, we would need to solve

$$\max_{\lambda,v} \sum_{t=1}^{T} \left[m_t \log\left(\frac{\lambda_t}{v_{0t} + \sum_{i\in S_t} v_i} \right) - \lambda_t \frac{\sum_{i\in S_t} v_i}{v_{0t} + \sum_{i\in S_t} v_i} + \sum_{i\in S_t} z_{it} \log(v_i) \right]$$

s.t.

$$v_{0t} = r\left[(1-\alpha)\sum_{i\in I_t} v_i + \alpha \sum_{i\in S_t} v_i \right]$$

$$\sum_{i=1}^{n} v_i = 1.$$

We will see in Sect. Solution using Frank–Wolfe method that this formulation with upper bound constraints on the arrival rate is easy to solve using the Frank–Wolfe method.

Extended EM algorithm

In this section we extend the EM algorithm of Vulcano et al. (2012) with non-homogeneous product sets (Sect. Non-homogeneous product set) and the ability to control the availability of outside alternative (Sect. Handling market share constraint and no-purchase option).

We incorporate I_t, the offer set of retailer at time t, and v_{0t}, the time-dependent preference weight for the outside alternative into the algorithm. We use

$$v_{0t} = r\left[(1-\alpha)\sum_{i\in I_t} v_i + \alpha \sum_{i\in S_t} v_i \right],$$

where $r = \frac{1-s}{s}$ and $\alpha \in [0,1]$ controls the availability of the outside alternative. Adding these into the formulation, we can modify the key E-step equations of the EM algorithm as

$$\hat{X}_{jt} = \begin{cases} \frac{v_j}{(1+r)\sum_{i\in I_t} v_i} \frac{\sum_{h\in S_t} v_h + v_{0t}}{\sum_{h\in S_t} v_h} \sum_{h\in S_t} z_{ht}, & \text{if} j \notin S_t \cup \{0\}, \\ \frac{\sum_{h\in S_t} v_h + v_{0t}}{(1+r)\sum_{i\in I_t} v_i} z_{jt}, & \text{if} j \in S_t, \end{cases}$$

$$\hat{Y}_{jt} = \frac{\sum_{h\notin S_t\cup\{0\}} v_h - v_{0t} + r\sum_{i\in I_t} v_i}{(1+r)\sum_{i\in I_t} v_i} z_{jt}, \quad j \in S_t,$$

$$\hat{X}_{0t} = r\sum_{i\in I_t} \hat{X}_{it},$$

$$\hat{Y}_{0t} = \frac{v_{0t}}{\sum_{i\in S_t} v_i + v_{0t}} \sum_{h\notin S_t\cup\{0\}} \hat{X}_{ht}.$$

For the M-step, we need to maximize the conditional expected, complete data log-likelihood function, which becomes

$$\mathcal{L}(\boldsymbol{v}) = \sum_{t=1}^{T}\sum_{j\in I_t} \hat{X}_{jt} \log\left(\frac{v_j}{(1+r)\sum_{i\in I_t} v_i} \right) + \sum_{t=1}^{T} \hat{X}_{0t} \log(s). \tag{13}$$

To find the maxima, the first-order conditions become

$$\frac{\partial \mathcal{L}}{\partial v_i} = \sum_{t=1}^{T} \mathbb{1}\left[i \in I_t\right] \left(\frac{\hat{X}_{it}}{v_i} - \frac{\sum_{k\in I_t} \hat{X}_{kt}}{\sum_{k\in I_t} v_k} \right) = 0, \quad i = 1, \ldots, n$$

which is a system of n nonlinear equations in v_i. The solution can be obtained with existing implementations of iterative methods, such as Newton–Raphson. We can also use a simple fixed point iteration algorithm by rewriting $\nabla\mathcal{L} = F(\boldsymbol{v}) = 0$ to $\boldsymbol{\Phi}(\boldsymbol{v}) = \boldsymbol{v}$ and then using the fixed point iteration $\boldsymbol{v}^{(l+1)} := \boldsymbol{\Phi}(\boldsymbol{v}^{(l)})$. In our case the update becomes

$$v_i^{(l+1)} := \frac{\sum_{t=1}^{T} \mathbb{1}\left[i \in I_t\right] X_{it}}{\sum_{t=1}^{T} \mathbb{1}\left[i \in I_t\right] \frac{\sum_{k\in I_t} X_{kt}}{\sum_{k\in I_t} v_k^{(l)}}}, \quad i = 1, \ldots, n.$$

In case of homogeneous product set, so that $I_t = \{1, \ldots, n\}$, $\forall t$, we get a closed form solution to the M-step, that is

$$v_i = \frac{\sum_{t=1}^{T} X_{it}}{\sum_{j=1}^{n}\sum_{t=1}^{T} X_{jt}}, \qquad i = 1, \ldots, n.$$

The result is easy to obtain by assuming $\sum_{i=1}^{n} v_i = 1$. Note that this closed form equation is different from the one derived in Vulcano et al. (2012), since they used a different parametrization by setting $v_0 = 1$.

Algorithm 2 presents the extended EM algorithm with non-homogeneous product set and the ability to control the availability of outside alternative.

Algorithm 2 EM algorithm with non-homogeneous product set and the control of availability of outside alternative (OA)

I_t: set of offered products at time t (product set)
S_t: set of products available for purchase at time t ($S_t \subset I_t$)
$r = \frac{1-s}{s}$, where s is market share of the retailer
z_{jt}: observed sales at time t for product j
α: control parameter ($\alpha = 1$: OA available as retailer; $\alpha = 0$: OA fully available)
E-step
for $t = 1, \ldots, T$ **do**
  $v_{0t} = r\left[(1-\alpha)\sum_{i\in I_t} v_i + \alpha \sum_{h\in S_t} v_h\right]$ ▷ OA preference weight at t
  for $j \in I_t$ **do**
    if $j \notin S_t$ **then**
      $X_{jt} = \frac{v_j}{(1+r)\sum_{i\in I_t} v_i} \frac{\sum_{h\in S_t} v_h + v_{0t}}{\sum_{h\in S_t} v_h} \sum_{h\in S_t} z_{ht}$
    else ($j \in S_t$)
      $Y_{jt} = \frac{\sum_{h\notin S_t} v_h - v_{0t} + r\sum_{i\in I_t} v_i}{(1+r)\sum_{i\in I_t} v_i} z_{jt}$
      $X_{jt} = z_{jt} - Y_{jt}$
    end if
  end for
  $X_{ot} = r\sum_{i\in I_t} X_{it}$
  $Y_{ot} = \frac{v_{0t}}{\sum_{h\in S_t} v_h + v_{0t}} \sum_{h\notin S_t} X_{ht}$
end for
M-step
Find v_i, $i = 1, \ldots, n$ as solution to the system of nonlinear equations $F(\boldsymbol{v}) = 0$:
$\sum_{t=1}^{T} \mathbb{1}\left[i \in I_t\right]\left(\frac{X_{it}}{v_i} - \frac{\sum_{k\in I_t} X_{kt}}{\sum_{k\in I_t} v_k}\right) = 0, \quad i = 1, \ldots, n$
Special case (homogeneous product set), if $I_t = \{1, \ldots, n\}$, $\forall t$:
$v_i = \frac{\sum_{t=1}^{T} X_{it}}{\sum_{j=1}^{n}\sum_{t=1}^{T} X_{jt}}, \quad i = 1, \ldots, n$

If, in practice, we observe data with partial availability, we recommend to split the sales to fully open and closed assortments (Algorithm 1), apply extended EM (Algorithm 2), and aggregate the solution back. It would be an interesting future research topic to further extend the EM algorithm and incorporate constraints on the arrival rates (Sect. Constrained parameter space).

Example: limited OA availability

In this section we apply Algorithm 2 on the simulated example from Vulcano et al. (2012) and demonstrate what happens when we limit the availability of the outside alternative simultaneously with the retailer's availability ($\alpha = 1$). The observed sales data are presented in Table 9.

First let us look at the solution, using $\alpha = 0$, presented in Table 10. We closely recover the results of Vulcano et al. (2012), since we make the same assumption that the outside alternative is always available.

Using $\alpha = 1$, the estimated primary demand at each time period is equal to the observed purchases, because the availability of outside alternative is restricted as the retailer's availability, preserving the model induced market share s at each time t. The results are presented in Table 11.

Notice that the estimated preference weights in Tables 10 and 11 are close to each other, the estimates are not sensitive to the value of α. However, the estimated arrival rates changed drastically, due to the change in the model-induced market shares.

Parameter α can be used as a tool for the practitioner to control the availability of the outside alternative between these two edge cases. We would like to emphasize, however, that competitor matching can be risky and the value of α should ideally be inferred from exogenous data. The conservative practice is to use $\alpha = 0$.

Table 9 Simulated example of Vulcano et al. (2012)

Sales	Period														
	15	**14**	**13**	**12**	**11**	**10**	**9**	**8**	**7**	**6**	**5**	**4**	**3**	**2**	**1**
1	10	15	11	14											
2	11	6	11	8	20	16									
3	5	6	1	11	4	5	14	7	11						
4	4	4	4	1	6	4	3	5	9	9	6	9			
5	0	2	0	0	1	0	1	3	0	3	3	5	2	3	3

Example: non-homogeneous product set

To demonstrate the extended EM algorithm on a non-homogeneous product set, let us revisit the example presented in Table 8 (Sect. Non-homogeneous product set). This is a hypothetical airline sales example with a schedule change. We observe 3 flights, where the sales of products of flight 1 are discontinued and the repeated as flight 2, and flight 3 has twice the observed sales of flights 1 and 2. The solution using $\alpha = 0$ is presented in Table 12.

For flights 1 and 2 we get the same estimated primary demand and preference weights as in Table 10, since we duplicated that example over the time horizon. For flight 3 we estimate twice the primary demand and preference weights, which was expected, since we artificially doubled the numbers. The method is consistent, and can handle non-homogeneous product set in a mathematically formal way.

It is interesting to note here that the EM algorithm of Vulcano et al. (2012) cannot distinguish between product i not being available for sale ($i \notin S_t$) as opposed to not being part of the product set ($i \notin I_t$). Because of this, naive application of the EM algorithm would estimate primary demand for non-existing products of flight 1 and 2, grossly overestimating the demand. The total estimated arrival rate is $\sum_{i=1}^{30} \lambda_t = 5324.10$, while using the extended EM algorithm we get $\sum_{i=1}^{30} \lambda_t = 4421.53$.

It is also interesting to mention that using the extended EM algorithm with $\alpha = 1$ we get $\sum_{i=1}^{30} \lambda_t = 2365.71$, and just like before, the total primary demand at time period t is equal to the observed purchases. It is easy to show in general that in case $\alpha = 1$ it follows that $\lambda_t = \left(\sum_{i=1}^{n} z_{it}\right)/s$. If we apply the algorithm with other values of $\alpha \in [0, 1]$ we observe a linear decrease of total estimated arrival rate as a function of α. The model induced market share of the retailer increases as α increases which decreases the estimated demand.

Algorithm 2 is a simple but yet powerful extension of the EM algorithm which can natively handle non-homogenous product set, and can control the availability of the outside alternative by a simple parameter. The price of this extension is that in the M-step we need to solve for the roots of a system of nonlinear equations, instead of having a closed form solution. However, we can use a simple fixed point iteration as the M-step.

Constrained optimization

In this section we develop algorithms to solve the estimation problem with constrained arrival rates (Sect. Constrained parameter space). We will extend the minorization–maximization (MM) algorithm of Abdallah and Vulcano (2016) and also present a solution utilizing the Frank–Wolfe algorithm. We will also incorporate partial availability, non-homogeneous product set, and the ability to control the availability of outside alternative into the model, and show how to estimate the parameters using iterative algorithms.

Consider the incomplete log-likelihood

Table 10 Estimated demand and parameters using EM algorithm with $\alpha = 0$

Estimates	Period															v_i
	15	**14**	**13**	**12**	**11**	**10**	**9**	**8**	**7**	**6**	**5**	**4**	**3**	**2**	**1**	
1	10	15	11	14	15.03	12.12	13.04	10.87	14.49	15.91	11.93	18.56	11.51	17.27	17.27	1.000
2	11	6	11	8	14.35	11.48	10.45	8.71	11.61	12.75	9.56	14.88	9.23	13.84	13.84	0.801
3	5	6	1	11	2.87	3.59	6.88	3.44	5.41	6.22	4.67	7.26	4.50	6.75	6.75	0.391
4	4	4	4	1	4.31	2.87	1.47	2.46	4.42	3.43	2.29	3.43	2.68	4.02	4.02	0.233
5	0	2	0	0	0.72	0.00	0.49	1.47	0.00	1.14	1.14	1.91	0.63	0.95	0.95	0.055
λ_t	42.86	47.14	38.57	48.57	53.26	42.95	46.19	38.50	51.33	56.37	42.28	65.76	40.78	61.18	61.18	

Table 11 Estimated demand and parameters using EM algorithm with $\alpha = 1$

Estimates	Period															v_i
	15	**14**	**13**	**12**	**11**	**10**	**9**	**8**	**7**	**6**	**5**	**4**	**3**	**2**	**1**	
1	10	15	11	14	12.50	10.08	7.26	6.05	8.06	4.84	3.63	5.65	0.81	1.21	1.21	1.000
2	11	6	11	8	11.94	9.55	5.75	4.79	6.39	3.83	2.87	4.47	0.64	0.96	0.96	0.792
3	5	6	1	11	2.39	2.98	3.88	1.94	3.05	1.92	1.44	2.24	0.32	0.48	0.48	0.396
4	4	4	4	1	3.58	2.39	0.83	1.39	2.50	1.06	0.71	1.06	0.20	0.30	0.30	0.245
5	0	2	0	0	0.60	0.00	0.28	0.83	0.00	0.35	0.35	0.59	0.04	0.06	0.06	0.046
λ_t	42.86	47.14	38.57	48.57	44.29	35.71	25.71	21.43	28.57	17.14	12.86	20.00	2.86	4.29	4.29	

$$l_I(\boldsymbol{v}, \boldsymbol{\lambda}) = \sum_{t=1}^{T} \left[m_t \log\left(\frac{\lambda_t}{v_{0t} + \sum_{i \in S_t} v_i \cdot o_{it}} \right) - \lambda_t \frac{\sum_{i \in S_t} v_i \cdot o_{it}}{v_{0t} + \sum_{i \in S_t} v_i \cdot o_{it}} + \sum_{i \in S_t} z_{it} \log(v_i \cdot o_{it}) \right] \tag{14}$$

which is an extension of the log-likelihood in Abdallah and Vulcano (2016) with partial availability, as discussed in Sect. Extending the attraction model. In case $o_{it} = 1$ when $i \in S_t$, the log-likelihood simplifies to the one considering only fully open and closed assortments. Note that we are using preference weights $\boldsymbol{v}$ in the model, but we could express the model in the utility space by using $\boldsymbol{v} = \exp(\boldsymbol{\beta})$.

The constrained optimization problem we need to solve is given by

$$\begin{aligned} &\max_{\boldsymbol{v}, \boldsymbol{\lambda}} \; l_I(\boldsymbol{v}, \boldsymbol{\lambda}) \\ &\text{s.t.} \\ &\lambda_t \leq L_t, \quad t = 1, \ldots, T, \end{aligned} \tag{15}$$

where L_t is an upper bound on the arrival rate at time t. The motivation behind putting an upper bound on the arrival rates was explained in Sect. Constrained parameter space, where we discussed two specific ways of constraining the arrival rates. To solve the constrained maximum likelihood estimation problem, we follow the idea in Abdallah and Vulcano (2016). We express the optimization problem as a function of $\boldsymbol{v}$ by applying part of the Karush–Kuhn–Tucker (KKT) conditions on the Lagrangian function, removing $\boldsymbol{\lambda}$ from the problem. The Lagrangian function of (15) becomes

$$\mathcal{L}(\boldsymbol{v}, \boldsymbol{\lambda}, \boldsymbol{\mu}) = l_I(\boldsymbol{v}, \boldsymbol{\lambda}) - \sum_{t=1}^{T} \frac{\mu_t}{v_{0t} + \sum_{i \in S_t} v_i \cdot o_{it}} (\lambda_t - L_t). \tag{16}$$

Note that we used a simple rescaling $(1/(v_{0t} + \sum_{i \in S_t} v_i \cdot o_{it})$ for time t) of the Lagrange multipliers $\boldsymbol{\mu}$. After we apply the KKT conditions to remove $\boldsymbol{\lambda}$ from the problem, and some algebra (see "Appendix"), the problem becomes

$$\begin{aligned} \max_{\boldsymbol{v}} \; & \sum_{i=1}^{n} K_i \log(v_i) - \sum_{t \notin \mathcal{B}(\boldsymbol{v})} m_t \log\left(\sum_{i \in S_t} v_i \cdot o_{it} \right) \\ & - \sum_{t \in \mathcal{B}(\boldsymbol{v})} m_t \log\left(v_{0t} + \sum_{i \in S_t} v_i \cdot o_{it} \right) - \sum_{t \in \mathcal{B}(\boldsymbol{v})} L_t \frac{\sum_{i \in S_t} v_i \cdot o_{it}}{v_{0t} + \sum_{i \in S_t} v_i \cdot o_{it}} \\ & + C_3(\mathcal{B}(\boldsymbol{v})), \end{aligned} \tag{17}$$

where $C_3(\mathcal{B}(\boldsymbol{v}))$ is defined in (26), $K_i = \sum_{t=1}^{T} z_{it}$ and $\mathcal{B}(\boldsymbol{v})$ represents the set of time periods where the upper bound constraints on the arrival rates are binding, that is

Table 12 Schedule change example: estimated demand and parameters ($\alpha = 0$)

Estimates	Period															$\mathbf{v}_i$
	1	**2**	**3**	**4**	**5**	**6**	**7**	**8**	**9**	**10**	**11**	**12**	**13**	**14**	**15**	
flt1-prod1	10	15	11	14	15.03	12.12	13.04	10.87	14.49	15.91	11.93	18.56	11.51	17.27	17.27	1.000
flt1-prod2	11	6	11	8	14.35	11.48	10.45	8.71	11.61	12.75	9.56	14.88	9.23	13.84	13.84	0.801
flt1-prod3	5	6	1	11	2.87	3.59	6.88	3.44	5.41	6.22	4.67	7.26	4.50	6.75	6.75	0.391
flt1-prod4	4	4	4	1	4.31	2.87	1.47	2.46	4.42	3.43	2.29	3.43	2.68	4.02	4.02	0.233
flt1-prod5	0	2	0	0	0.72	0.00	0.49	1.47	0.00	1.14	1.14	1.91	0.63	0.95	0.95	0.055
flt2-prod1																1.000
flt2-prod2																0.801
flt2-prod3																0.391
flt2-prod4																0.233
flt2-prod5																0.055
flt3-prod1	20	30	22	28	30.07	24.25	26.08	21.73	28.98	31.82	23.87	37.12	23.02	34.54	34.54	2.000
flt3-prod2	22	12	22	16	28.71	22.97	20.90	17.42	23.22	25.50	19.13	29.75	18.45	27.68	27.68	1.603
flt3-prod3	10	12	2	22	5.74	7.18	13.76	6.88	10.81	12.44	9.33	14.52	9.00	13.51	13.51	0.782
flt3-prod4	8	8	8	2	8.61	5.74	2.95	4.92	8.85	6.86	4.57	6.86	5.36	8.04	8.04	0.465
flt3-prod5	0	4	0	0	1.44	0.00	0.98	2.95	0.00	2.29	2.29	3.81	1.26	1.89	1.89	0.110
λ_t	128.57	141.43	115.71	145.71	159.79	128.86	138.58	115.49	153.98	169.10	126.83	197.29	122.35	183.53	183.53	
	16	**17**	**18**	**19**	**20**	**21**	**22**	**23**	**24**	**25**	**26**	**27**	**28**	**29**	**30**	
flt1-prod1																1.000
flt1-prod2																0.801
flt1-prod3																0.391
flt1-prod4																0.233
flt1-prod5																0.055
flt2-prod1	10	15	11	14	15.03	12.12	13.04	10.87	14.49	15.91	11.93	18.56	11.51	17.27	17.27	1.000
flt2-prod2	11	6	11	8	14.35	11.48	10.45	8.71	11.61	12.75	9.56	14.88	9.23	13.84	13.84	0.801
flt2-prod3	5	6	1	11	2.87	3.59	6.88	3.44	5.41	6.22	4.67	7.26	4.50	6.75	6.75	0.391
flt2-prod4	4	4	4	1	4.31	2.87	1.47	2.46	4.42	3.43	2.29	3.43	2.68	4.02	4.02	0.233
flt2-prod5	0	2	0	0	0.72	0.00	0.49	1.47	0.00	1.14	1.14	1.91	0.63	0.95	0.95	0.055
flt3-prod1	20	30	22	28	30.07	24.25	26.08	21.73	28.98	31.82	23.87	37.12	23.02	34.54	34.54	2.000
flt3-prod2	22	12	22	16	28.71	22.97	20.90	17.42	23.22	25.50	19.13	29.75	18.45	27.68	27.68	1.603
flt3-prod3	10	12	2	22	5.74	7.18	13.76	6.88	10.81	12.44	9.33	14.52	9.00	13.51	13.51	0.782
flt3-prod4	8	8	8	2	8.61	5.74	2.95	4.92	8.85	6.86	4.57	6.86	5.36	8.04	8.04	0.465
flt3-prod5	0	4	0	0	1.44	0.00	0.98	2.95	0.00	2.29	2.29	3.81	1.26	1.89	1.89	0.110
λ_t	128.57	141.43	115.71	145.71	159.79	128.86	138.58	115.49	153.98	169.10	126.83	197.29	122.35	183.53	183.53	

$$\mathcal{B}(\boldsymbol{v}) = \left\{ t \middle| L_t < m_t \frac{v_{0t} + \sum_{i \in S_t} v_i \cdot o_{it}}{\sum_{i \in S_t} v_i \cdot o_{it}} \right\}. \tag{18}$$

Through the definitions of v_{0t} and market share constraint we can incorporate non-homogeneous product set in the model, and the ability to control the availability of outside alternative. These details were discussed in Sects. Non-homogeneous product set and Handling market share constraint and no-purchase option. We will consider two formulations here. In the first one we use market share constraint

$$\sum_{t=1}^{T} \left[(1-\alpha) \sum_{i \in I_t} v_i + \alpha \sum_{i \in S_t} v_i \cdot o_{it} \right] = \tilde{s} \sum_{t=1}^{T} v_{0t} \tag{19}$$

which is equivalent to the aggregate constraint (11) discussed in Sect. Handling market share constraint and no-purchase option, with using $\tilde{s} = s/(1-s)$. We assume that OA preference weights v_{0t} are known, and we can use standard scaling $v_{0t} = 1$. In Sect. Solution using MM algorithm we will show how to use the MM algorithm to solve this problem.

In the second formulation we incorporate market share constraint (10) into the objective function and constrain the preference weights by using

$$v_{0t} = r \left[(1-\alpha) \sum_{i \in I_t} v_i + \alpha \sum_{i \in S_t} v_i \cdot o_{it} \right]$$
$$\sum_{\forall i} v_i = 1. \tag{20}$$

This first equation removes v_{0t} from the problem, while the second equation avoids having multiple solutions by rescaling $\boldsymbol{v}$. In Sect. Solution using Frank–Wolfe method we will show how the Frank–Wolfe method will lead to a simple coordinate descent algorithm to solve this problem.

Solution using MM algorithm

In this section we present a solution to the optimization problem (15) with constraint (19) using the MM algorithm, building on Abdallah and Vulcano (2016). The idea behind MM algorithms is to find a surrogate function that minorizes the original objective function, maximize the surrogate function, and continue this iteratively. For more information on the MM algorithm, in general, see Hunter and Lange (2000a, 2004).

After removing $\boldsymbol{\lambda}$ from the problem using the KKT conditions, we arrive to (17). If we rearrange the last term of the objective to aid the minorization, the optimization problem becomes

$$\max_{\boldsymbol{v}} \sum_{i=1}^{n} K_i \log(v_i) - \sum_{t \notin \mathcal{B}(\boldsymbol{v})} m_t \log\left(\sum_{i \in S_t} v_i \cdot o_{it} \right)$$
$$- \sum_{t \in \mathcal{B}(\boldsymbol{v})} m_t \log\left(v_{0t} + \sum_{i \in S_t} v_i \cdot o_{it} \right) + \sum_{t \in \mathcal{B}(\boldsymbol{v})} L_t \frac{v_{0t}}{v_{0t} + \sum_{i \in S_t} v_i \cdot o_{it}}$$
$$+ C_3(\mathcal{B}(\boldsymbol{v}))$$

s.t.

$$\sum_{t=1}^{T} \left[(1-\alpha) \sum_{i \in I_t} v_i + \alpha \sum_{i \in S_t} v_i \cdot o_{it} \right] = \tilde{s} \sum_{t=1}^{T} v_{0t}, \tag{21}$$

where v_{0t} are known. A common technique to find a minorizer is to use supporting hyperplanes (Hunter and Lange 2004). Since the second, third, and fourth terms in (21) are convex, we can use first-order Taylor approximation to the convex functions $-\log(x)$ and $\frac{1}{x}$, that is, for all $x, y > 0$

$$-\log(y) \geq -\log(x) - \frac{1}{x}(y - x)$$
$$\frac{1}{y} \geq \frac{1}{x} - \frac{1}{x^2}(y - x).$$

In our specific case we will use

$$-\log(y) \geq 1 - \log(x) - y/x$$
$$\frac{1}{y} \geq -\frac{y}{x^2} + \frac{2}{x}$$

with $y = \sum_{i \in S_t} v_i \cdot o_{it}$ and $x = \sum_{i \in S_t} v_i^{(k)} \cdot o_{it}$ or $y = v_{0t} + \sum_{i \in S_t} v_i \cdot o_{it}$ and $x = v_{0t} + \sum_{i \in S_t} v_i^{(k)} \cdot o_{it}$, where $\boldsymbol{v}^{(k)}$ is the value of $\boldsymbol{v}$ at iteration k. Therefore, the minorizer function becomes

$$g(\boldsymbol{v}|\boldsymbol{v}^{(k)}) - \sum_{i=1}^{n} K_i \log(v_i) - \sum_{t \notin \mathcal{B}(\boldsymbol{v}^{(k)})} m_t \frac{\sum_{i \in S_t} v_i \cdot o_{it}}{\sum_{i \in S_t} v_i^{(k)} \cdot o_{it}}$$
$$- \sum_{t \in \mathcal{B}(\boldsymbol{v}^{(k)})} m_t \frac{\sum_{i \in S_t} v_i \cdot o_{it}}{v_{0t} + \sum_{i \in S_t} v_i^{(k)} \cdot o_{it}} \tag{22}$$
$$- \sum_{t \in \mathcal{B}(\boldsymbol{v}^{(k)})} \frac{L_t v_{0t} \left(\sum_{i \in S_t} v_i \cdot o_{it} \right)}{\left(v_{0t} + \sum_{i \in S_t} v_i^{(k)} \cdot o_{it} \right)^2} + C_0,$$

where C_0 contains the constant terms independent of $\boldsymbol{v}$. Although the minorizer $g(\boldsymbol{v}|\boldsymbol{v}^{(k)})$ is developed locally for fixed $\boldsymbol{v}^{(k)}$, it can be shown to be globally dominated by the $\mathcal{L}(\boldsymbol{v}, \boldsymbol{\lambda}, \boldsymbol{\mu})$. Computational results show that it is actually a fairly tight minorizer as well. We now need to solve a single market share constraint optimization problem:

$$\max_{v} \sum_{i=1}^{n} K_i \log(v_i) - \sum_{t \notin \mathcal{B}(v^{(k)})} m_t \frac{\sum_{i \in S_t} v_i \cdot o_{it}}{\sum_{i \in S_t} v_i^{(k)} \cdot o_{it}}$$
$$- \sum_{t \in \mathcal{B}(v^{(k)})} m_t \frac{\sum_{i \in S_t} v_i \cdot o_{it}}{v_{0t} + \sum_{i \in S_t} v_i^{(k)} \cdot o_{it}}$$
$$- \sum_{t \in \mathcal{B}(v^{(k)})} \frac{L_t v_{0t} \left(\sum_{i \in S_t} v_i \cdot o_{it} \right)}{\left(v_{0t} + \sum_{i \in S_t} v_i^{(k)} \cdot o_{it} \right)^2}$$

s.t.

$$\sum_{t=1}^{T} \left[(1-\alpha) \sum_{i \in I_t} v_i + \alpha \sum_{i \in S_t} v_i \cdot o_{it} \right] = \tilde{s} \sum_{t=1}^{T} v_{0t}.$$

Since the objective function is concave, the constraint is convex, and we are maximizing, there exists a single scalar η such that the first-order optimality condition holds for Lagrangian

$$\mathcal{L}(v, \eta) = g(v|v^{(k)}) - \eta \left(\sum_{t=1}^{T} \left[(1-\alpha) \sum_{i \in I_t} v_i + \alpha \sum_{i \in S_t} v_i \cdot o_{it} \right] - \tilde{s} \sum_{t=1}^{T} v_{0t} \right).$$

The first-order optimality condition is

$$K_j = v_j \left(\sum_{t \notin \mathcal{B}(v^{(k)})} m_t \frac{\mathbb{1}(j \in S_t) \cdot o_{jt}}{\sum_{i \in S_t} v_i^{(k)} \cdot o_{it}} + \sum_{t \in \mathcal{B}(v^{(k)})} m_t \frac{\mathbb{1}(j \in S_t) \cdot o_{jt}}{v_{0t} + \sum_{i \in S_t} v_i^{(k)} \cdot o_{it}} + \sum_{t \in \mathcal{B}(v^{(k)})} L_t \frac{v_{0t} \mathbb{1}(j \in S_t) \cdot o_{jt}}{\left(v_{0t} + \sum_{i \in S_t} v_i^{(k)} \cdot o_{it} \right)^2} + \eta \left[(1-\alpha) \sum_{t=1}^{T} \mathbb{1}(j \in I_t) + \alpha \sum_{t=1}^{T} \mathbb{1}(j \in S_t) \cdot o_{jt} \right] \right) \quad (23)$$

which leads to the MM update of v summarized in Algorithm 3. In the update step we use Newton's method to find η in every MM iteration, summarized in Algorithm 4. For more details, please see "Appendix".

Algorithm 3 MM algorithm to estimate $(\boldsymbol{\lambda}, \boldsymbol{v})$ in (15) with market share constraint (19)

Input: $\{(z_t, o_t, S_t, I_t, v_{0t}, L_t)\}_{t=1}^{T}$, s, α
Let $K_j = \sum_{t=1}^{T} z_{jt}$, $n_j = \sum_{t=1}^{T} \mathbb{1}(j \in I_t)$ and $o_j = \sum_{t=1}^{T} o_{jt}$, $j = 1, \ldots, n$
Let $m_t = \sum_{j=1}^{n} z_{jt}$, $t = 1, \ldots, T$ and $\tilde{s} = s/(1-s)$
Let $k = 0$, initialize $v^{(0)}$
while Stopping criteria is not satisfied **do**
  $\mathcal{B}(v^{(k)}) = \left\{ t \middle| L_t < m_t \frac{v_{0t} + \sum_{i \in S_t} v_i^{(k)} \cdot o_{it}}{\sum_{i \in S_t} v_i^{(k)} \cdot o_{it}} \right\}$
  for $j = 1, \ldots, n$ **do**
    $A_j = \sum_{t \notin \mathcal{B}(v^{(k)})} m_t \frac{\mathbb{1}(j \in S_t) \cdot o_{jt}}{\sum_{i \in S_t} v_i^{(k)} \cdot o_{it}} + \sum_{t \in \mathcal{B}(v^{(k)})} m_t \frac{\mathbb{1}(j \in S_t) \cdot o_{jt}}{v_{0t} + \sum_{i \in S_t} v_i^{(k)} \cdot o_{it}} + \sum_{t \in \mathcal{B}(v^{(k)})} L_t \frac{v_{0t} \mathbb{1}(j \in S_t) \cdot o_{jt}}{\left(v_{0t} + \sum_{i \in S_t} v_i^{(k)} \cdot o_{it} \right)^2}$
  end for
  Find η using **Algorithm 4**
  for $j = 1, \ldots, n$ **do**
    $v_j^{(k+1)} = \frac{K_j}{A_j + \eta[(1-\alpha) n_j + \alpha o_j]}$
  end for
  $k = k + 1$
end while
for $t = 1, \ldots, T$ **do**
  $\lambda_t = \min \left(L_t, m_t \frac{v_{0t} + \sum_{i \in S_t} v_i^{(k)} \cdot o_{it}}{\sum_{i \in S_t} v_i^{(k)} \cdot o_{it}} \right)$
end for

Algorithm 4 Newton's method to find η

Let $k = 0$, $\eta^0 = 0$, and $f^0 = 1$
while $f^k > \epsilon$ **do**
  $f^k = \sum_{j=1}^{n} \frac{K_j[(1-\alpha)n_j+\alpha o_j]}{A_j+\eta^k[(1-\alpha)n_j+\alpha o_j]} - \tilde{s}\sum_{t=1}^{T} v_{0t}$
  $g^k = -\sum_{j=1}^{n} \frac{K_j[(1-\alpha)n_j+\alpha o_j]^2}{(A_j+\eta^k[(1-\alpha)n_j+\alpha o_j])^2}$
  $\eta^{k+1} = \eta^k - \frac{f^k}{g^k}$
  $k = k+1$
end while

Solution using Frank–Wolfe method

In this section we present a solution to the optimization problem (15) with constraint (20) using the Frank–Wolfe algorithm. The Frank–Wolfe, or conditional gradient method is an iterative optimization algorithm for constrained convex optimization, where in each iteration, it considers a linear approximation of the objective function, and moves towards a minimizer of this linear function. For more information on the Frank–Wolfe algorithm, see Frank and Wolfe (1956) and Bertsekas (1999).

After removing λ from the problem using the KKT conditions, we arrive to (17). Therefore, we need to solve the constrained optimization problem

$$\max_{\boldsymbol{v}} \sum_{i=1}^{n} K_i \log(v_i) - \sum_{t\notin\mathcal{B}(\boldsymbol{v})} m_t \log\left(\sum_{i\in S_t} v_i \cdot o_{it}\right) - \sum_{t\in\mathcal{B}(\boldsymbol{v})} m_t \log\left(v_{0t} + \sum_{i\in S_t} v_i \cdot o_{it}\right) - \sum_{t\in\mathcal{B}(\boldsymbol{v})} L_t \frac{\sum_{i\in S_t} v_i \cdot o_{it}}{v_{0t} + \sum_{i\in S_t} v_i \cdot o_{it}} + C_3(\mathcal{B}(\boldsymbol{v})) \quad (24)$$

s.t.

$$\sum_{i=1}^{n} v_i - 1,$$

where we plug in $v_{0t} = r\left[(1-\alpha)\sum_{i\in I_t} v_i + \alpha\sum_{i\in S_t} v_i \cdot o_{it}\right]$ to include the market share constraints into the objective function.

The first step of the Frank–Wolfe algorithm is the direction finding subproblem, which becomes

$$\max_{\boldsymbol{y}} \nabla f\left(\boldsymbol{v}^{(k)}\right)^T \boldsymbol{y}$$

s.t. $\quad (25)$

$$\sum_{i=1}^{n} y_i = 1,$$

where $\nabla f\left(\boldsymbol{v}^{(k)}\right)$ is the gradient vector of the objective function (24) evaluated at the solution of iteration k. The elements of $\nabla f(\boldsymbol{v})$ are calculated as

$$\frac{\partial f}{\partial v_j} = \frac{K_j}{v_j} - \sum_{t\notin\mathcal{B}(\boldsymbol{v})} m_t \frac{\mathbb{1}(j\in S_t)o_{jt}}{\sum_{i\in S_t} v_i \cdot o_{it}} - \sum_{t\in\mathcal{B}(\boldsymbol{v})} m_t \frac{\mathbb{1}(j\in S_t)o_{jt}(r\alpha+1)}{v_{0t} + \sum_{i\in S_t} v_i \cdot o_{it}} - \sum_{t\in\mathcal{B}(\boldsymbol{v})} L_t \frac{\mathbb{1}(j\in S_t)o_{jt} r(1-\alpha)\left(\sum_{i\in I_t} v_i\right)}{\left(v_{0t} + \sum_{i\in S_t} v_i \cdot o_{it}\right)^2} - \sum_{t\in\mathcal{B}(\boldsymbol{v})} m_t \frac{\mathbb{1}(j\in I_t) r(1-\alpha)}{v_{0t} + \sum_{i\in S_t} v_i \cdot o_{it}} + \sum_{t\in\mathcal{B}(\boldsymbol{v})} L_t \frac{\mathbb{1}(j\in I_t) r(1-\alpha)\left(\sum_{i\in S_t} v_i \cdot o_{it}\right)}{\left(v_{0t} + \sum_{i\in S_t} v_i \cdot o_{it}\right)^2}$$

where

$$v_{0t} = r\left[(1-\alpha)\sum_{i\in I_t} v_i + \alpha\sum_{i\in S_t} v_i \cdot o_{it}\right].$$

For detailed derivation and computational formula, please see "Appendix". The direction finding subproblem in (25) is a fractional knapsack problem (Korte and Vygen 2012), which solution becomes

$$y_i = \begin{cases} 1, & \text{if } i = \arg\max\left\{ \left.\frac{\partial f}{\partial v_i}\right|_{\boldsymbol{v}=\boldsymbol{v}^{(k)}}, i = 1, \ldots, n\right\}, \\ 0, & \text{otherwise.} \end{cases}$$

We take a step in the direction of the maximum element of the gradient, only changing that variable in the current step. The Frank–Wolfe algorithm reduces to a coordinate descent algorithm, successively maximizing along coordinate directions determined by the largest value of the gradient vector. The update step of Frank–Wolfe is

$$\boldsymbol{v}^{(k+1)} = \boldsymbol{v}^{(k)} + \gamma_k\left(\boldsymbol{y} - \boldsymbol{v}^{(k)}\right)$$

which simplifies to

$$v_i^{(k+1)} = \begin{cases} \gamma_k + (1-\gamma_k)v_i^{(k)}, & \text{if } i = \arg\max\left\{ \left.\frac{\partial f}{\partial v_i}\right|_{\boldsymbol{v}=\boldsymbol{v}^{(k)}}, i = 1, \ldots, n\right\}, \\ v_i^{(k)}, & \text{otherwise.} \end{cases}$$

For step size γ_k we can use the default choice $\gamma_k = 2/(k+2)$ or perform a line search to find γ_k that minimizes $f\left(\mathbf{v}^{(k)} + \gamma_k\left(\mathbf{s} - \mathbf{v}^{(k)}\right)\right)$ subject to $0 \leq \gamma_k \leq 1$. In practice, we implemented a backtracking linesearch using the Armijo's rule (Nocedal and Wright 2006). The Frank–Wolfe algorithm is summarized in Algorithm 5, while the backtracking linesearch is presented in Algorithm 6.

Algorithm 5 Frank-Wolfe algorithm to estimate $(\boldsymbol{\lambda}, \boldsymbol{v})$ in (15) with constraint (20)

Input: $\{(\boldsymbol{z}_t, \boldsymbol{o}_t, S_t, I_t, L_t)\}_{t=1}^{T}$, s, α
Let $K_j = \sum_{t=1}^{T} z_{jt}$, $j = 1, \ldots, n$
Let $m_t = \sum_{j=1}^{n} z_{jt}$, $t = 1, \ldots, T$ and $r = (1-s)/s$
Let $k = 0$, initialize $\boldsymbol{v}^{(0)}$
while Stopping criteria is not satisfied **do**
  for $t = 1, \ldots, T$ **do**
    $v_{0t}^{(k)} = r\left[(1-\alpha)\sum_{i \in I_t} v_i^{(k)} + \alpha \sum_{i \in S_t} v_i^{(k)} \cdot o_{it}\right]$
  end for
  $\mathcal{B}(\boldsymbol{v}^{(k)}) = \left\{ t \,\middle|\, L_t < m_t \frac{v_{0t}^{(k)} + \sum_{i \in S_t} v_i^{(k)} \cdot o_{it}}{\sum_{i \in S_t} v_i^{(k)} \cdot o_{it}} \right\}$
  for $j = 1, \ldots, n$ **do**

$$
\begin{aligned}
g_j = \; & \frac{K_j}{v_j^{(k)}} - \sum_{t \notin \mathcal{B}(\boldsymbol{v}^{(k)})} m_t \frac{\mathbb{1}(j \in S_t) o_{jt}}{\sum_{i \in S_t} v_i^{(k)} \cdot o_{it}} - \\
& \sum_{t \in \mathcal{B}(\boldsymbol{v}^{(k)})} m_t \frac{\mathbb{1}(j \in S_t) o_{jt} (r\alpha + 1)}{v_{0t}^{(k)} + \sum_{i \in S_t} v_i^{(k)} \cdot o_{it}} - \sum_{t \in \mathcal{B}(\boldsymbol{v}^{(k)})} L_t \frac{\mathbb{1}(j \in S_t) o_{jt} r (1-\alpha) \left(\sum_{i \in I_t} v_i^{(k)}\right)}{\left(v_{0t}^{(k)} + \sum_{i \in S_t} v_i^{(k)} \cdot o_{it}\right)^2} - \\
& \sum_{t \in \mathcal{B}(\boldsymbol{v}^{(k)})} m_t \frac{\mathbb{1}(j \in I_t) r (1-\alpha)}{v_{0t}^{(k)} + \sum_{i \in S_t} v_i^{(k)} \cdot o_{it}} + \sum_{t \in \mathcal{B}(\boldsymbol{v}^{(k)})} L_t \frac{\mathbb{1}(j \in I_t) r (1-\alpha) \left(\sum_{i \in S_t} v_i^{(k)} \cdot o_{it}\right)}{\left(v_{0t}^{(k)} + \sum_{i \in S_t} v_i^{(k)} \cdot o_{it}\right)^2}
\end{aligned}
$$

  end for
  $l = \arg\max\{g_j,\ j = 1, \ldots, n\}$ ▷ Find direction
  $\boldsymbol{y} = \boldsymbol{e}_l$ ▷ $\boldsymbol{e}_i$ is ith unit vector
  Use $\gamma_k = 2/(k+2)$ or find γ_k using Algorithm 6 ▷ Find step size
  $\boldsymbol{v}^{(k+1)} = \boldsymbol{v}^{(k)} + \gamma_k\left(\boldsymbol{y} - \boldsymbol{v}^{(k)}\right)$ ▷ Frank-Wolfe update
  $k = k + 1$
end while
for $t = 1, \ldots, T$ **do**
  $v_{0t}^{(k)} = r\left[(1-\alpha)\sum_{i \in I_t} v_i^{(k)} + \alpha \sum_{i \in S_t} v_i^{(k)} \cdot o_{it}\right]$
  $\lambda_t = \min\left(L_t, m_t \frac{v_{0t}^{(k)} + \sum_{i \in S_t} v_i^{(k)} \cdot o_{it}}{\sum_{i \in S_t} v_i^{(k)} \cdot o_{it}}\right)$
end for

Algorithm 6 Backtracking linesearch by Armijo's rule to find step size

Initialize β, τ, $\gamma^{(0)}$ $\triangleright$ e.g. $\beta = 0.001$, $\tau = 0.5$, $\gamma^{(0)} = 1$
Let $h = 0$, $a = 1$, $b = 2$
Let $\boldsymbol{s} = \boldsymbol{y} - \boldsymbol{v}^{(h)}$
Let

$$f_v = \sum_{i=1}^{n} K_i \log(v_i^{(k)}) - \sum_{t \notin \mathcal{B}(\boldsymbol{v}^{(k)})} m_t \log\left(\sum_{i \in S_t} v_i^{(k)} \cdot o_{it}\right) - \sum_{t \in \mathcal{B}(\boldsymbol{v}^{(k)})} m_t \log\left(v_{0t}^{(k)} + \sum_{i \in S_t} v_i^{(k)} \cdot o_{it}\right) - \sum_{t \in \mathcal{B}(\boldsymbol{v}^{(k)})} L_t \frac{\sum_{i \in S_t} v_i^{(k)} \cdot o_{it}}{v_{0t}^{(k)} + \sum_{i \in S_t} v_i^{(k)} \cdot o_{it}} + C_3\left(\mathcal{B}(\boldsymbol{v}^{(k)})\right)$$

Let $g_v s = \sum_{i=1}^{n} g_i s_i$
while $a < b$ **do**
  $\boldsymbol{w} = \boldsymbol{v}^{(k)} + \gamma^{(h+1)} \boldsymbol{s}$
  $w_{0t} = r\left[(1-\alpha)\sum_{i \in I_t} w_i + \alpha \sum_{i \in S_t} w_i \cdot o_{it}\right],\ t = 1, \ldots, T$

$$a = \sum_{i=1}^{n} K_i \log(w_i) - \sum_{t \notin \mathcal{B}(\boldsymbol{w})} m_t \log\left(\sum_{i \in S_t} w_i \cdot o_{it}\right) - \sum_{t \in \mathcal{B}(\boldsymbol{w})} m_t \log\left(w_{0t} + \sum_{i \in S_t} w_i \cdot o_{it}\right) - \sum_{t \in \mathcal{B}(\boldsymbol{w})} L_t \frac{\sum_{i \in S_t} w_i \cdot o_{it}}{w_{0t} + \sum_{i \in S_t} w_i \cdot o_{it}} + C_3\left(\mathcal{B}(\boldsymbol{w})\right)$$

  $b = f_v + \gamma^{(h+1)} \beta g_v s$
  $\gamma^{(h+1)} = \tau \gamma^{(h)}$
  $h = h + 1$
end while

Example: non-homogeneous product set with constraint

To demonstrate Algorithms 3 and 5, let us revisit the example presented in Table 8 by adding constraints to the problem. We constrain the arrival rates to be less than twice the observed sales, that is $L_t = 2m_t$. We use $\alpha = 0$, assuming that the OA product is always available, and set $v_{0t} = 1$ for the MM algorithm. Note that $o_t = S_t$, since we have fully open and closed assortments. The results, using the MM algorithm, are presented in Table 13.

We observe the expected symmetry in the results, that is the estimated demands and preference weights are same for flights 1 and 2, and twice for flight 3. We can see that the estimated value of λ_t is equal to the upper bound L_t for time periods 7–15 and 22–30. Using the Frank–Wolfe algorithm, we converge to the same solution. Note that Frank–Wolfe is solving a problem with different constraints, but for this symmetric example and in the case of $\alpha = 0$ the two different formulations are equivalent. It is also interesting to note that using a built-in nonlinear optimization routine in R (sequential quadratic programming), we can solve the problem in 141 iterations, taking 2.53 s, converging to the same solution. The Frank–Wolfe algorithm with Armijo's rule converges in 267 iterations taking 0.12 s, and the MM algorithm converges in 11 iterations taking only 0.006 s. The convergence of the Frank–Wolfe algorithm is sublinear, in general, and using the default step size $\gamma_k = 2/(k+2)$ results in even much slower convergence. Note, however, that the Frank–Wolfe algorithm solves an optimization problem with market share constraint at each time period t substituted into the objective function. For this problem we were not able to develop the MM algorithm due to the difficulty of finding a suitable minorizer. The formulation of the MM algorithm uses an aggregate market share constraint, where we assume that v_{0t} are known. The solutions are not equivalent, in general.

Conclusions and future research

In this paper, we discussed some of the practical limitations of the standard choice-based demand models used in the literature to estimate demand from sales transaction data. We presented modifications and extensions of the models and discussed data preprocessing and solution techniques which can be useful for practitioners dealing with sales transaction data. We hope that these discussions could facilitate further

Table 13 Schedule change example with constraint: estimated demand and parameters using MM algorithm

Estimates	Period															v_i
	1	2	3	4	5	6	7	8	9	10	11	12	13	14	15	
flt1-prod1	10.41	11.45	9.37	11.80	12.47	10.06	8.74	7.29	9.72	5.83	4.37	6.80	0.97	1.46	1.46	1.000
flt1-prod2	9.40	10.34	8.46	10.65	11.26	9.08	7.89	6.58	8.77	5.26	3.95	6.14	0.88	1.32	1.32	0.903
flt1-prod3	5.11	5.62	4.60	5.79	6.12	4.94	4.29	3.58	4.77	2.86	2.15	3.34	0.48	0.72	0.72	0.491
flt1-prod4	3.71	4.08	3.33	4.20	4.44	3.58	3.11	2.59	3.46	2.07	1.56	2.42	0.35	0.52	0.52	0.356
flt1-prod5	1.38	1.52	1.24	1.56	1.65	1.33	1.16	0.97	1.29	0.77	0.58	0.90	0.13	0.19	0.19	0.133
flt2-prod1																1.000
flt2-prod2																0.903
flt2-prod3																0.491
flt2-prod4																0.356
flt2-prod5																0.133
flt3-prod1	20.82	22.90	18.74	23.59	24.94	20.11	17.49	14.57	19.43	11.66	8.74	13.60	1.94	2.91	2.91	2.000
flt3-prod2	18.79	20.67	16.91	21.30	22.52	18.16	15.79	13.16	17.54	10.52	7.89	12.28	1.75	2.63	2.63	1.806
flt3-prod3	10.22	11.24	9.20	11.58	12.24	9.87	8.58	7.15	9.54	5.72	4.29	6.68	0.95	1.43	1.43	0.982
flt3-prod4	7.41	8.15	6.67	8.40	8.88	7.16	6.22	5.19	6.92	4.15	3.11	4.84	0.69	1.04	1.04	0.712
flt3-prod5	2.76	3.03	2.48	3.13	3.31	2.67	2.32	1.93	2.57	1.54	1.16	1.80	0.26	0.39	0.39	0.265
λ_t	128.57	141.43	115.71	145.71	154.04	124.22	108.00	90.00	120.00	72.00	54.00	84.00	12.00	18.00	18.00	
	16	**17**	**18**	**19**	**20**	**21**	**22**	**23**	**24**	**25**	**26**	**27**	**28**	**29**	**30**	
flt1-prod1																1.000
flt1-prod2																0.903
flt1-prod3																0.491
flt1-prod4																0.356
flt1-prod5																0.133
flt2-prod1	10.41	11.45	9.37	11.80	12.47	10.06	8.74	7.29	9.72	5.83	4.37	6.80	0.97	1.46	1.46	1.000
flt2-prod2	9.40	10.34	8.46	10.65	11.26	9.08	7.89	6.58	8.77	5.26	3.95	6.14	0.88	1.32	1.32	0.903
flt2-prod3	5.11	5.62	4.60	5.79	6.12	4.94	4.29	3.58	4.77	2.86	2.15	3.34	0.48	0.72	0.72	0.491
flt2-prod4	3.71	4.08	3.33	4.20	4.44	3.58	3.11	2.59	3.46	2.07	1.56	2.42	0.35	0.52	0.52	0.356
flt2-prod5	1.38	1.52	1.24	1.56	1.65	1.33	1.16	0.97	1.29	0.77	0.58	0.90	0.13	0.19	0.19	0.133
flt3-prod1	20.82	22.90	18.74	23.59	24.94	20.11	17.49	14.57	19.43	11.66	8.74	13.60	1.94	2.91	2.91	2.000
flt3-prod2	18.79	20.67	16.91	21.30	22.52	18.16	15.79	13.16	17.54	10.52	7.89	12.28	1.75	2.63	2.63	1.806
flt3-prod3	10.22	11.24	9.20	11.58	12.24	9.87	8.58	7.15	9.54	5.72	4.29	6.68	0.95	1.43	1.43	0.982
flt3-prod4	7.41	8.15	6.67	8.40	8.88	7.16	6.22	5.19	6.92	4.15	3.11	4.84	0.69	1.04	1.04	0.712
flt3-prod5	2.76	3.03	2.48	3.13	3.31	2.67	2.32	1.93	2.57	1.54	1.16	1.80	0.26	0.39	0.39	0.265
λ_t	128.57	141.43	115.71	145.71	154.04	124.22	108.00	90.00	120.00	72.00	54.00	84.00	12.00	18.00	18.00	

methodological progress and even more rigorous theoretical research in this domain. We presented an algorithm to split sales transaction data observed under partial availability, and we extended an EM algorithm for the case where we observe a non-homogeneous product set. We developed two iterative optimization algorithms which incorporated partial availability information, non-homogeneous product set, ability to control the availability of outside alternative, and an upper bound on the arrival rates. In one formulation, we used market share constraint at each time period and incorporated them into the objective function through the preference weights of the outside alternative. This formulation was solved using the Frank–Wolfe algorithm, leading to a simple coordinate descent algorithm. We discussed another formulation, which used a single, aggregate market share constraint over the time horizon, and assumed the knowledge of preference weights of the outside alternative. Using this formulation, we could develop a very fast, iterative minorization-maximization algorithm building on the work in Abdallah and Vulcano (2016).

Future extension of these methods are possible. For instance, after using the sales splitting algorithm, it would be interesting to group the arrival rates in the EM algorithm and avoid having too many parameters to be estimated. Similarly, it would be of practical interest to extend the EM algorithm to the constrained optimization case or introduce other regularization on the parameters. The MM and the Frank–Wolfe algorithms developed in this paper could be extended to include covariates and additional preference weights for the product with lowest fare. The Frank–Wolfe algorithm could be further improved by taking the gradient descent direction instead of coordinate descent. Finally, in practical applications, we often encounter sparse demand distribution with excess number of zeros and heavy tails. A natural extension of the currently used models would be to use a zero-inflated Poisson or negative binomial distribution to describe the customer arrival process.

Appendix

Using KKT conditions to remove λ

The KKT condition of Lagrangian (15) are

$$\frac{\partial \mathcal{L}(\boldsymbol{v}, \boldsymbol{\lambda}, \boldsymbol{\mu})}{\partial \lambda_t} = \frac{m_t}{\lambda_t} - \frac{\mu_t + \sum_{i \in S_t} v_i \cdot o_{it}}{v_{0t} + \sum_{i \in S_t} v_i \cdot o_{it}} = 0, \quad t = 1, \dots, T$$

$$\mu_t(\lambda_t - L_t) = 0, \quad t = 1, \dots, T$$

$$\mu_t \geq 0, \quad t = 1, \dots, T$$

$$\lambda_t \leq L_t, \quad t = 1, \dots, T.$$

Expressing λ_t and μ_t in first equation results in

$$\lambda_t = m_t \frac{v_{0t} + \sum_{i \in S_t} v_i \cdot o_{it}}{\mu_t + \sum_{i \in S_t} v_i \cdot o_{it}},$$

$$\mu_t = \frac{m_t}{\lambda_t}\left(v_{0t} + \sum_{i \in S_t} v_i \cdot o_{it}\right) - \sum_{i \in S_t} v_i \cdot o_{it}.$$

Plugging back μ_t, the complementary slackness condition becomes

$$\left(\frac{m_t}{\lambda_t}\left(v_{0t} + \sum_{i \in S_t} v_i \cdot o_{it}\right) - \sum_{i \in S_t} v_i \cdot o_{it}\right)(\lambda_t - L_t) = 0.$$

Therefore, we have that one of the following conditions should hold:

$$\lambda_t^1 = L_t,$$

$$\lambda_t^2 = m_t \frac{v_{0t} + \sum_{i \in S_t} v_i \cdot o_{it}}{\sum_{i \in S_t} v_i \cdot o_{it}}.$$

So, the partial KKT condition stated earlier reduces to the following conditions:

If $L_t < m_t \frac{v_{0t} + \sum_{i \in S_t} v_i \cdot o_{it}}{\sum_{i \in S_t} v_i \cdot o_{it}}$ then

$$\lambda_t = L_t$$

$$\mu_t = \frac{m_t}{L_t}\left(v_{0t} + \sum_{i \in S_t} v_i \cdot o_{it}\right) - \sum_{i \in S_t} v_i \cdot o_{it}$$

else

$$\lambda_t = m_t \frac{v_{0t} + \sum_{i \in S_t} v_i \cdot o_{it}}{\sum_{i \in S_t} v_i \cdot o_{it}},$$

$$\mu_t = 0.$$

Let us define $\mathcal{B}(\boldsymbol{v}) = \left\{ t \middle| L_t < m_t \frac{v_{0t} + \sum_{i \in S_t} v_i \cdot o_{it}}{\sum_{i \in S_t} v_i \cdot o_{it}} \right\}$ The Lagrangian function can be simplified to

$$\begin{aligned}
\mathcal{L}(\boldsymbol{v}, \boldsymbol{\lambda}, \boldsymbol{\mu}) &= l_I(\boldsymbol{v}, \boldsymbol{\lambda}) - \sum_{t=1}^{T}\left(\frac{\mu_t}{v_{0t} + \sum_{i \in S_t} v_i \cdot o_{it}}\right)(\lambda_t - L_t) \\
&= \sum_{t=1}^{T}\sum_{i \in S_t} z_{it} \log(v_i \cdot o_{it}) - \sum_{t \notin \mathcal{B}(\boldsymbol{v})} m_t \log\left(\sum_{i \in S_t} v_i \cdot o_{it}\right) \\
&\quad + \sum_{t \in \mathcal{B}(\boldsymbol{v})} m_t \log\left(\frac{L_t}{v_{0t} + \sum_{i \in S_t} v_i \cdot o_{it}}\right) \\
&\quad - \sum_{t \in \mathcal{B}(\boldsymbol{v})} L_t \frac{\sum_{i \in S_t} v_i \cdot o_{it}}{v_{0t} + \sum_{i \in S_t} v_i \cdot o_{it}} + C_1(\mathcal{B}(\boldsymbol{v})) \\
&= \sum_{t=1}^{T}\sum_{i \in S_t} z_{it} \log(v_i \cdot o_{it}) - \sum_{t \notin \mathcal{B}(\boldsymbol{v})} m_t \log\left(\sum_{i \in S_t} v_i \cdot o_{it}\right) \\
&\quad - \sum_{t \in \mathcal{B}(\boldsymbol{v})} m_t \log\left(v_{0t} + \sum_{i \in S_t} v_i \cdot o_{it}\right) \\
&\quad - \sum_{t \in \mathcal{B}(\boldsymbol{v})} L_t \frac{\sum_{i \in S_t} v_i \cdot o_{it}}{v_{0t} + \sum_{i \in S_t} v_i \cdot o_{it}} + C_2(\mathcal{B}(\boldsymbol{v})) \\
&= \sum_{i=1}^{n} K_i \log(v_i) - \sum_{t \notin \mathcal{B}(\boldsymbol{v})} m_t \log\left(\sum_{i \in S_t} v_i \cdot o_{it}\right) \\
&\quad - \sum_{t \in \mathcal{B}(\boldsymbol{v})} m_t \log\left(v_{0t} + \sum_{i \in S_t} v_i \cdot o_{it}\right) \\
&\quad - \sum_{t \in \mathcal{B}(\boldsymbol{v})} L_t \frac{\sum_{i \in S_t} v_i \cdot o_{it}}{v_{0t} + \sum_{i \in S_t} v_i \cdot o_{it}} \\
&= \sum_{i=1}^{n} K_i \log(v_i) - \sum_{t \notin \mathcal{B}(\boldsymbol{v})} m_t \log\left(\sum_{i \in S_t} v_i \cdot o_{it}\right) - \\
&\quad \sum_{t \in \mathcal{B}(\boldsymbol{v})} m_t \log\left(v_{0t} + \sum_{i \subset S_t} v_i \cdot o_{it}\right) \\
&\quad + \sum_{t \in \mathcal{B}(\boldsymbol{v})} L_t \frac{v_{0t}}{v_{0t} + \sum_{i \in S_t} v_i \cdot o_{it}} \\
&\quad + C_3(\mathcal{B}(\boldsymbol{v})),
\end{aligned}$$

where

$$C_1(\mathcal{B}(\boldsymbol{v})) = \sum_{t \notin \mathcal{B}(\boldsymbol{v})} \left(m_t \log m_t - m_t\right),$$
$$C_2(\mathcal{B}(\boldsymbol{v})) = \sum_{t \notin \mathcal{B}(\boldsymbol{v})} \left(m_t \log m_t - m_t\right) + \sum_{t \in \mathcal{B}(\boldsymbol{v})} m_t \log L_t,$$
$$C_3(\mathcal{B}(\boldsymbol{v})) = \sum_{t \notin \mathcal{B}(\boldsymbol{v})} \left(m_t \log m_t - m_t\right) + \sum_{t \in \mathcal{B}(\boldsymbol{v})} \left(m_t \log L_t - L_t\right). \tag{26}$$

Note that

$$\sum_{t=1}^{T} \sum_{i \in S_t} z_{it} \log(v_i \cdot o_{it}) = \sum_{t=1}^{T} \sum_{i \in S_t} z_{it} \log(v_i) + C_4 = \sum_{i=1}^{n} K_i \log(v_i) + C_4$$

for some constant C_4, because $z_{it} = 0$ when $i \notin S_t$.

Newton's method to find η

To simplify notation, let us define

$$A_j = \sum_{t \notin \mathcal{B}_j(\boldsymbol{v}^{(k)})} m_t \frac{\mathbb{1}(j \in S_t) \cdot o_{jt}}{\sum_{i \in S_t} v_i^{(k)}} + \sum_{t \in \mathcal{B}_j(\boldsymbol{v}^{(k)})} m_t \frac{\mathbb{1}(j \in S_t) \cdot o_{jt}}{v_{0t} + \sum_{i \in S_t} v_i^{(k)}} + \sum_{t \in \mathcal{B}_j(\boldsymbol{v}^{(k)})} L_t \frac{v_{0t} \mathbb{1}(j \in S_t) \cdot o_{jt}}{\left(v_{0t} + \sum_{i \in S_t} v_i^{(k)}\right)^2},$$

$$n_j = \sum_{t=1}^{T} \mathbb{1}(j \in I_t),$$

$$o_j = \sum_{t=1}^{T} \mathbb{1}(j \in S_t) \cdot o_{jt} = \sum_{t=1}^{T} o_{jt},$$

where $\mathcal{B}_j(\boldsymbol{v}^{(k)}) = \mathcal{B}(\boldsymbol{v}^{(k)}) \cap \{t | j \in S_t\}$, n_j represents the number of times product j is in the offer set I_t over time horizon T, and o_j represents the proportion of time product j is available S_t over time horizon T. Using the notation above, Eq. (23) simplifies to

$$K_j = v_j A_j + v_j \eta \left[(1 - \alpha) n_j + \alpha o_j\right].$$

Expressing v_j in the above equation and plugging it into the market share constraint (19) leads to

$$(1 - \alpha) \sum_{t=1}^{T} \sum_{j \in I_t} \frac{K_j}{A_j + \eta\left[(1 - \alpha) n_j + \alpha o_j\right]} + \alpha \sum_{t=1}^{T} \sum_{j \in S_t} \frac{K_j \cdot o_{jt}}{A_j + \eta\left[(1 - \alpha) n_j + \alpha o_j\right]} = \tilde{s} \sum_{t=1}^{T} v_{0t}$$

and finally to

$$\sum_{j=1}^{n} \frac{K_j\left[(1 - \alpha) n_j + \alpha o_j\right]}{A_j + \eta\left[(1 - \alpha) n_j + \alpha o_j\right]} = \tilde{s} \sum_{t=1}^{T} v_{0t},$$

where we used general equalities

$$\sum_{t=1}^{T} \sum_{j \in I_t} K_j = \sum_{j=1}^{n} n_j K_j,$$
$$\sum_{t=1}^{T} \sum_{j \in S_t} K_j \cdot o_{jt} = \sum_{j=1}^{n} o_j K_j.$$

This leads to

$$f = \sum_{j=1}^{n} \frac{K_j\left[(1 - \alpha) n_j + \alpha o_j\right]}{A_j + \eta\left[(1 - \alpha) n_j + \alpha o_j\right]} - \tilde{s} \sum_{t=1}^{T} v_{0t}$$

and

$$g = \frac{\mathrm{d}f}{\mathrm{d}\eta} = -\sum_{j=1}^{n} \frac{K_j\left[(1 - \alpha) n_j + \alpha o_j\right]^2}{\left(A_j + \eta\left[(1 - \alpha) n_j + \alpha o_j\right]\right)^2}$$

and to Newton's method summarized in Algorithm (4).

Computation of gradient in Frank–Wolfe algorithm

Here we will derive the gradient vector of the objective function (24), and show a computational formula. The elements of $\nabla f(\boldsymbol{v})$ can be derived as

$$
\begin{aligned}
\frac{\partial f}{\partial v_i} = \frac{K_i}{v_i} - \sum_{t \notin \mathcal{B}(v)} m_t \frac{\mathbb{1}(i \in S_t) o_{it}}{\sum_{i \in S_t} v_i \cdot o_{it}} \\
- \sum_{t \in \mathcal{B}(v)} m_t \frac{r(1-\alpha)\mathbb{1}(i \in I_t) + (r\alpha + 1)\mathbb{1}(i \in S_t) o_{it}}{v_{0t} + \sum_{i \in S_t} v_i \cdot o_{it}} \\
- \sum_{t \in \mathcal{B}(v)} L_t \frac{\mathbb{1}(i \in S_t) o_{it}\left(v_{0t} + \sum_{i \in S_t} v_i \cdot o_{it}\right) - \left(\sum_{i \in S_t} v_i \cdot o_{it}\right)\left(r(1-\alpha)\mathbb{1}(i \in I_t) + (r\alpha + 1)\mathbb{1}(i \in S_t) o_{it}\right)}{\left(v_{0t} + \sum_{i \in S_t} v_i \cdot o_{it}\right)^2} \\
= \frac{K_i}{v_i} - \sum_{t \notin \mathcal{B}(v)} m_t \frac{\mathbb{1}(i \in S_t) o_{it}}{\sum_{i \in S_t} v_i \cdot o_{it}} \\
- \sum_{t \in \mathcal{B}(v)} m_t \frac{r(1-\alpha)\mathbb{1}(i \in I_t) + (r\alpha + 1)\mathbb{1}(i \in S_t) o_{it}}{v_{0t} + \sum_{i \in S_t} v_i \cdot o_{it}} \\
- \sum_{t \in \mathcal{B}(v)} L_t \frac{r(1-\alpha)\left(\sum_{i \in I_t} v_i\right)\mathbb{1}(i \in S_t) o_{it} - r(1-\alpha)\left(\sum_{i \in S_t} v_i \cdot o_{it}\right)\mathbb{1}(i \in I_t)}{\left(v_{0t} + \sum_{i \in S_t} v_i \cdot o_{it}\right)^2} \\
= \frac{K_i}{v_i} - \sum_{t \notin \mathcal{B}(v)} m_t \frac{\mathbb{1}(i \in S_t) o_{it}}{\sum_{i \in S_t} v_i \cdot o_{it}} \\
- \sum_{t \in \mathcal{B}(v)} m_t \frac{\mathbb{1}(i \in S_t) o_{it}(r\alpha + 1)}{v_{0t} + \sum_{i \in S_t} v_i \cdot o_{it}} \\
- \sum_{t \in \mathcal{B}(v)} L_t \frac{\mathbb{1}(i \in S_t) o_{it} r(1-\alpha)\left(\sum_{i \in I_t} v_i\right)}{\left(v_{0t} + \sum_{i \in S_t} v_i \cdot o_{it}\right)^2} \\
- \sum_{t \in \mathcal{B}(v)} m_t \frac{\mathbb{1}(i \in I_t) r(1-\alpha)}{v_{0t} + \sum_{i \in S_t} v_i \cdot o_{it}} \\
+ \sum_{t \in \mathcal{B}(v)} L_t \frac{\mathbb{1}(i \in I_t) r(1-\alpha)\left(\sum_{i \in S_t} v_i \cdot o_{it}\right)}{\left(v_{0t} + \sum_{i \in S_t} v_i \cdot o_{it}\right)^2}.
\end{aligned}
$$

To implement the gradient with matrix-vector operations let us define

$$\boldsymbol{K} \in \mathbb{R}_+^n - K_i \text{ is the total purchases of product } i \text{ over the selling horizon}$$
$$\boldsymbol{v} \in \mathbb{R}_+^n - v_i \text{ is the preference weight for product} i$$
$$\boldsymbol{v}_0 \in \mathbb{R}_+^T - v_{0t} \text{ is the preference weight for outside alternative at time} t$$
$$r \in \mathbb{R}_+ - r = (1-s)/s \text{ , where } s \text{ is the host market share}$$
$$\boldsymbol{I} \in \{0,1\}^{nxT} - I_{it} = 1 \text{ if product } i \text{ is in the offer set at time} t$$
$$\boldsymbol{S} \in \{0,1\}^{nxT} - S_{it} = 1 \text{ if product } i \text{ is available for sale at time} t$$
$$\boldsymbol{O} \in [0,1]^{nxT} - o_{it} \text{ is percentage of time product} i \text{is available for sale during time period } t$$
$$\boldsymbol{m} \in \mathbb{R}_+^T - m_t \text{ is the total purchases of all products at time } t$$
$$\boldsymbol{L} \in \mathbb{R}_+^T - L_t \text{ is the upper bound on} \lambda_t$$
$$\boldsymbol{B} \in \{0,1\}^T - B_t = 1 \text{ if bound} L_t \text{is violated at time} t \left(L_t < m_t \frac{v_{0t} + \sum_{i \in S_t} v_i \cdot o_{it}}{\sum_{i \in S_t} v_i \cdot o_{it}} \right)$$
$$\alpha \in [0,1] - \text{parameter to control availability of outside alternative}$$

Then

$$\begin{aligned}
\nabla f(\boldsymbol{v}) = \boldsymbol{K} \oslash \boldsymbol{v} + (\boldsymbol{S} \circ \boldsymbol{O}) \big[-\boldsymbol{m} \oslash \left((\boldsymbol{S} \circ \boldsymbol{O})^T \boldsymbol{v} \right) \circ (\boldsymbol{1} - \boldsymbol{B}) \\
-(r\alpha + 1)\boldsymbol{m} \oslash \left(\boldsymbol{v}_0 + (\boldsymbol{S} \circ \boldsymbol{O})^T \boldsymbol{v} \right) \circ \boldsymbol{B} \\
-(r\alpha + 1)\boldsymbol{L} \circ \left(\boldsymbol{I}^T \boldsymbol{v} \right) \oslash \left(\boldsymbol{v}_0 + (\boldsymbol{S} \circ \boldsymbol{O})^T \boldsymbol{v} \right)^{\circ 2} \big] \\
+ \boldsymbol{I} \big[-r(1-\alpha)\boldsymbol{m} \oslash \left(\boldsymbol{v}_0 + (\boldsymbol{S} \circ \boldsymbol{O})^T \boldsymbol{v} \right) \circ \boldsymbol{B} \\
-(r\alpha + 1)\boldsymbol{L} \circ \left((\boldsymbol{S} \circ \boldsymbol{O})^T \boldsymbol{v} \right) \oslash \left(\boldsymbol{v}_0 + (\boldsymbol{S} \circ \boldsymbol{O})^T \boldsymbol{v} \right)^{\circ 2} \big],
\end{aligned}$$

where $\oslash$, $\circ$, and $\circ 2$ denote the elementwise subtraction, addition, and square operations.

Acknowledgements The authors would like to gratefully acknowledge Guillermo Gallego who suggested the idea of splitting the sales in case of partial availability.

References

Abdallah, T., and G. Vulcano. 2016. Demand estimation under the multinomial logit model from sales transaction data. Working Paper. https://www.researchgate.net/publication/303408073. Accessed 21 Nov 2019.

Ben-Akiva, M., and S. Lerman. 1994. *Discrete choice analysis: Theory and applications to travel demand*, 6th ed. Cambridge: The MIT Press.

Bertsekas, D. 1999. *Nonlinear programming*. Belmont: Athena Scientific.

Cao, Y., A. Kleywegt, and H. Wang. 2019. Network revenue management under a spiked multinomial logit choice model. Working Paper. SSRN:https://ssrn.com/abstract=3200531. Accessed 13 Aug 2019.

Dai, J., W. Ding, A. Kleywegt, X. Wang, and Y. Zhang. 2014. *Choice-based revenue management for parallel flights*. Working paper. Georgia Institute of Technology.

Frank, M., and P. Wolfe. 1956. An algorithm for quadratic programming. *Naval Research Logistics Quarterly* 3: 95–110.

Gallego, G., R. Ratliff, and S. Shebalov. 2015. A general attraction model and sales-based linear program for network revenue management under customer choice. *Operations Research* 63 (1): 212–232.

Hunter, D.R., and K. Lange. 2000a. Rejoinder to discussion of "optimization transfer using surrogate objective functions". *Journal of Computational and Graphical Statistics* 9: 52–59.

Hunter, D.R., and K. Lange. 2004. A tutorial on MM algorithms. *The American Statistician* 58 (1): 30–37.

Korte, B.H., and J. Vygen. 2012. *Combinatorial optimization: Theory and algorithms*. New York: Springer.

Nocedal, J., and S.J. Wright. 2006. *Numerical optimization, Springer series in operations research and financial engineering*, 2nd ed. New York: Springer.

Sharif Azadeh, S., P. Marcotte, and G. Savard. 2014. A taxonomy of demand uncensoring methods in revenue management. *Journal of Revenue and Pricing Management* 13: 440–456.

Talluri, K., and G. van Ryzin. 2004. Revenue management under a general discrete choice model of consumer behavior. *Management Science* 50 (1): 15–33.

Train, K. 2003. *Discrete choice methods with simulation*. New York: Cambridge University Press.

Vulcano, G., G. van Ryzin, and R. Ratliff. 2012. Estimating primary demand for substitutable products from sales transaction data. *Operations Research* 60 (2): 313–334.

Norbert Remenyi is a Statistician at Sabre Research, working on statistical modeling, machine learning, forecasting, and revenue management. He received his Ph.D. Degree in Statistics (Industrial and Systems Engineering) from Georgia Institute of Technology in 2012. Before joining Sabre his research focus was on Bayesian modeling in the wavelet domain with applications to nonparametric regression, image denoising, and estimation of partially linear models.

Xiaodong Luo holds a Ph.D. in Operations Research from MIT (1995) and an M.S. in Computer Science from McMaster University (1991). He has been a senior research scientist at Shenzhen Research Institute of Big Data since November of 2020. Before that, he worked at Sabre for 18 years. He has worked on many airline problems such as fleet assignment, crew scheduling, Ops Recovery, as well as pricing and revenue management. He has also worked for i2 Technologies for 6 years on Supply Chain management. His interest lies in modeling and using advanced solution techniques for complex practical problems.

Journal of Revenue and Pricing Management (2021) 20:301–315
https://doi.org/10.1057/s41272-021-00313-2

PRACTICE ARTICLE

How recommender systems can transform airline offer construction and retailing

Amine Dadoun[1,2] · Michael Defoin-Platel[2] · Thomas Fiig[3] · Corinne Landra[2] · Raphaël Troncy[1]

Received: 15 July 2020 / Accepted: 18 September 2020 / Published online: 20 March 2021

Abstract
Recommender systems have already been introduced in several industries such as retailing and entertainment, with great success. However, their application in the airline industry remains in its infancy. We discuss why this has been the case and why this situation is about to change in light of IATA's New Distribution Capability standard. We argue that recommender systems, as a component of the Offer Management System, hold the key to providing customer centricity with their ability to understand and respond to the needs of the customers through all touchpoints during the traveler journey. We present six recommender system use cases that cover the entire traveler journey and we discuss the particular mind-set and needs of the customer for each of these use cases. Recent advancements in Artificial Intelligence have enabled the development of a new generation of recommender systems to provide more accurate, contextualized and personalized offers to customers. This paper contains a systematic review of the different families of recommender system algorithms and discusses how the use cases can be implemented in practice by matching them with a recommender system algorithm.

Keywords Recommender systems · Artificial intelligence · Dynamic offer construction · NDC

Introduction

A recommender system can be seen as an algorithm to compute the probability that a user (customer) would like to interact with an item (product or service). These systems were originally introduced to overcome the problem of information overload that customers face when exposed to a large catalog of products or services. By providing the customers with contextualized and personalized recommendations, recommender systems aim at narrowing down the search to a manageable subset of products that are relevant to the customer.

Recommender systems have proven to be popular for both customers and sellers, particularly for online retail (Resnick and Varian 1997). The most representative example is Amazon that has become one of the largest retailers in the world because, among other important things such as a large selection of products and a fast and reliable delivery chain, it offers best-of-breed customer experience as a result of an extensive use of recommender systems. Recommender systems result in a more personalized shopping experience, giving customers the feeling of being understood and recognized which contributes in building trust and in maintaining loyalty.

From the seller's point of view, recommender systems offer the possibility to control and to increase the exposure of their catalog by driving customers toward products lacking visibility. Recommender systems are also notoriously good at decreasing bounce rate and at increasing average time spent on a web page for online selling (Taghipour and Kardan 2008). Finally, recommender systems have also proved to be very effective offline in email marketing

✉ Amine Dadoun
amine.dadoun@eurecom.fr

Michael Defoin-Platel
michael.defoinplatel@amadeus.com

Thomas Fiig
thomas.fiig@amadeus.com

Corinne Landra
clandra@amadeus.com

Raphaël Troncy
raphael.troncy@eurecom.fr

1 Eurecom, Sophia Antipolis, Biot, France

2 Amadeus, Sophia Antipolis, Biot, France

3 Amadeus, Kastrup, Denmark

Chapter 7 was originally published as Dadoun, A., Defoin-Platel, M., Fiig, T., Landra, C. & Troncy, R. Journal of Revenue and Pricing Management (2021) 20:301–315. https://doi.org/10.1057/s41272-021-00313-2.

campaigns allowing sellers to run so-called "one-to-one marketing" at scale (Jannach and Jugovac 2019).

Recommender systems are growing in popularity in the travel industry to address the complex set of decisions customers face when booking a flight, selecting a hotel or finding relevant events and activities at their destination. For example, Airbnb is now offering real-time personalization of search rankings within its marketplace (Grbovic and Cheng 2018). Travel agencies or brokers have recently called upon the research community to work further on the particularities of making recommendations in the context of travel. The online travel agency Trivago sponsored the 2019 Recommender Systems Challenge as part of the Association for Computing Machinery (ACM) RecSys conference in order to improve their current recommender system for hotels.

However, despite the successful application of recommender systems across many industries, airline offer construction and retailing remains quite rudimentary with little or no differentiation in how products and services are selected, retailed, or priced across customers. There are several reasons for this. First, in the current airline distribution model, airlines have delegated control of the offer construction to content aggregators, such as global distribution systems (GDSs). Real-time interactions with the airline systems are quite limited, and the pricing function which is used to create offers on behalf of the airline is governed by industry standards that only enable very few parameters to differentiate the content based on who the traveler is. Therefore, airlines cannot provide personalized and contextualized offers in a meaningful way. Second, the responsibility of the offer construction and retailing has historically been managed across separate departments within the airline organization. Offer construction and retailing were therefore never part of a broader and holistic customer experience management strategy.

We believe the current approach is inadequate and that the key to profitability is to manage offers consistently in an integrated Offer Management System (OMS) serving the customer throughout the traveler journey from inspiration to post-trip. However, realizing this vision will require significant advancements in both the science of offer construction and in the distribution capabilities employed across all distribution channels, being direct as well as intermediated.

On the distribution side, this advancement will happen as part of IATA's New Distribution Capability (NDC), which will allow airlines to move toward customer centric airline retailing. NDC is an enabler for the application of airline OMS including recommender systems. Industry adoption of NDC has continued to grow in recent years. As of August 2020, 40 airlines, 20 aggregators and 10 sellers are NDC certified level 4 (the highest level) covering booking of NDC content as well as supporting changes of the order IATA (2020).

On the science side, the airline industry literature is still underdeveloped in terms of how dynamic offer construction could be designed and implemented. The key contributions of this paper are therefore to detail illustrative examples of recommender system use cases in the airline industry context and to discuss how these use cases could be implemented in practice with the benefits for both airlines and travelers.

The remainder of this paper is structured as follows. In the next section, we present the traveler journey and we identify use cases for recommender systems. Next, we describe the traditional airline distribution model, the new distribution model enabled by NDC, and the airline's Offer Management System, which will dramatically influence airline offer construction and retailing. We then review the scientific concepts behind each family of recommender systems. Subsequently, we match the use cases with the most appropriate families of algorithms. Finally, we provide some conclusions and we outline some future research directions.

Recommender system use cases throughout the traveler journey

The traveler journey is a key consideration to understand the customer needs and intents (Fig. 1). In their report, Frost & Sullivan (2014) indicate that there "are certain moments when the customer is in a purchasing mind-set and thinking about his trip and what he will need". For example, at the booking stage, the customer is in a "planning" mind-set. At this stage, the airline can approach the customer with more "expensive" offers such as cabin upgrade, or flexibility options. Close to departure (48 h/24 h), the customer has a different mind-set—making the final preparations for his trip. At this moment, airlines could propose the customer with extra baggage, airport transfer, parking, priority check-in, or fast track access. In this section, we detail some use cases for recommender systems along different phases of the traveler journey.

To provide more in-depth discussion, we focus on recommender systems that are under airline control. These use cases cover customers that actively search and book travel products through the standard distribution channels enabled by NDC—both direct and indirect channels. Thus, use cases for recommender systems regarding customer acquisition through the Internet giants' web interfaces, social media, and search engines will not be covered, since in these cases, the recommender systems reside outside the airline's control.

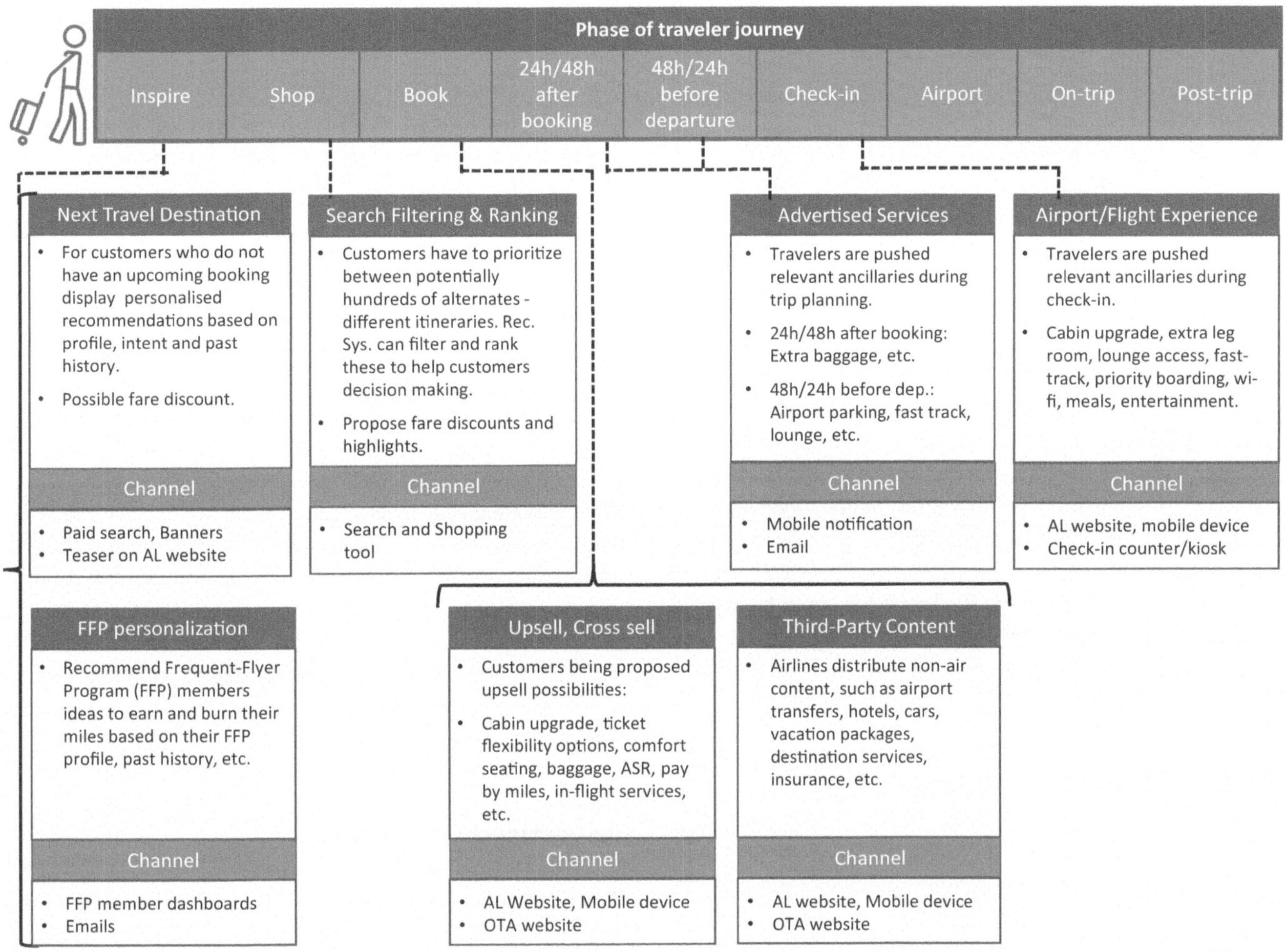

Fig. 1 Recommender system use cases throughout the traveler journey

Next travel destination

The inspiration phase is a key opportunity to influence the customer decision-making process. We distinguish between *passive inspiration* and *interactive inspiration*. The former represents the case when a customer (typically anonymously) lands on a web page and receives travel inspiration simply because some routes are popular in general, while the latter corresponds to the case where the customer interacts with the recommender system by providing personalized search criteria. In the following, to be concrete, we take the assumption that the customer stays anonymous and is engaged in interactive inspiration, providing the recommender system with more leverage.

Affinity shopping tools can be employed to create a personalized shopping experience. Rather than selecting the traditional criteria of origin/destination and calendar dates, these tools enable inspiration based on personalized criteria, such as customers' budget and interests (events or destination type such as beach, city, etc.). A recommender system with access to information of upcoming events (e.g., jazz festivals, sport events, exhibitions, etc.), and real-time information about flight prices and promotional fares (campaigns) could be used to recommend the most appropriate destinations and dates that match the customers criteria. Further, it could also recommend how the offers should be retailed using rich format such as infographics, photos and videos. For example, a trip during the summer to Nice Côte d'Azur in France, should have a very different presentation depending on if the customer is interested in beach, nightlife or a culinary experience.

FFP personalization

The frequent-flyer program (FFP) business model is dependent on FFP members having sufficient incentive to earn and burn their points. However, in reality, this may not be so easy. Premium-tier members with large point balances may not be able to find availability on attractive flights or premium classes due to blackouts or lack of award availability, while low-tier members with small point balances often cannot afford a redemption ticket and see no value in the program.

Recommender systems are in a good position to increase the number of points burned using information about both the member's point balance and the availability of award tickets. For example, the premium-tier member may be offered to burn points for upgrades for his/her family on their annual vacation trip (to mitigate the dilution risk of the award ticket substituting a commercial ticket) or non-air content not readily accessible for purchase on the open market (e.g., backstage passes to concerts, games, etc.). For the low-tier member, recommender systems could offer a "discount" toward the fare of a commercial ticket.

Several other use cases for recommender systems can also be identified, such as incentivizing members to earn points to reach the next tier level or burn points that are close to expiration. In all these cases, the system may be able to increase the value of the program by sending personalized emails to members with the right offer at the right time.

Search filtering and ranking

For a customer who makes searches by comparison shopping, booking air travel can be a daunting experience. He or she must prioritize among potentially hundreds of itineraries, with different prices and product characteristics across multiple partner airlines. As a result, it becomes almost impossible for the customer to make a purchase decision. Today, most search algorithms aim at finding the lowest fares but, in doing so, create irrelevant or unattractive itineraries that distract or overwhelm the customer.

A recommender system can filter the choice set into a manageable number of alternatives and rank them in order of relevancy based on an understanding of the customer's stated criteria. In this way, the recommender system both guides the customer in his decision process and benefits the airline through improved conversion rates. We may also add new customized criteria beyond the usual origin destination, date range, flying time, ground time and overnight stay criteria to incorporate product attributes such as cabin, ticket flexibility, seat reservation and baggage allowance that are not typically considered in comparison shopping requests today.

Upsell, cross-sell and third-party content

When the customer has decided on his preferred itinerary, he enters the booking stage. During the booking stage, the recommender system has ideal information about the customer and his travel party—not only the current trip destination, duration, and already-selected ancillary services, but also the customer's profile and historic purchases. At the booking stage, the customer is in a planning mind-set and this is an ideal opportunity to both increase ancillary revenues for the airlines as well as offer a one-stop shopping experience that covers the customer's full journey.

Examples of products that could be recommended at this stage include upsell offers such as cabin upgrades or ticket flexibility options, as well as cross-sell offers such as baggage, advance seat reservations or in-flights services (e.g., meals). In addition, the airline can also offer third-party content. Based on the customer needs, the commercial relation with the third parties, the prices and availabilities for the relevant resources, the recommender system can propose simple products such as insurance, airport transfers, etc., or even more complex bundled travel such as vacation packages that include hotels and rental cars.

Advertised services

During the post-shopping period, the airline has an opportunity to push offers to customers through unsolicited mail or via notification on a mobile device. This period is a critical phase for the customers' last-minute decisions and preparations for their trip. Customers can be approached with ancillary services such as extra luggage, airport parking, seat selection, priority check-in, etc., and also be informed of availability of cabin upgrades that are aligned with their preferences. Again, the offer and communication would be very different between a family of four traveling long-haul from Frankfurt to New York City in economy class for a two weeks' vacation, versus a business purpose customer traveling the same itinerary and cabin, but staying only for two days. A recommender system would propose not only the most relevant offers but also the most relevant channel and time to push these offers with the benefit of increased adoption rates and customer satisfaction.

Airport/flight experience

During check-in, the customers actively interact with the airline via employees at the check-in counter, the kiosk, or on mobile devices. During this phase, the customer is focusing on the practicalities before takeoff. This may regard logistics of how to navigate through the airport, but the customer may also wish to indulge themselves with restaurants, lounge access, or cabin upgrades, which could be paid for example using FFP points.

Considering the personas mentioned before, the family of four returning from their vacation in New York City may have excess baggage, while the business purpose customer returning from New York City on a red-eye flight may be looking for an upgrade to the business cabin. These examples serve to illustrate that customers' needs may vary significantly and that the airline has an opportunity to approach the customers with relevant offers based on a deep understanding of their needs, preferences and intent.

Toward a new distribution capability for the airline industry

In this section, we first detail the traditional airline distribution model. This will provide the necessary background for understanding the objectives behind the new distribution standards, known as the new distribution capability (NDC), which we discuss subsequently. We demonstrate that NDC is an enabler for the application of the airline OMS including recommender systems.

Traditional distribution model

Figure 2 shows how a customer's request for an itinerary is passed from a retailing platform (airline retailing platform, or other retailing platforms), possibly through a distributor, and to the airline's Inventory system for evaluation, using the distribution model in place today. For the direct channel (direct connect), the airline fully controls the shopping and pricing flow. However, for the indirect channels, the current distribution paradigm relies on a two-step process. First, the airline files fares with data distributors such as ATPCO or SITA. These filed fares drive the construction and pricing of the products that can be offered to the customers. Then, the availability computation within the airline's Inventory system (flight execution) determines which of the filed fares are made available for sale. The airlines control the availability computation via their Revenue Management Systems (RMS), which essentially can be performed using offline optimization (airline planning).

Other retailing platforms may interact directly with the airline's flight execution layer via proprietary interfaces. Distributors such as the GDSs acquire the filed fares content and have the authorization to build offers on behalf of the airlines (delegated shopping & pricing). The distributors then poll the airline's availability to determine which fare products are available for sale. Consistency across indirect channels is enabled by highly standardized content and associated processing logic that the GDSs adopt and implement when accepting airline content and developing their shopping and pricing engines. This means that there is a limited ability for customer-specific information to be used in the indirect distribution channel. In principle, even if the airlines could create contextualized and personalized offers in the direct channel, this would create inconsistency that cannot be resolved among the distribution channels.

New distribution capability (NDC)

The new distribution capability (NDC) is a set of new technical communication standards that was initiated almost a decade ago by the International Air Transport Association (IATA). The vision with NDC is to modernize airline distribution and enable airlines to have better control of their offers and their retailing. We list below the most important benefits for airlines that are adopting NDC, which are of

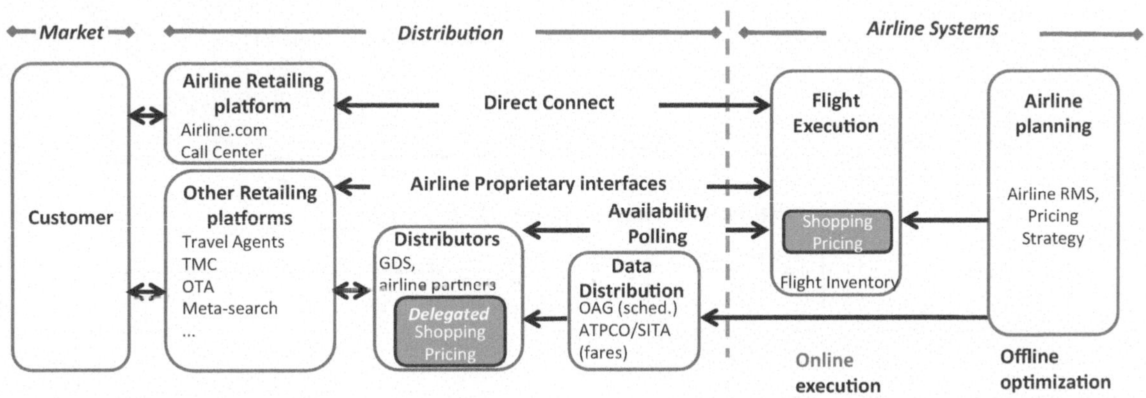

Fig. 2 Traditional distribution model

particular relevance for this paper. For further information on the objectives and benefits of NDC, we refer the reader to (Hoyles 2015).

- **Personalized and contextualized offers**. The airlines will have access to customer and contextual information in a shopping or booking request, which will allow for personalized and contextualized offers.
- **Dynamic offers**. The airlines will be able to create, distribute, and fulfill dynamic offers as described in the next section.
- **Dynamic pricing**. The airlines can employ dynamic pricing using a continuous price.
- **Retailing**. The airlines can provide the retailing platforms with product description that encompasses retailing preferences and information. For instance, rich media content that further complements their offers using visual elements, such as infographics, photos, videos, etc.
- **Merchandising**. The airlines will be able to employ merchandising techniques to affect customers purchase behavior.

Figure 3 shows how airlines are aspiring to take control of the offer creation, at scale and across all distribution channels.

In the NDC environment, airlines still make the decision of distributing via direct channels and/or via indirect channels with third-party intermediation. However, delegation of the offer creation to intermediaries no longer exists. Instead, each customer shopping request in an agent's front-office system is passed to the airline OMS, either directly in the case of NDC direct connect distribution, or via an aggregator in the case of NDC Intermediated distribution. Note that the airline proprietary interfaces and availability polling arrows in Fig. 2 have been replaced by NDC direct connect and NDC intermediated arrows in Fig. 3, enabling a cost efficient deployment at scale for the distribution network actors. The airline's OMS creates a set of one or more offers that are returned to the customer. Each offer is individually tagged with an offer ID that can be used in any subsequent request on that offer. If the customer accepts an offer, the offer is converted into an order and the contract with the customer is established.

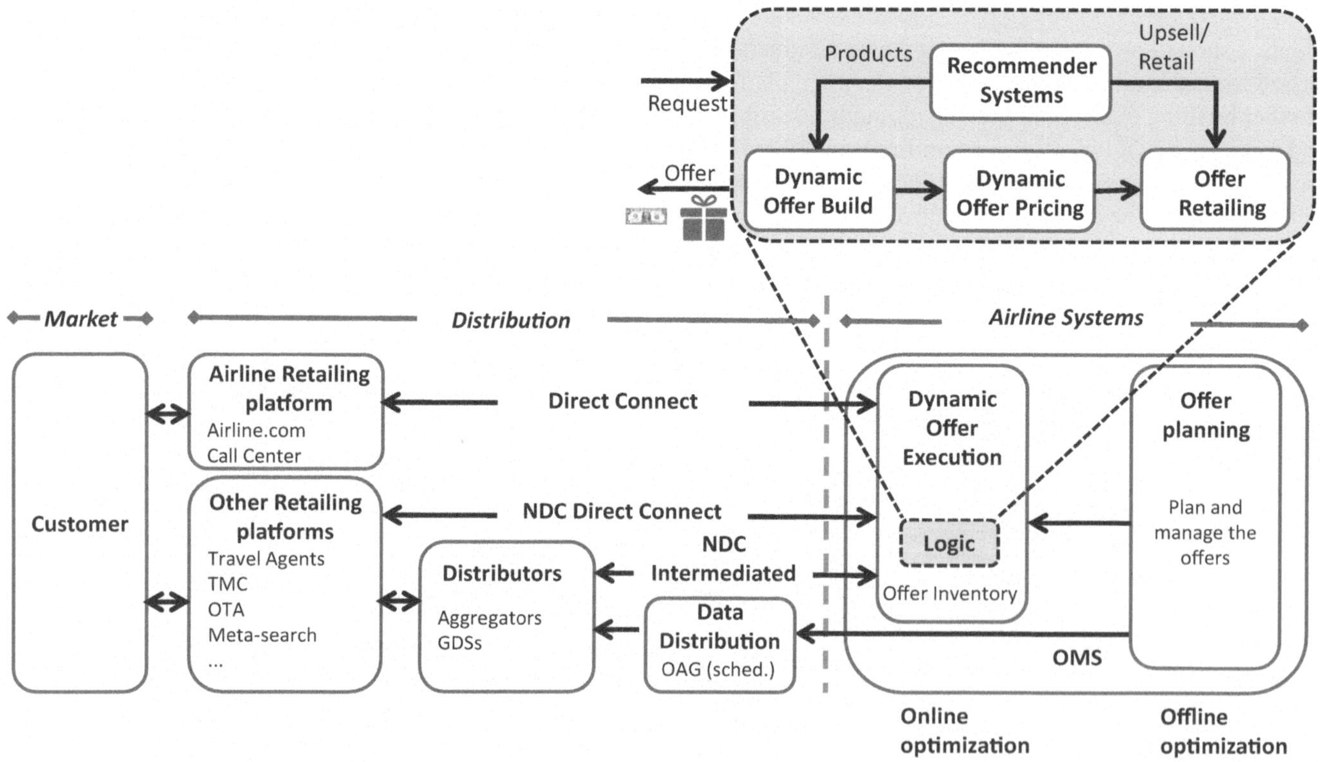

Fig. 3 Distribution model using NDC

The offer management system (OMS)

As seen in Fig. 3, the airline OMS controls the offer construction and retailing for both the direct channel and the indirect channel in NDC. We can think about OMS as an extension of the airline's RMS in several dimensions.

The main extensions are as follows. First, RMS optimizes only the prices (actually the availabilities) of the pre-filed flight products, while OMS optimizes both product components (flight products, ancillaries, third-party content) and prices. Second, unlike RMS which provides the same price to all customers for a given flight and fare product, OMS may differentiate among customers and construct personalized and contextualized offers. Third, and not considered by RMS, OMS may construct one or multiple offers in a so-called *offer set* that will be displayed together as options. For further information, we direct readers to (Fiig et al. 2018).

Finally, because RMS does not differentiate among customers, the price computation can essentially be precomputed during the offline optimization processes and the online process is a lightweight execution logic. For OMS, this is not the case, as computing personalized and contextual offers is designed to be a real-time decision and the optimization logic must be moved to the online domain. This has significant ramifications for the IT system design of the OMS, which we will discuss below.

The online optimization logic of the OMS is comprised of the following components, which is illustrated in the inset in Fig. 3. In particular, we would like to draw attention to the role of recommender systems in guiding both the Dynamic Offer Build and the Offer Retailing, which has also been exemplified with the recommender system use cases presented.

- **Dynamic Offer Build**. This module makes the determination of the relevant set of products (flights, ancillaries, and third-party content) to be returned at the individualized customer level.
- **Dynamic Offer Pricing**. This module takes as input the offers that were built by "Dynamic Offer Build" and determines for each of these the selling price that maximizes the contribution considering both customer and contextual information.
- **Offer Retailing**. This module aims to increase conversion rates by applying merchandizing techniques to affect the customer's purchasing behavior.

In the description above, we have seen the different functional steps of an OMS to dynamically construct, price and retail an offer. However, we also need to consider the ecosystem that will trigger and support this process. In particular, online search engines have strict performance requirements. As these engines generate thousands of search transactions per booking, these IT systems need to be extremely cost-effective, scalable and resilient, to provide real-time dynamic offer construction and retailing while providing consistency across all distribution channels. Recent advancements in technology and infrastructure capabilities can enable airlines and system providers to accomplish these goals. For example, cloud infrastructure and real-time worldwide data synchronization and processing power allow data centers across continents to host and run local instances of the online optimization logic, accessible to any distribution channel, while continuously being under airline control.

The science of recommender systems

Introduction to recommender systems

In the terminology of recommender systems, the customers are referred to as *users* and the products in the catalog are referred to as *items*. Hence, a recommender system can be seen as a way to compute the probability that a user would like to interact with an item and use this probability to recommend the most relevant subset of items to him. Depending on the context, an interaction would correspond to the act of searching, buying, visiting, watching, etc.

In its most simple form, a recommender system is typically built in three consecutive steps: *information collection*, *learning* and *recommendation* (Isinkaye et al. 2015). The information collection phase consists in building a weighted graph $G = (U, I, E, w)$, where U, the set of users, and I, the set of items, are the nodes in the graph and E corresponds to the set of edges. These edges represent the past interactions between users and items. There are no edges between the users nor the items, hence the graph is bipartite. The strength of these past interactions is given by the function $w{:}E \mapsto [0, 1]$.

In the learning phase, a machine learning (ML) algorithm is used to train a model $\mathcal{W}$ that approximates w in G. Finally, in the recommendation phase, the trained model is used to predict, for every possible pair $(u, i) \in (U \times I)$, the strength of the interaction between user u and item i. From these predictions, it is then possible to derive the list of items that could be recommended to the users.

From tapestry (Goldberg et al. 1992), introduced in the early 90s that is considered as the first example of a working collaborative filtering algorithm, to the massive usage of deep learning algorithms (Zhang et al. 2019), the research on recommender systems is now one of the most prolific topics in the artificial intelligence (AI) literature.

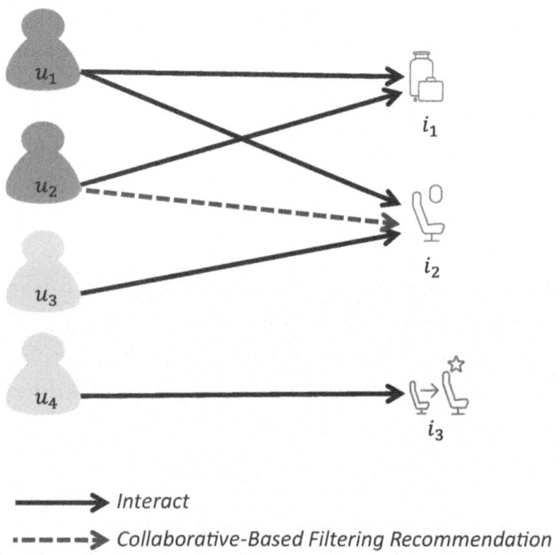

Fig. 4 CF recommender systems: bipartite graph between users and items showing how item i_2 is recommended to user u_2 through a CF algorithm

Machine learning models to predict user-item interactions have evolved from using simple models such as linear and logistic regression to deep neural network models that endow them non-linearity, and thus allow them to find non-linear patterns in the data. However, each of these approaches has its own specificities and it is important to understand their strengths and limitations when addressing a particular use case. In this section, we review the main families of recommender systems.

Collaborative filtering recommender systems (CF)

Collaborative filtering (CF) algorithms are among the most widely used algorithms in the field of recommender systems (Sarwar et al. 2001) and have been applied in industries such as e-commerce or online entertainment to recommend the most relevant products or movies to their customers. In the original formulation, a CF algorithm relies only on the interactions present in the graph *G* without any additional knowledge or information about the items or the users.

Figure 4 shows an illustrative example of the bipartite user-item graph *G* for ancillary products. The graph contains interactions between users (travelers) and items (seat, baggage, etc.) represented by the solid arrows, while the dashed arrows represent the recommendations obtained from CF algorithms. Let us consider the item i_1 (baggage) for example. Users u_1 and u_2 both purchased this item. Furthermore, user u_1 also purchased item i_2, thus item i_2 is recommended to user u_2.

We can divide CF algorithms into two different classes of methods, the first one relying on *Matrix Factorization* techniques (Hu et al. 2008) and the second one, named *Neighborhood Methods* (Sarwar et al. 2001), relying on computing the similarity between users or items.

Over the years, significant progress has been made to improve CF algorithms, for example in terms of learning speed (He et al. 2016) or accuracy (Rendle et al. 2009; He et al. 2017). Nevertheless, despite their proven overall effectiveness and usability, CF algorithms are still limited especially when users interact with a restricted number of items (*data sparsity*) or when new users or items frequently enter the system and, consequently, past interactions are not available (the user or item *cold start problem*).

Content-based filtering recommender system (CB)

The content-based (CB) filtering method (Lieberman 1995) aims at building user preference profiles based not only on historical user-to-item interactions but also on a form of description of these items that is often represented by a set of keywords or properties. Conversely, it is also possible to associate items to user profiles by looking at the description of the users interacting with them.

In Fig. 5, we present the graph *G* enriched with the item properties needed for the use of CB recommender systems. Each item (ancillary product) is characterized by a set of properties: for example, the baggage item has the value "C" for the Reason for Issuance Code (RFIC) and the value "A"

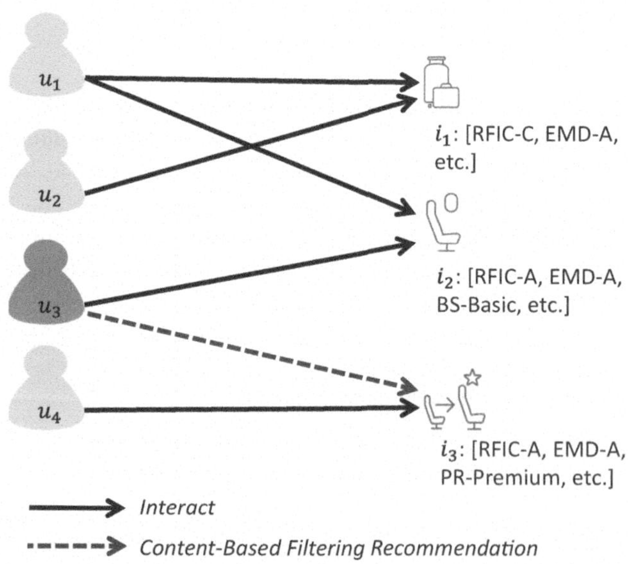

Fig. 5 CB recommender systems: bipartite graph between users and items enriched with item descriptions showing how item i_3 is recommended to user u_3 through CB algorithm

for the Electronic Miscellaneous Document (EMD) category, as it is a flight-associated product. In this example, the CB algorithm recommends item i_3 (premium seat) to user u_3 because item i_3 has the same characteristics of item i_2 which user u_3 has interacted with (added in his cart) in the past.

With CB filtering, even new items without any previously observed interactions will have at least a description that can be used by the system to provide recommendations. Hence, the problem of item *cold start* is mitigated. Nevertheless, CB filtering methods also have some shortcomings. For example, building and maintaining relevant representations for every item can turn into a heavy feature engineering task. Also, introducing novelty into what is being recommended to a given user is not possible since the system works only by looking at content associated with the user's past interactions.

One of the alternatives to deal with the above mentioned limitations such as the lack of novelty consists in mixing CB and CF techniques in what is referred to as Hybrid recommender systems in the literature (Melville et al. 2002; Khrouf and Troncy 2013).

Context-aware recommender system (CA)

CF or CB algorithms model the users' behavior by relying on past user-item interactions or on the content of the items. However, to better capture the complex decision-making process that the users are following when exposed to a selection of items (e.g., the offer set construction by OMS), it is

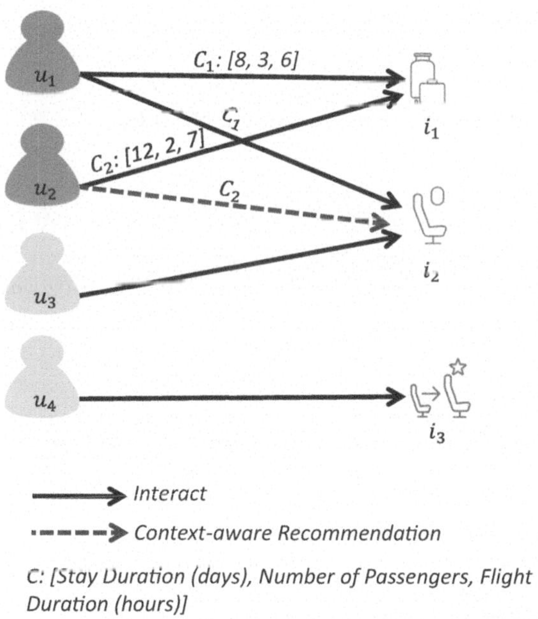

Fig. 6 CA recommender systems: bipartite graph between users and items enriched with contextual information showing how item i_2 is recommended to user u_2 through CA algorithm

crucial to consider the overall context of this process. For instance, a user who wants to travel during summer with four people for two weeks (likely leisure travel) will not have the same needs when traveling alone for two days during a winter week (likely business travel).

A context-aware (CA) recommender system should first be able to collect contextual information and then make use of it to better tailor the offers depending on the circumstances. In Fig. 6, we present the graph G enriched with contextual information. As an illustration, let us consider that the user u_1 who purchased both items i_1 (baggage) and i_2 (seat) for his trip to Paris which will last 8 days with a flight duration of 6 hours. On the other hand, we consider the user u_2 that will travel from New York to Paris on a similarly long flight (7 hours) for 12 days and purchased item i_1 in addition to the flight ticket. Item i_2 is being recommended to user u_2, as contexts C_1 & C_2 are closely related.

Several initiatives have been conducted to enrich existing recommendation approaches with contextual information. We can categorize them into three different groups (Adomavicius and Tuzhilin 2015): (i) Contextual Pre-filtering (Adomavicius and Tuzhilin 2005) where the contextual information is used only to filter out the graph of user-item interactions to keep only the data pertaining to a particular context; (ii) Contextual Post-filtering (Panniello et al. 2009) where the context is used to produce contextualized recommendations on top of what a traditional recommender system suggests; and finally (iii) Contextual Modeling (Karatzoglou et al. 2010; Rendle 2010; Xiao et al. 2017) where the context itself is considered by the model as input information together with the user-item interaction graph.

Knowledge-aware recommender system (KA)

According to Paulheim (2017), a Knowledge Graph (KG) (i) mainly describes real-world entities and their interrelations, organized in a graph, (ii) defines possible classes and relations of entities in a schema, (iii) allows for potentially interrelating arbitrary entities with each other and (iv) covers various topical domains.

KGs became an increasingly popular research direction toward cognition and human-level intelligence, and are now used in many AI applications such as semantic search or automatic fraud detection. In recent years, KGs have also been introduced in Knowledge-Aware (KA) recommender systems (Palumbo et al. 2017) in order to enrich the graph of user-item interactions with more complex and structured information about the users, the items, and the interactions themselves.

In Fig. 7, an example of a KA recommender system is shown. Beyond the simple lists of properties already

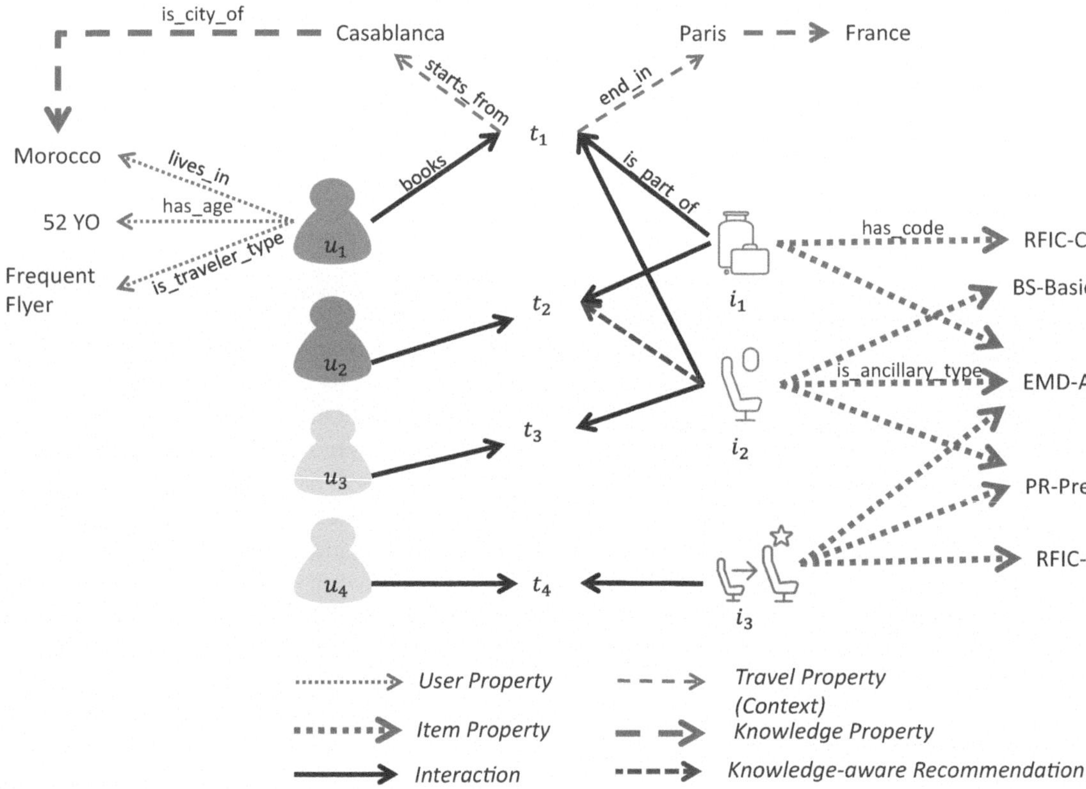

Fig. 7 KA recommender systems: knowledge graph representing user-item interactions in addition to information about users, items and the context of each interaction showing how item i_2 is recommended to the user u_2 via KA algorithm over the knowledge graph

managed by previous versions of recommender systems, KGs represent and leverage semantically rich relations between entities. We see that travel t_1 booked by user u_1 starts from Casablanca, a city in Morocco, which is also the country where user u_1 lives. By construction, KGs can easily be linked between each other. For example, it would be straightforward to extend the graph from Fig. 7 to include cities' main Points of Interest (Monti et al. 2018).

One remarkable thing about KA recommender systems is their ability to make use of the KG structure to provide better recommendations (Sun et al. 2018). Using deep learning, and in particular, graph embedding techniques (Zhang et al. 2016; Palumbo et al. 2017), it is now possible to turn virtually any type of information into a vector that the system can learn. Dadoun et al. (2019) proposed to use the Semantic Trail Dataset (Monti et al. 2018) that contains users' check-ins in many cities around the world to build location embeddings for travel recommendations.

Session-based recommender system (SB)

Recommender system approaches based on historical user-item interactions are very powerful because they are able to exploit long-term user profiles (Ludewig and Jannach 2018). However, in real-world applications such as e-commerce platforms, many new users visit the system every day for

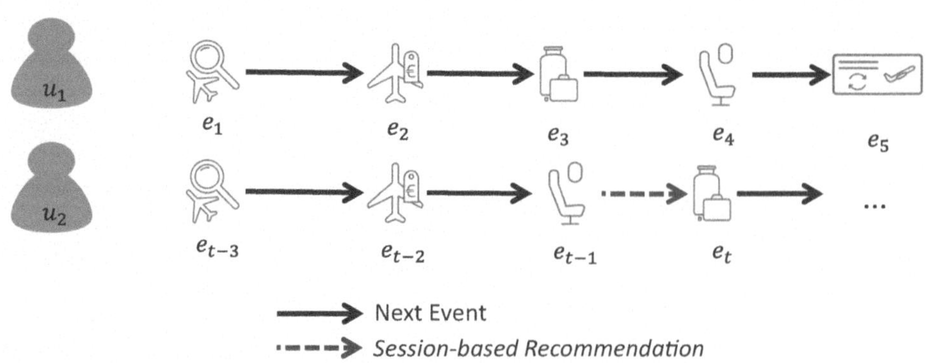

Fig. 8 Session-based recommender systems: sequence of user events (interaction with the catalog), user u_2 is being recommended a bag at t through SB algorithm

which no historical information is available (the user *cold start* problem).

It is therefore necessary to analyze users' live sequence of actions (for instance, their sequence of clicks) to identify patterns and generate recommendations (Linden et al. 2003). This approach can range from simply detecting frequently co-occurring actions (Agrawal et al. 1993) to a more in-depth modeling of the sequence itself with deep learning techniques (Hidasi et al. 2016).

In Fig. 8, user u_1 starts a browsing session looking for a flight (event e_1), then chooses his flight (e_2) and adds it to the shopping cart, and he decides to add two ancillaries (seat and baggage) which represent events e_3 and e_4, to finally make his booking e_5. On the other hand, user u_2 follows the same path as u_1 for his first two events and decides at $t-1$ to add a seat to his shopping cart. Since adding seat and baggage in the same shopping cart are two co-occurring events, the session-based (SB) recommender system will propose to user u_2 to add baggage to his cart.

Beyond the different families of algorithms described in this section, the field of recommender systems is in constant evolution with more and more complex approaches being regularly proposed to address the limitations of the previous generations. As an example, a promising research direction mixing reinforcement learning (Sutton and Barto 2018) and recommender systems (Rohde et al. 2018; Zhao et al. 2018) is being explored with the ambition to focus on long-term returns and break the pernicious feedback loop of recommendation as described in (Chaney et al. 2018).

Adapting recommender systems for offer construction and retailing

In this section, we revisit the use cases introduced in the Section "Recommender system use cases throughout the traveler journey" and we discuss how they can be implemented in practice using the families of recommender system algorithms described in the previous section. We identify the most appropriate algorithms given the non-functional requirements, such as (i) the available input data, (ii) the output data, (iii) the chosen objectives, and (iv) the operational constraints (e.g., response times). For each use case, we also provide relevant metrics that could be used to assess the quality of each recommender system. Figure 9 provides a summary of this analysis.

		Next Travel Destination	FFP Personalization	Search Filtering & Ranking	Upsell, Cross sell & Third Party content	Advertised Services	Airport/Flight Experience
Input Data	Past user-item interactions	-	✔	-	✔	✔	✔
	User information	-	✔	-	✔	✔	✔
	Item Information	✔	✔	✔	✔	✔	✔
	Context Information	✔	✔	✔	✔	✔	✔
	Knowledge Graph	-	✔	-	✔	✔	✔
	Live interactions	✔	-	✔	✔	-	-
	Extra information	RMS, Interests, Budget, Upcoming events	FFP, RMS	RMS	Third Party	-	FFP, RMS
Output Data	Offer build	destination, date range	action to burn points	ranking of offers	ranking of offers	ancillary proposition	ranking of offers
	Offer retail	presentation, infographics, description	offers timing	offers highlight	offers highlight	offer timing	presentation, infographics, description
Objectives	Travelers' Loyalty	-	✔	-	✔	✔	✔
	Air product conversion	✔	✔	✔	-	-	-
	Ancillary product Conversion	-	-	-	✔	✔	✔
	Third Party Conversion	-	-	-	✔	-	-
	Miles burned	-	✔	-	-	-	-
Specifics	Response Time	-	-	✔	✔	-	✔
	Data Acquisition	-	-	✔	✔	-	✔
Algorithms Family		CB, CA	CA, KA	CB, CA, SB	SB, CA, KA	KA	CA, KA

Fig. 9 Summary of recommender system algorithms for each use case given the input data, outputs, objectives and constraints. Algorithms in brackets are feasible, while the algorithms without bracket are preferred

Next travel destination

We take the assumption that the customer (user) is anonymous at this stage of the traveler journey. Hence, for this use case, we cannot rely on the past interactions of the user and we discard the use of sophisticated algorithms such as KA that are most effective with this information. Instead, we consider using CA algorithms in a post-filtering fashion starting with CB or SB algorithms to rank destinations based on either the content of the destinations (CB) or the user's clicks through his live interactions (SB). The outputs of the CB/SB algorithms can then be filtered according to the criteria specified by the user from the search tool. Metrics used to evaluate the recommendations could be click-through rate and conversion rate.

FFP personalization

In this use case, the customer identity is known and we can therefore leverage on individual FFP data—such as tier level, point balance, point expiration dates, recency, frequency, and monetary value—but also on price/point conversion rates for the recommended itineraries and services in order to produce meaningful recommendations. The algorithm must also be able to mix this information with a variety of other data from different sources, ranging from the product catalog of air and non-air products, the customer travel history, and the product availability and prices provided by the RMS.

Hence, because of their data integration capabilities, KA algorithms appear to be the natural choice for this complex use case. Moreover, as demonstrated in Yao et al. (2015), KA can be extended to include contextual information allowing the algorithm to capture the travel intent of the user. Metrics used to evaluate the recommendations could be conversion rate and FFP points burned.

Search filtering & ranking

We take the assumption that the customer (user) is anonymous during this stage. In this situation, the recommender system will have to rely on stated criteria (origin-destination, date range, stops, etc.), the context of the search (search time and date, type of the device being used, etc.), product attributes (cabin, flexibility, baggage allowance, etc.), and possible extended criteria depending on the capabilities of the search tool. The recommender system may also employ user navigation behavior to better understand the travel intent. Given the input data available, CA/SB recommender systems (Rendle 2010; Sarwar et al. 2001) seem to be judicious choices provided that session data can be acquired and response time kept within acceptable limits. Metrics used to evaluate the recommendations could be click-through rate, conversion rate, and sales.

Upsell, cross-sell and third-party content

At this stage, the customer identity is known. However, the customer travel history will, in many cases, still be absent or rather limited. In this case, SB/CA algorithms could be considered. On the other hand, when customer travel history is present, hybrid approaches integrating personalized recommendations could be investigated using for example the KA algorithms. Response time and data acquisition are important specifics of this use case and must be taken into consideration before the preferred algorithm is chosen. Of note, the SB algorithms have a very fast execution time compared to CA and KA, which may impact the choice. Metrics used to evaluate the recommendations could be Conversion Rate, ancillary/third-party revenue and adoption rates.

Advertised services

Targeting customers with unsolicited notifications can be counter-productive and lead to adversarial effects on customer loyalty if done incorrectly. It is therefore critical to identify the customers that we expect to react positively to an advertised service. This problem can be seen as an inverse recommendation scenario—recommending a user to an item.

This problem is well-suited for the KA algorithm. Indeed, in this use case where the customer identity is known, the algorithm can take advantage of a diverse set of data: collaborative information (e.g., historical ancillary purchases), user-related information (e.g., number in party), item-related information (e.g., product descriptions), and context-related information (e.g., attributes of the current order). Additionally, other ML approaches such as contextual multi-armed bandits (Li et al. 2010) could also be employed to find the best timing and channel for sending the notifications. Metrics used to evaluate the recommendations could be click-through rate, conversion rate, and incremental revenue.

Airport/flight experience

The time period spent at the airport or during the flight itself is a particularly favorable window of opportunity for the airlines to approach the traveler with personalized and contextualized offers. The algorithms of choice could be CF or CB given their ability to learn the preferences of the travelers and provide near real-time recommendations, especially when the product catalog is rather limited. Alternatively, the CA algorithm should be also considered, since this algorithm is able to capture travel intent which may well be of importance in this use case. The conversion rate, incremental revenue, FFP points burned are the most appropriate metrics to evaluate how these algorithms perform.

Conclusion and future research directions

Recommender systems have already been introduced in several industries such as retailing and entertainment, where their capability to display personalized and contextualized recommendations have provided benefits to customers and sellers alike. However, their application in the airline industry remains in its infancy. In this paper, we explain that this is primarily a result of the limitations of IT systems that delegate airline control of offer creation to content aggregators. The traditional distribution paradigm relies on a two-step process—fare filing which drives the product and price construction, followed by the availability computation—which provides airlines with limited control over offer construction and retailing. Further, the airlines are unaware of the customer's identity and therefore unable to generate personalized recommendations.

NDC is an enabler for the airlines to provide contextualized and personalized offers, thereby opening the door for the application of recommender systems via the airlines offer management systems (OMS). We believe that recommender systems hold the key to customer centricity with their ability to understand and respond to the needs of the customers throughout all touchpoints during the traveler journey, which we have exemplified with airline-specific recommender system use cases.

We have explained how recent advances in AI have enabled the development of a new generation of recommender systems to provide more accurate, contextualized and personalized offers to users. However, choosing one family of algorithms over another can be a complex task for a travel industry expert because of the large number of algorithms described in the literature and the particularities of the travel domain. Therefore, we have for each of the use cases, provided guidance by identifying the preferred algorithms.

While we have discussed how the application of recommender systems can provide "short-term" (or transactional) benefit to the airline through increased ancillary adoption rates and revenue, we believe that recommender systems may have an even greater opportunity for improving customer experience and increasing customer loyalty by enabling airlines to understand their customers' needs, preferences and intent. The impacts of effective recommendations and retailing on customer loyalty in the airline industry have yet to be explored.

We propose three main areas for our future research directions.

- **Empirical Study**. The next logical step is to perform an empirical study of the performance of the algorithms using actual airline data. This requires to partner with airlines in order to acquire real life data.
- **Explainability**. One of the main challenges for the AI community is to bring explainability to decision-making algorithms. Indeed, it is crucial to understand why an algorithm has recommended a specific item. One popular method of explainability arises from Neighborhood Methods that can state, for example, that "a customer that bought this item, also bought these items". The KA recommender systems are also ideally suited for this purpose, as this algorithm constructs an explainable path within the knowledge graph that lead to the item recommendation (Song et al. 2019). Moreover, performing an ablation study on algorithm inputs, where an input of a model is removed to assess the effect on algorithm performance, would allow us to understand what input data are the most beneficial for an accurate prediction.
- **Industry disruptions**. The Covid-19 pandemic has disrupted the airline industry in an unprecedented way. The industry may not experience a smooth recovery but rather in waves as different countries open/close for air traffic in response to pandemic evolution. This raises questions of the performance and robustness of the different algorithms in the presence of sparse, scattered, and constantly evolving data.

References

Adomavicius, G., and A. Tuzhilin. 2005. Toward the next generation of recommender systems: A survey of the state-of-the-art and possible extensions. *IEEE Transaction on Knowledge and Data Engineering* 17 (6): 734–749.

Adomavicius, G., and A. Tuzhilin. 2015. *Context-Aware Recommender Systems*, 191–226. Berlin: Springer.

Agrawal, R., T. Imieliundefinedski, and A. Swami. 1993. Mining association rules between sets of items in large databases. In ACM SIGMOD International Conference on Management of Data (ICDM), pp. 207–216. Washington, D.C., USA.

Chaney, A. J. B., B. M. Stewart, and B. E. Engelhardt. (2018). How algorithmic confounding in recommendation systems increases homogeneity and decreases utility. In 12^h ACM Conference on Recommender Systems (RecSys), pp. 224—232, Vancouver, British Columbia, Canada.

Dadoun, A., R. Troncy, O. Ratier, and R. Petitti. (2019). Location embeddings for next trip recommendation. In 9th International Workshop on Location on the Web (LocWeb), pp. 896–903, San Francisco, USA.

Fiig, T., R. Guen, and M. Gauchet. 2018. Dynamic pricing of airline offers. *Journal of Revenue and Pricing Management* 17: 04.

Frost and Sullivan. 2014. *Thinking like a retailer*. https://amadeus.com/documents/en/blog/pdf/2014/12/report-thinking-like-a-retailer-airline-merchandising.pdf.

Goldberg, D., D. Nichols, B.M. Oki, and D. Terry. 1992. Using collaborative filtering to weave an information tapestry. *Communication of the ACM* 35 (12): 61–70.

Grbovic, M., and H. Cheng. 2018. Real-time personalization using embeddings for search ranking at Airbnb. In 24th ACM SIGKDD international conference on knowledge discovery and data mining (KDD).

He, X., H. Zhang, M.-Y. Kan, and T.-S. Chua. 2016. Fast matrix factorization for online recommendation with implicit feedback. In 39th International ACM conference on research and development in information retrieval (SIGIR), pp. 549–558.

He, X., L. Liao, H. Zhang, L. Nie, X. Hu, and T. Chua. 2017. Neural collaborative filtering. In 26th International conference on world wide web (WWW), pp. 173–182.

Hidasi, B., A. Karatzoglou, L. Baltrunas, and D. Tikk. (2016). Session-based recommendations with recurrent neural networks. arXiv 1511.06939. https://arxiv.org/abs/1511.06939.

Hoyles, Y. 2015. New distribution capability (NDC)—together, let's build airline retailing. https://www.iata.org/contentassets/6de4dce5f38b45ce82b0db42acd23d1c/guide-ndc-registration-certification-program.pdf.

Hu, Y., Y. Koren, and C. Volinsky. 2008. Collaborative filtering for implicit feedback datasets. In 8th IEEE international conference on data mining (ICDM), pp. 263–272.

IATA. 2020. Together, let's build airline retailing ndc program update january. https://www.iata.org/contentassets/6de4dce5f38b45ce82b0db42acd23d1c/ndc-standard-presentation.pdf.

Isinkaye, F., Y. Folajimi, and B. Ojokoh. 2015. Recommendation systems: Principles, methods and evaluation. *Egyptian Informatics Journal*. https://doi.org/10.1016/j.eij.2015.06.005.

Jannach, D., and M. Jugovac. 2019. Measuring the business value of recommender systems. *ACM Transactions on Management Information Systems*. https://doi.org/10.1145/3370082.

Karatzoglou, A., X. Amatriain, L. Baltrunas, and N. Oliver. 2010. Multiverse recommendation: N-dimensional tensor factorization for context-aware collaborative filtering. In 4th ACM conference on recommender systems (RecSys), pp. 79–86, Barcelona, Spain.

Khrouf, H., and R. Troncy. 2013. Hybrid event recommendation using linked data and user diversity. In 7th ACM conference on recommender systems (RecSys), pp. 185–192.

Li, L., W. Chu, J. Langford, and R. E. Schapire. 2010. A contextual-bandit approach to personalized news article recommendation. In 19th International conference on world wide web, pp. 661–670, Raleigh, North Carolina, USA.

Lieberman, H. 1995. Letizia: An agent that assists web browsing. In 14th International joint conference on artificial intelligence (IJCAI).

Linden, G., B. Smith, J. York. 2003. Amazon.Com recommendations: Item-to-item collaborative filtering. *IEEE Internet Computing, 7*(1):76-80

Ludewig, M., and D. Jannach. 2018. Evaluation of session-based recommendation algorithms. *User Modeling and User-Adapted Interaction* 28 (4–5): 331–390.

Melville, P., R. Mooney, and R. Nagarajan. 2002. Content-boosted collaborative filtering for improved recommendations. In 18th National conference on artificial intelligence (AAAI), pp. 187–192.

Monti, D., E. Palumbo, G. Rizzo, R. Troncy, and M. Morisio. 2018. Semantic trails of city explorations: How do we live a city. arXiv 1812.04367. http://arxiv.org/abs/1812.04367.

Palumbo, E., G. Rizzo, and R. Troncy. 2017. Entity2Rec: Learning user-item relatedness from knowledge graphs for top-N item recommendation. In 11th ACM conference on recommender systems (RecSys), pp. 32–36, Como, Italy.

Panniello, U., A. Tuzhilin, M. Gorgoglione, C. Palmisano, and A. Pedone. 2009. Experimental comparison of pre- vs. post-filtering approaches in context-aware recommender systems. In 3rd ACM conference on recommender systems (RecSys), pp. 265–268.

Paulheim, H. 2017. Knowledge graph refinement: A survey of approaches and evaluation methods. *Semantic Web* 8 (3): 489–508.

Rendle, S. 2010. Factorization machines. In IEEE international conference on data mining (ICDM), pp. 995–1000.

Rendle, S., C. Freudenthaler, Z. Gantner, and L. Schmidt-Thieme. 2009. BPR: Bayesian personalized ranking from implicit feedback. In 25th Conference on uncertainty in artificial intelligence (UAI), pp. 452–461.

Resnick, P., and H.R. Varian. 1997. Recommender systems. *Communication of the ACM* 40 (3): 56–58.

Rohde, D., S. Bonner, T. Dunlop, F. Vasile, and A. Karatzoglou. 2018. RecoGym: A reinforcement learning environment for the problem of product recommendation in online advertising. In International workshop on offline evaluation for recommender systems (REVEAL).

Sarwar, B., G. Karypis, J. Konstan, and J. Riedl. 2001. Item-based collaborative filtering recommendation algorithms. In 10th International conference on world wide web (WWW), pp. 285–295.

Song, W., Z. Duan, Z. Yang, H. Zhu, M. Zhang, and J. Tang. 2019. Explainable knowledge graph-based recommendation via deep reinforcement learning. arXiv 1906.09506. https://arxiv.org/abs/1906.09506.

Sun, Z., J. Yang, J. Zhang, A. Bozzon, L.-K. Huang, and C. Xu. 2018. Recurrent Knowledge Graph Embedding for Effective Recommendation. In 12th ACM conference on recommender systems (RecSys), pp. 297–305, Vancouver, British Columbia, Canada.

Sutton, R.S., and A.G. Barto. 2018. *Reinforcement Learning: An Introduction. A Bradford Book*. Cambridge: The MIT Press.

Taghipour, N., and A. Kardan. 2008. A hybrid web recommender system based on q-learning. In ACM symposium on applied computing (SAC), pp. 1164–1168, Fortaleza, Ceara, Brazil.

Xiao, J., H. Ye, X. He, H. Zhang, F. Wu, and T.-S. Chua. 2017. Attentional factorization machines: Learning the weight of feature interactions via attention networks. In 26th International joint conference on artificial intelligence (IJCAI), pp. 3119–3125.

Yao, W., J. He, G. Huang, J. Cao, and Y. Zhang. 2015. A Graph-based model for context-aware recommendation using implicit feedback data. *World Wide Web* 18 (5): 1351–1371.

Zhang, F., N. J. Yuan, D. Lian, X. Xie, and W.-Y. Ma. 2016. Collaborative Knowledge Base Embedding for Recommender Systems. In 22nd ACM SIGKDD international conference on knowledge discovery and data mining, pp. 353–362.

Zhang, S., L. Yao, A. Sun, and Y. Tay. 2019. Deep learning based recommender system: A survey and new perspectives. *ACM Computing Surveys* 52: 1. https://doi.org/10.1145/3285029.

Zhao, X., L. Xia, L. Zhang, Z. Ding, D. Yin, and J. Tang. 2018. Deep reinforcement learning for page-wise recommendations. In 12th ACM conference on recommender systems (RecSys), pp. 95–103, Vancouver, British Columbia, Canada.

Amine Dadoun is an industrial PhD student in Data Science at EURECOM in collaboration with Amadeus. His research focuses on recommender systems applied to the travel domain, including Deep Learning based Recommender Systems, Knowledge Graph & Word

Embeddings for Recommender Systems, and Session Based Recommender Systems.

Michael Defoin-Platel is Head of Machine Learning Services at Amadeus. He received a PhD in Artificial Intelligence from the University of Nice. Michael has been conducting academic research in fundamental and applied AI during 10 years before moving to the industry. He is now responsible for the development of AI in Amadeus products.

Thomas Fiig is the Director, Chief Scientist at Amadeus. He holds a Ph.D. in theoretical physics and mathematics from the University of Copenhagen, Denmark. He has more than 20 years of revenue management experience and has published several papers on forecasting, offer optimization, reinforcement learning.

Corinne Landra is Associate Director Expert at Amadeus. She has more than 10 years's experience on shopping and pricing solutions serving both the direct and indirect channels. In her current role, Corinne is focusing on designing the future of Airline Offer Management System, defining as part of it, the architecture around pricing and product dynamicity. She represents Amadeus at ATPCO and IATA New Distribution Capability (NDC) industry fora.

Raphaël Troncy is an Associate Professor at EURECOM. He got his PhD from Grenoble University (France) in 2004 and had research positions at INA (France), CNR (Italy) and CWI (The Netherlands). He published nearly 200 scientific papers in various journals, top tier conferences and associated workshops and held key organisational roles in multiple world-leading conferences. He will be General Chair of The Web Conference in 2022 He is an expert in knowledge engineering, information extraction and recommender systems. He is the primary investigator of many National and European projects where semantic technologies and information extraction are used together to build knowledge graphs, exploratory search engines and recommender systems.

Journal of Revenue and Pricing Management (2021) 20:316–321
https://doi.org/10.1057/s41272-021-00314-1

PRACTICE ARTICLE

A note on the advantage of context in Thompson sampling

Michael Byrd[1] · Ross Darrow[2]

Received: 12 May 2020 / Accepted: 8 September 2020 / Published online: 24 March 2021

Abstract
Personalization has become a focal point of modern revenue management. However, it is often the case that minimal data are available to appropriately make suggestions tailored to each customer. This has led to many products making use of reinforcement learning-based algorithms to explore sets of offerings to find the best suggestions to improve conversion and revenue. Arguably the most popular of these algorithms are built on the foundation of the multi-arm bandit framework, which has shown great success across a variety of use cases. A general multi-arm bandit algorithm aims to trade-off adaptively exploring available, but under observed, recommendations, with the current known best offering. While much success has been achieved with these relatively understandable procedures, much of the airline industry is losing out on better personalized offers by ignoring the context of the transaction, as is the case in the traditional multi-arm bandit setup. Here, we explore a popular exploration heuristic, Thompson sampling, and note implementation details for multi-arm and contextual bandit variants. While the contextual bandit requires greater computational and technical complexity to include contextual features in the decision process, we illustrate the value it brings by the improvement in overall expected

Keywords Bandit algorithms · Online learning · E-commerce

Introduction

Real-time, model-driven insights become increasingly important for maintaining a competitive edge in the marketplace. Many upcoming applications of revenue management require or will be enhanced by real-time experimentation. Some examples:

1. *Ancillary pricing* While common to initialize ancillary prices with customer surveys and conjoint analysis Green et al. (2001), after initial prices are set, it's often useful to do live experiments to further optimize the prices. Even well-designed experiments may not accurately capture the true willingness-to-pay of an item and, worse, can not adapt to changing behaviors over time without further studies.
2. *Personalized offers* Recently, under mild assumptions, it was estimated that profits could be improved over 10% by using personalized pricing Dubé and Misra (2019). For an introduction to offer management for airlines, see Vinod et al. (2018).
3. *Online shopping and booking* Airlines and online travel agencies want to experiment with different display approaches to improve conversion.

Offline analysis, such as traditional A/B/n testing, has shown to be too slow to adapt to the current rapidly changing marketplace. The main criticisms being that an A/B/n test is expected to run for a predetermined amount of time while also uniformly allocating traffic allocation regardless of recently observed findings. To address these problems, reinforcement learning algorithms have been introduced to quickly learn toward an optimal solution in real-time. Many problems fall into the bandit paradigm, which balances between recommending new offerings with the current known best offering. The most well known of these approaches is the multi-arm bandit Whittle (1980). The travel industry has begun making extensive use of multi-arm bandit algorithms to address many web-scale problems.

✉ Michael Byrd
mcbyrd100@gmail.com

Ross Darrow
ross_darrow@prodigy.net

1 Yum! Brands, 7100 Corporate Drive, Plano, TX 75024, USA

2 Bedford, TX 76021, USA

Chapter 8 was originally published as Byrd, M. & Darrow, R. Journal of Revenue and Pricing Management (2021) 20:316–321. https://doi.org/10.1057/s41272-021-00314-1.

While the multi-arm bandit works well, the traditional algorithms ignore the available contextual data. Incorporating each unique context into business decisions has been shown to help adaptively address similar issues with minimal data. For example, a well-known empirical experiment with the contextual bandit illustrated substantial gains in performance for personalizing news articles Li et al. (2010). To illustrate the advantage of using context, we explore a multi-arm bandit algorithm employed by many in the travel industry, Thompson sampling, and show how recent advancements extended it to handle context.

Overview

In "Background" section, we present the foundation for bandit algorithms. This includes reviewing Thompson sampling. "Thompson Sampling with Context" section then further inspects a contextual bandit algorithm using Thompson sampling. Here, we review the algorithm and inspect improvements to ease computation. In "Simulation" section, we implement a simulation study to investigate the potential performance improvements for the contextual algorithm over bandit algorithms that ignore the context. Finally "Conclusion" section offers final thoughts and future improvements.

Background

We consider the bandit problem with binary reward. For time step $t = 1, 2, \ldots$ suppose a recommendation is to be made from one of K possible options. Upon being received, the recommendation is either selected or not selected; we note that these concepts can easily be adapted to incorporate the profitability into the decision process. The goal is to recommend the option that has the highest chance of success. Suppose, of the K options available to recommend, that each option has p_A features. We refer to each option as an *arm*, where $\boldsymbol{a}_i \in \mathbb{R}^{p_A}$ denotes the set of features associated with arm i. Further suppose each instance has an additional p_I features, denoted $\boldsymbol{w}_t \in \mathbb{R}^{p_I}$. We use $\boldsymbol{x}_{t,i}$ to refer generally to the ith *context* at time t as the collection of features composed of $\boldsymbol{w}_t$ and $\boldsymbol{a}_i$. The contextual features include the arm features, instance features, and any function of the two, such as interactions, and are assumed to be of dimension p.

Let $y_t \in \{0, 1\}$ be the reward from the recommendation at time t. We assume that

$$\mathbb{E}(y_t|\boldsymbol{x}_{t,i}) = f(\boldsymbol{x}_{t,i};\ \boldsymbol{\Theta}), \quad (1)$$

where $\boldsymbol{\Theta}$ are learnable parameters. We wish to recommend the arm

$$\alpha_t^* = \arg\max_i \left\{\mathbb{E}(y_t|x_{t,i})\right\}_{i=1}^K, \quad (2)$$

which corresponds to the arm with the highest expected reward; in the case of a binary reward, (2) corresponds to the arm with the highest probability of success. Let α_t denote the index corresponding to the arm selected at time t and $\boldsymbol{\Theta}_t$ be the parameters learned at time t. Define the cumulative regret at time t by

$$r_t = \sum_{j=1}^t f(x_{j,\alpha_j^*}, \boldsymbol{\Theta}) - f(x_{j,\alpha_j}, \boldsymbol{\Theta}). \quad (3)$$

We assume the parameters $\boldsymbol{\Theta}_t$ are updated as frequently as possible, preferably every iteration, though this may not always be the case in practice Joulani et al. (2013).

For instance, a common problem in airline offer management is offering personalized bundles to different travelers. Suppose each traveler arrives sequentially at time $t = 1, 2, \ldots,$ where it is desired to offer the bundle which has the highest probability of being purchased. There are K possible bundles, each having various add-ons like offering a checked bag or allowing for refunds on the ticket. Suppose each offering is one-hot encoded to encapsulate if the bundle contains that offering, then the offerings can represent the arm's feature set. The complete context can include an indicator if the traveler is a business or leisure traveler, and all interactions of the bundle's offerings with the traveler type. Naturally, if the bundle was created sequentially by the traveler, each new suggestion to be potentially added into the bundle could incorporate the outcome of prior suggestions.

While tempting to always select the item that is estimated to give the highest expected reward, this could often lead to suboptimal results. Typically, bandit algorithms are used when historical data are not available. Hence, it is reasonable that the optimal solution has yet to have been learned due to having no data to incorporate into the estimated $\boldsymbol{\Theta}_t$. This has inspired the exploration–exploitation trade-off of Audibert et al. (2009), which aims to incorporate the uncertainty about undersampled arms when choosing the arm to recommend. While many strategies exist in the study of Chapelle and Li (2011), Garivier and Cappé (2011), and Karnin et al. (2013), the general notion is that the more an arm is sampled, the more certain we are about the outcome. Hence, arms we are confident to perform worse than other arms will begin to never be picked with time, ultimately converging to the arm that minimizes the regret per each given instance.

Thompson sampling for bandit problems

Many exploration policies exist in practice, but we focus on Thompson sampling of Chapelle and Li (2011) Agrawal and Goyal (2012). Thompson sampling puts a prior distribution on $\boldsymbol{\Theta}$, and then aims to iteratively update the posterior distribution with each newly observed example. Access to the posterior distribution quantifies the uncertainty of the estimate of $\boldsymbol{\Theta}$ at time t. When making a recommendation, Thompson sampling algorithms generate $\boldsymbol{\Theta}_t^*$ from the current posterior distribution and use it in calculating the probability of success for each arm. This process incorporates the uncertainty of the estimate into the decision making to allow better exploration. As time progresses, the estimate of $\boldsymbol{\Theta}$ will become more certain and the posterior distribution will contract around $\boldsymbol{\Theta}$, which leads to $\boldsymbol{\Theta}_t^*$ often being close to $\boldsymbol{\Theta}$. Thompson sampling has been shown to efficiently trade-off exploring the set of available arms with offering arms more probable to be successful by Kaufmann et al. (2012). There has been considerable success achieved by Thompson sampling of Ferreira et al. (2018), and it has empirically outperformed other multi-arm bandit approaches of Chapelle and Li (2011).

Formally, we assume that $\boldsymbol{\Theta} \sim \pi(\boldsymbol{\Theta})$ and are interested in iteratively computing

$$\pi(\boldsymbol{\Theta}|\boldsymbol{X},\boldsymbol{y}) = \frac{\pi(\boldsymbol{\Theta})\pi(\boldsymbol{X}_t,\boldsymbol{y}_t|\boldsymbol{\Theta})}{\pi(\boldsymbol{X}_t,\boldsymbol{y}_t)}, \tag{4}$$

where $\boldsymbol{X}_t = (\boldsymbol{x}_{1,\alpha_1},\dots,\boldsymbol{x}_{t,\alpha_t})^T$ and $\boldsymbol{y}_t = (y_1,\dots,y_t)^T$ denote the historical contexts for the arms recommended and their outcomes, respectively. The denominator is often referred to the normalizing constant, where

$$\pi(\boldsymbol{X},\boldsymbol{y}) = \int_{\boldsymbol{\Theta}} \pi(\boldsymbol{X}_t,\boldsymbol{y}_t|\boldsymbol{\Theta})\pi(\boldsymbol{\Theta})d\boldsymbol{\Theta}. \tag{5}$$

Often (5) can be intractable; however, $\pi(\boldsymbol{\Theta})$ and $\pi(\boldsymbol{X}_t,\boldsymbol{y}_t|\boldsymbol{\Theta})$ can be chosen to be conjugate which gives closed form updates for $\pi(\boldsymbol{\Theta}|\boldsymbol{X},\boldsymbol{y})$ Gelman et al. (2013). Upon observing a new instance at time step $t+1$, $\boldsymbol{\Theta}_{t+1}^*$ is drawn at random from $\pi(\boldsymbol{\Theta}|\boldsymbol{X}_t,\boldsymbol{y}_t)$ and used to select the arm to be recommended as

$$\alpha_t = \arg\max_i \left\{ f(x_{t,i}; \Theta_{t+1}^*) \right\}_{i=1}^K.$$

Finally, the posterior is recomputed with the new recommendation and the outcome as in (4).

Thompson sampling for the multi-arm bandit

To first illustrate Thompson sampling, we show it in the classic setting of the multi-arm bandit problem. The multi-arm bandit problem ignores the available context when making a recommendation at each instance. In a sense, the multi-arm bandit is exploring the marginal success rate for each available arm. A Thompson sampler for a multi-arm bandit problem with binary reward puts a prior distribution on the probability of success and incorporates the empirical successes into the posterior update. This can easily be achieved using a beta prior and binomial likelihood, which is conjugate and gives closed form updates for (4). Then, at each iteration, each arm is randomly sampled from its corresponding posterior distribution, and the arm with the highest sample is recommended.

Formally, consider unknown success probabilities for each arm $\boldsymbol{\Theta} = (\theta_1,\dots,\theta_K)^T$, where $\theta_k \sim Beta(a,b)$ for all $k = 1,\dots,K$. Letting $\pi(y|\theta_k) \sim Bern(\theta_k)$, then, at time t, the posterior distribution for the success of arm k is

$$\pi(\theta_k|\boldsymbol{y}_t) \sim Beta(a + s_{t,k}, b + n_{t,k} - s_{t,k}), \tag{6}$$

where $n_{t,k} = \sum_{i=1}^t \mathbf{1}(\alpha_i = k)$ is the total number of times arm k was recommended at time t and $s_{t,k} = \sum_{i=1}^t \mathbf{1}(\alpha_i = k)y_i$ is the total number of successes for arm k at time t. Then, for instance, at time step $t+1$, possible arm success probabilities, $\boldsymbol{\Theta}_{t+1}^* = (\theta_1^*,\dots,\theta_K^*)$, are generated according to (6). The recommended arm is selected by

$$\alpha_t^* = \arg\max_i \{\theta_1^*,\dots,\theta_K^*\}. \tag{7}$$

As time progresses, the more the posterior distribution will reflect the potential success rate for each arm. With more observations, the tighter the *Beta* posterior will contract about the true success probability. Note that each arm's success probability is completely independent from one another. Hence, if contextual information gives insight into multisets of arms, then the regret could better be minimized by more quickly finding optimal arm traits.

Thompson sampling with context

To illustrate the benefits of including contextual information, we inspect a contextual bandit algorithm using Thompson sampling. While incorporating additional information can help reduce the overall regret, it comes at the cost of complexity as conjugacy is not easily achieved for (4). For the sake of complexity, we assume a linear relationship between the context and success probability, such that

$$\mathbb{E}(y_t|\boldsymbol{x}_{t,i}) = h(\boldsymbol{x}_{t,i}^T\boldsymbol{\theta}) \tag{8}$$

for regression coefficients $\boldsymbol{\theta} \in \mathbb{R}^p$. The function h is assumed to be the logit function, $h(x) = 1/(1+\exp(-x))$. Notably, any model that gives a posterior distribution, like a Bayesian neural network Riquelme et al. (2018), can be used, but this adds additional complexity to the update process.

With the assumed likelihood $y_t \sim Bern(h(\boldsymbol{x}_{t,i}^T\boldsymbol{\theta}))$, the choice for the prior on the regression coefficients is still needed. This, however, causes problems due to the intractability of (5) for most reasonable priors. Such issues are well known in Bayesian statistics, where heavy use of Gibbs samplers is common place. A recent advance showed a data augmentation technique can facilitate efficient sampling algorithms for (8) when $\pi(\boldsymbol{\theta}) \sim N(\boldsymbol{\mu}, \boldsymbol{\Sigma})$ for mean vector $\boldsymbol{\mu}$ and covariance matrix $\boldsymbol{\Sigma}$ Polson et al. (2013). This technique was adapted by Dumitrascu et al. (2018) in the context of Thompson sampling and illustrated great improvements over other contextual bandit algorithms.

Pólya-gamma augmentation for binomial regression

Here, we briefly explore the data augmentation technique that facilitates the sampling of the posterior distribution for the linear contextual bandit with binary reward. A random variable z is distributed according to a Pólya-Gamma distribution with parameters $b \in \mathbb{R}^+$ and $c \in \mathbb{R}$ if

$$z = \frac{1}{2\pi^2} \sum_{j=1}^{\infty} \frac{g_k}{(j-1/2)^2 + c^2/(4\pi^2)}, \tag{9}$$

where $g_j \sim Gamma(b, 1)$ for $j \in \mathbb{N}$; if z is a Pólya-Gamma random variable with parameters b and c, then $z \sim PG(b, c)$. A useful identity was identified by Polson et al. (2013), showing

$$\frac{(e^{\psi})^a}{(1+e^{\psi})^b} = 2^{-b} e^{\kappa\psi} \int_0^{\infty} e^{-z\psi^2/2} p(z) dz \tag{10}$$

for $z \sim PG(b, 0)$. Naturally (10) bears resemblance to the logit function, where we can set $\psi = \boldsymbol{x}_{t,i}^T \boldsymbol{\theta}$, and was adapted to Bayesian MCMC procedures for posterior estimation.

With some manipulation, it can be shown that

$$\pi(\boldsymbol{\theta}|\boldsymbol{X}_t, \boldsymbol{y}_t, \boldsymbol{z}_t) \propto \pi(\boldsymbol{\theta}) \prod_{i=1}^{t} \exp(\frac{z_i}{2}(\boldsymbol{x}_{i,a_i} - \kappa_i/z_i)^2) \tag{11}$$

for $\kappa_i = y_i - 1/2$ and $PG(1,\ 0)$ latent variables $\boldsymbol{z}_t = (z_1, \ldots, z_t)^T$. When $\pi(\boldsymbol{\theta}) \sim N(\boldsymbol{\mu}, \boldsymbol{\Sigma})$ for mean vector $\boldsymbol{\mu}$ and covariance matrix $\boldsymbol{\Sigma}$, then (11) is proportional to a Normal distribution. Further, it can be shown that the full-conditional distribution for the latent variables $\boldsymbol{\theta}$ and $\boldsymbol{z}_t$ can be found in closed form

$$\boldsymbol{\theta}|\boldsymbol{X}_t, \boldsymbol{y}_t, \boldsymbol{z}_t \sim N(\boldsymbol{\mu}^*, \boldsymbol{\Sigma}^*) \tag{12}$$

$$z_i|\boldsymbol{X}_t, \boldsymbol{y}_t, \boldsymbol{\theta} \sim PG(1, \boldsymbol{x}_{i,a_i}^T \boldsymbol{\theta}), \tag{13}$$

where $\boldsymbol{\Sigma}^* = (\boldsymbol{X}_t^T \boldsymbol{Z}_t \boldsymbol{X}_t + \boldsymbol{\Sigma}^{-1})^{-1}$, $\boldsymbol{Z} = diag(\boldsymbol{z}_t)$, $\boldsymbol{\mu}^* = \boldsymbol{\Sigma}^*(\boldsymbol{X}_t \boldsymbol{\kappa} + \boldsymbol{\Sigma}^{-1}\boldsymbol{\mu})$, and $\boldsymbol{\kappa} = (\kappa_1, \ldots, \kappa_t)^T$ By iteratively updating between (12) and (13), the posterior distributions can be estimated with help from quick accept–reject sampling strategies of Makalic and Schmidt (n.d.).

Pólya-gamma augmentation for Thompson sampling

The Pólya-Gamma distribution augmentation was incorporated into a Thompson sampler by Dumitrascu et al. (2018) in a similar fashion to the presentation in Polson et al. (2013) for a Gibbs sampler to estimate binomial regression. Upon receiving each instance, the Pólya-Gamma Thompson sampler (PT-TS) approximates the posterior distribution of the regression coefficients, $\boldsymbol{\theta}$, by drawing M iterative samples of (12) and (13). This allows incorporating the most recent data into the shape of the posterior and progressively contracts the posterior distribution about the true regression coefficients. The Mth draw of $\boldsymbol{\theta}$ is used as $\boldsymbol{\theta}^*$ and used in calculating (8) for each arm, which then suggests the arm with largest expected success rate.

The guidance in Dumitrascu et al. (2018) advocates for using $M = 100$, giving a large burn-in time to better traverse the posterior distribution. Having large burn-in times is ideal, but difficult to compute in practice. This is further complicated by the difficult operations in (12) and (13), like matrix inversion, that are computationally difficult and must be performed in each iteration of the burn-in for each instance. The use of $M = 1$ was found by Dumitrascu et al. (2018) to give competitive results. This result suggests that the algorithm can quickly move toward ideal regions of the posterior.

As t grows, the posterior distribution will shrink toward the $\boldsymbol{\theta}$, giving little mass to be traversed. This, of course, is ideal, as we desired to find the true underlying distribution; however, the incremental value of each data point will decrease as the uncertainty, as measured by $\boldsymbol{\Sigma}^*$, increasingly shrinks in magnitude. We further propose a rolling window over the data used, taking only the $min(L, t)$ most recent observations and outcomes when estimating the posterior at time t. This (1) reduces the amount of computation and (2) allows for gradual adaption to underlying changes in the underlying process. We find in our experiments that, for a reasonable value of L, there is minimal difference in the observed regret between using all data or a rolling window.

Simulation

We implement a small simulation study to illustrate the gain in using PG-TS over Thompson sampling in the MAB setting. Each study is conducted assuming a linear relationship between the arm features and the interaction of the instance and arm features as in (8); note we do not include the instance features as they provide no information to the preference of an arm by themselves. For a horizon of $T = 1000$ steps, we observe randomly generated instances and recommend an arm for a random sampler, MAB Thompson sampler, and variants of the PG-TS sampler. Each instance is generated at random, where $\boldsymbol{w}_t \sim N(\boldsymbol{0}_{p_I}, diag(\boldsymbol{1}_{p_I}))$ for $t = 1, \ldots, T$. We generate the arm feature values $a_i \sim N(\boldsymbol{0}_{p_A}, diag(\boldsymbol{1}_{p_A}))$ for $i = 1, \ldots, K$ and regression coefficients $\theta \sim N(\boldsymbol{0}_p, diag(\boldsymbol{1}_p))$ at time $t = 1$, which persist for the entire time horizon. We consider two settings:

1. $K = 10$ arms with 2 instance features and 2 arm features,
2. $K = 100$ arms with 2 instance features and 2 arm features.

With 2 instance features and 2 arm features, then the model has a total of $p = 6$ features. Each setting inspected is replicated 25 times.

The MAB Thompson sampler is set with hyperparameters $a = 1$ and $b = 1$. For the PG-TS prior, we assume $\theta \sim N(\boldsymbol{0}_p, diag(\boldsymbol{1}_p))$ for each version; we note that, while we assume a somewhat strong prior, the Gibbs sampler is valid under vague priors if unwilling to make such an assumption Choi et al. (2017). We inspect PG-TS with a rolling window and without a rolling window, where, for the rolling window, we set $L = 100$. Additionally, we inspect the difference in changing the number of Gibbs sampler iterations at each instance, with $M = 25$ and $M = 1$. We permute each of the rolling window and Gibbs sampler iteration settings, giving 4 variants of PG-TS. The performance is measured by the average cumulative regret, as calculated in (3), at each time $t = 1, \ldots, T$ for each method; additionally, we show the 5% and 95% percentiles for the cumulative regret at each time.

We plot the results of Setting 1 and Setting 2 in Fig. 1. We opt to display the results for $M = 25$, but note that the $M = 1$ results were about 2.5 worse than the displayed PG-TS results; however, even with $M = 1$, both PG-TS outperformed the MAB Thompson sampler. The performance improvement from incorporating context is quite stark, where both the rolling window and standard PG-TS procedures upper percentile bound is much less than the MAB algorithm. Additionally, while the standard PG-TS does perform better than the rolling window version, the difference is relatively small. The rolling window then seems to reduce the amount of computation at a minimal cost to the overall model.

Conclusion

The need for adaptive exploration of online retail continues to grow. While many have addressed this issue with multi-arm bandit algorithms with great success, there is still room for improvement over simply estimating marginal success rates. Here, we have explored a recent contextual extension of the well-known Thompson sampler, called Pólya-Gamma Thompson Sampling, and empirically illustrated the potential gains in terms of minimizing the regret of the recommendations. While improvements to the regret are realized, it does come at the cost of more computation. These computations can be somewhat mitigated by the rolling window we imposed. Alternatively, computation could be reduced by computing updates in batches, rather than each iteration,

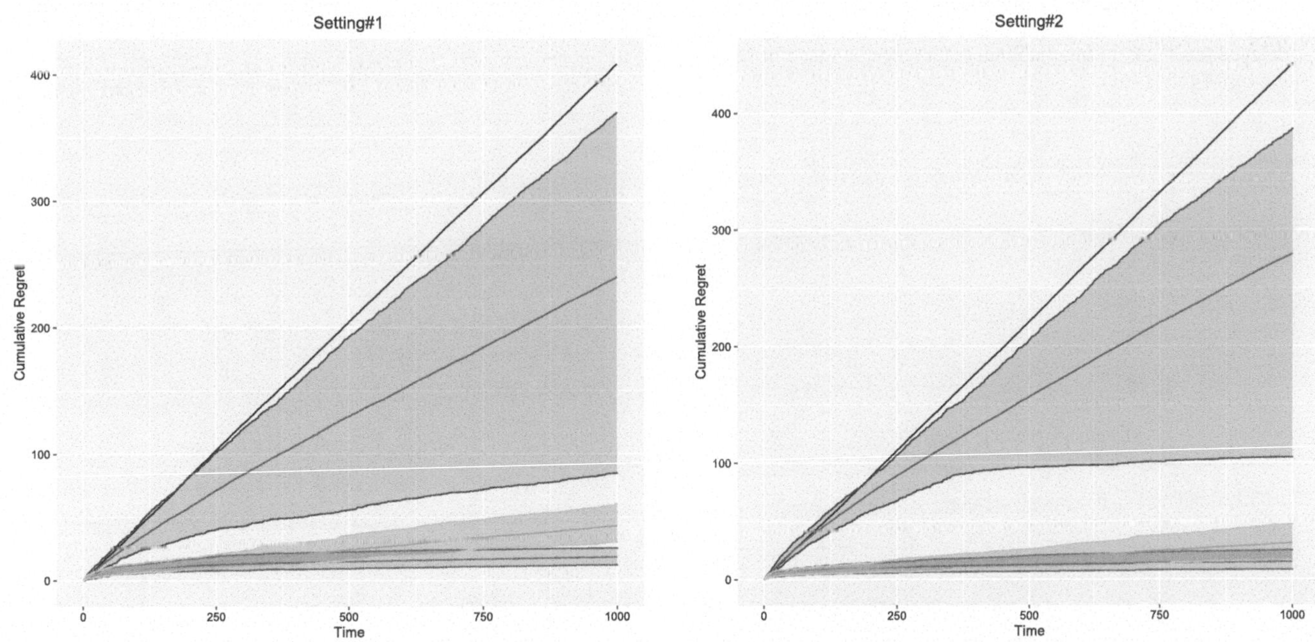

Fig. 1 The average cumulative regret with the 5 and 95% percentile bands for the MAB Thompson sampler and PG-TS with and without a rolling window for $M = 25$. Both settings use $p = 6$ total context features with $K = 10$ and $K = 100$ arms

or assuming the posterior covariance of the regression coefficients to be diagonal, simplifying the inverse operation. Regardless, as computational platforms continue to illustrate low-cost scalability, the improvement to the recommendation platform seems deserving of consideration for production systems.

References

Agrawal, S., Goyal, N. 2012. Analysis of thompson sampling for the multi-armed bandit problem. In Conference on learning theory (PP. 39–1).

Audibert, J.Y., R. Munos, and C. Szepesvári. 2009. Exploration-exploitation tradeoff using variance estimates in multi-armed bandits. *Theoretical Computer Science* 410 (19): 1876–1902.

Chapelle, O., Li, L. 2011. An empirical evaluation of thompson sampling. In Advances in neural information processing systems (pp. 2249–2257).

Choi, H.M., J.C. Román, et al. 2017. Analysis of polya-gamma GIBBS sampler for Bayesian logistic analysis of variance. *Electronic Journal of Statistics* 11 (1): 326–337.

Dubé, J- P., Misra, S. 2019. Personalized pricing and customer welfare. Chicago Booth School of Business Working Paper.

Dumitrascu, B., Feng, K., Engelhardt, B. 2018. Pg-ts: Improved thompson sampling for logistic contextual bandits. In Advances in neural information processing systems (pp. 4624–4633).

Ferreira, K.J., D. Simchi-Levi, and H. Wang. 2018. Online network revenue management using Thompson sampling. *Operations Research* 66 (6): 1586–1602.

Garivier, A., Cappé, O. 2011. The KL-UCB algorithm for bounded stochastic bandits and beyond. In Proceedings of the 24th annual conference on learning theory (pp. 359–376).

Gelman, A., J.B. Carlin, H.S. Stern, D.B. Dunson, A. Vehtari, and D.B. Rubin. 2013. *Bayesian data analysis*. Boca Raton: CRC Press.

Green, P.E., A.M. Krieger, and Y. Wind. 2001. Thirty years of conjoint analysis: Reflections and prospects. *Interfaces* 31 (3): S56–S73.

Joulani, P., Gyorgy, A., Szepesvári, C. 2013. Online learning under delayed feedback. In International conference on machine learning (pp. 1453–1461).

Karnin, Z., Koren, T., Somekh, O. 2013. Almost optimal exploration in multi-armed bandits. In International conference on machine learning (pp. 1238–1246).

Kaufmann, E., Korda, N., Munos, R. 2012. Thompson sampling: An asymptotically optimal finite-time analysis. In International conference on algorithmic learning theory (pp. 199–213).

Li, L., Chu, W., Langford, J., Schapire, R. E. 2010. A contextual-bandit approach to personalized news article recommendation. In Proceedings of the 19th international conference on world wide web (pp. 661–670).

Makalic, E., Schmidt, D. n.d. High-dimensional Bayesian regularised regression with the bayesreg package. arXiv:1611.06649v3

Polson, N.G., J.G. Scott, and J. Windle. 2013. Bayesian inference for logistic models using pólya-gamma latent variables. *Journal of the American Statistical Association* 108 (504): 1339–1349.

Riquelme, C., Tucker, G., Snoek, J. 2018. Deep bayesian bandits showdown: An empirical comparison of bayesian deep networks for thompson sampling. arXiv preprint arXiv:1802.09127.

Vinod, B., R. Ratliff, and V. Jayaram. 2018. An approach to offer management: Maximizing sales with fare products and ancillaries. *Journal of Revenue and Pricing Management* 17 (2): 91–101.

Whittle, P. 1980. Multi-armed bandits and the Gittins index. *Journal of the Royal Statistical Society* 42 (2): 143–149.

Michael Byrd is a Senior Data Scientist for Yum! and formerly a Senior Data Scientist at Sabre Research. He is responsible for building and deploying machine learning enabled applications at scale, particularly in fast moving, dynamic environments. His interests include high-dimensional variable selection, online machine learning algorithms, and Bayesian methods for machine learning.

Ross Darrow was a Director at Sabre Research, responsible for advanced statistical modeling and decision support. After three decades at American Airlines and Sabre, he recently joined Charter and Go, an innovation start-up for charter airlines, as Chief Scientist.

Journal of Revenue and Pricing Management (2021) 20:322–329
https://doi.org/10.1057/s41272-021-00315-0

PRACTICE ARTICLE

Shelf placement optimization for air products

Tomasz Szymanski[1] · Ross Darrow[2,3]

Received: 12 May 2020 / Accepted: 14 September 2020 / Published online: 17 March 2021

Abstract
Airlines are increasingly trying to differentiate their offers, so one airline's bare bones low fare may, for some travelers, be equivalent to another airline's high fare, if the low fare bundles some ancillaries a traveler wants. Because of differentiation of airline offers the air traveler must simultaneously compare air fares and the ancillaries included. Price comparison is also difficult as the same cabin offers vary by the ancillaries included across the competing airlines. We are proposing a shelf product assortment method for categorizing airline offers into utility levels, thus facilitating the choice task of air travelers. We describe possible known approaches, describe Sabre's proposal, provide information on the optimization methods used, and propose future work in the field.

Keywords Offer optimization · Airlines offer management · Bundles · Electronic retail · Shelf space product assortment · Constraint programming

Problem introduction

With airline deregulation and much increased competition in the industry, airlines are trying to differentiate their products by including in their offer goods and services that are something more than just the fare for carrying passengers from origin to destination. This creates additional revenue and makes the airline product more than just a commodity differentiated only by price. Common extras include number of bags included before an extra fee is imposed, seat comfort, check-in services (such as priority check-in), and in-flight services like premium meal. This applies both to the basic economy offer and to higher cabins offer (such as business class). The resulting fares, packaged with such ancillary goods, can be perceived as bundles and are usually marketed under some specific names called *brands*, and are generally called *branded fares*.

As particular airlines create their bundled offers in any way that optimizes total revenue for them, then the offers from different airlines are typically not directly comparable, as they contain services and services levels that are different across the airlines. This makes the customer choice difficult, made even more difficult by using different names that different airlines call their products (Gilbertson 2019; Waldek 2019).

Sabre, as a Global Distribution System (GDS), provides a marketplace for search and purchase of air offers. Sabre distributes air products for most airlines around the world, and travel agents around the world depend on Sabre to display (in seconds) up to 200 available offers on any origin & destination market. To compare different airline offers and make the marketplace more transparent for customers, Sabre has proposed the Storefront product. The product idea is to arrange airline bundled offers into value levels, called *shelves*, thus making them directly comparable and offering clarity of choice for customers.

Problem description

First let's introduce the terminology that we will use. An *attribute* is the name of the good or a service, different than just the base fare for transportation, that is included in the offer. For example, seat comfort level (in terms of seat pitch,

✉ Tomasz Szymanski
tomek.szymanski@gmail.com

Ross Darrow
ross_darrow@prodigy.net

1 ul. Spacerowa 23, Ciężkowice 33-190, Poland

2 3917 Cedar Ridge Dr, Bedford, TX 76021, USA

3 Sabre Research, Southlake, TX, USA

Chapter 9 was originally published as Szymanski, T. & Darrow, R. Journal of Revenue and Pricing Management (2021) 20:322–329. https://doi.org/10.1057/s41272-021-00315-0.

seat width, seat reclining) is an attribute. Another attribute is the number of free bag pieces or its weight included for free (called *baggage allowance*). The right to refund the unused ticket is another attribute called *refundability*. Both refundability, along with the right to exchange the ticket for other day or route (*exchangeability*), are called *voluntary changes*. The other popular ancillaries included are advance seat selection (also known as pre reserved seats), priority check-in and priority boarding. Traditionally the meal and free drinks are another attribute, while the example of attributes that were added in last few years are on-demand streaming (video) services, connectivity to internet on board thru Wi-fi network and USB power sockets. Providing the comprehensive catalog of all possible ancillaries (attributes) that may be included in airline offers is not practical here, as airlines keep differentiating their offer.

The attributes are rarely *binary* (meaning they are either just included or not included) but they are offered at different *levels* instead. The attribute *level* is the amount of value of the attribute that is included. For example, for seat comfort the possible values might be: basic seat with up to 32″ seat pitch, the standard seat with up to 34″ seat pitch, the extra legroom seat with 34″+seat pitch, the comfortable business class seat, ending up with a separate suite seat.

Example of levels for exchanges and refunds may be: not permitted, permitted for charge, permitted for free.

A *shelf* is the designation of the selected attributes and their minimum levels required for given airline offer to be classified in that shelf. An example of *shelf definition* can be: exchanges must be permitted, for free or for charge, and the seat pitch must be at least 34″. Shelves are typically given ordinal numbers: 0, 1, 2, and so on. The *shelf 0* typically contains the most basic fares with most basic attribute levels, while the higher shelves are offering more attributes and at higher levels. All the shelves with numbers above 0 are called *upsell shelves*. The term *upsell* reflects an attempt to sell more goods to the customers while presenting the higher shelves, and not just the cheapest fares (shelf 0). The distribution channel works with airlines and standards organizations (such as IATA) on how to name or depict the shelves (there is ongoing debate on this). Examples can be: number of stars (like hotel stars), icons for the attributes included, some new crafted names like virtual brands or new semi-cabins.

Possible solutions

So how can the offers from different airlines, created based on different product attributes and their levels, become comparable to each other?

First let's consider what are the desirable properties of shelf placement revealed during our market analysis:

1. Shelf definitions must be easy to understand: it must be clearly communicated what is offered within given shelf.
2. The attribute levels must only increase going from lower into higher shelves (strictly increasing, will be explained in detail later).
3. Possibly whole airlines offer is presented and all airline brands in particular.
4. Shelves are as full as possible.
5. Shelved display must have good conversion and the total revenue from shelved display must be higher that from the traditional lowest fares display.

Clustering

The task of grouping different objects into similar groups is traditionally done by *clustering*, which is a form of *unsupervised machine learning*. The clustering approach seems appealing because of the two characteristics of airline offers: first the attributes included are not always fully intersecting (you cannot compare), second the levels of attributes are not directly comparable (different airlines are using different scale or variants). Then the clustering approach would theoretically solve the problem of some attributes missing and the attribute levels being on the continuous scale. As attractive from theoretical perspective the solution was immediately turned down during product market validation. The reason is straightforward: the customers will not understand an abstract "goodness score", they will not understand what is included the groups (shelves). When clustering based on continuous distance metric it is not possible to provide *guarantees*, that a group (shelf) always contains some attributes with some specific minimum levels.

An example of the problem is an attempt to cluster offers into two clusters: it may easily occur that cluster 1 will contain fares that are *typically* exchangeable and have one piece of checked bag, with varying seat comfort, while cluster 2 will typically contain fares that are both exchangeable and refundable, would contain two pieces of checked free bags, and will usually have extra legroom seat comfort. However, the first cluster may also contain some fares that are fully exchangeable and refundable but have no free baggage allowance, while cluster 2 may contain fares that have very comfortable seat, on board video streaming, two bags allowance but which are not refundable at all.

Attribute based definitions

All these deficiencies can be solved by selecting some closed set of attributes that are ubiquitous in all airline offerings, assuming some fixed attribute subsets and their levels that are required for particular shelves and assigning the offers to the shelves based on the levels of the selected attributes.

For example, assuming we want to have three upsell shelves (shelves higher than the most basic shelf 0), then their attributes and levels might be defined as:

1. *Shelf 0* most basic fare, no attribute level guarantees.
2. *Shelf 1* exchangeable, for free or for fee.
3. *Shelf 2* exchangeable and refundable, for free or for fee.
4. *Shelf 3* exchangeable and refundable, for free or for fee, plus one piece of checked bag.

Please notice a very important aspect of such definitions of the shelves—the attribute levels are *strictly increasing*—all the attributes and their levels on shelf N + 1 are at least the same as on shelf N, plus there is something else added—a new attribute is added or an attribute level already existing on lower shelf, directly preceding this shelf, is increased.

Another example could be:

1. *Shelf 0* most basic fare, no attribute level guarantees.
2. *Shelf 1* exchangeable for free or for fee, one piece of checked bag.
3. *Shelf 1* exchangeable for free or for fee, 2 pieces of checked bag.
4. *Shelf 3* exchangeable and refundable for free or for fee, 2 pieces of checked bag, business seat.

Please note that if an offer is placed on shelf then it must fulfill all the required attribute levels requirements, while it may still offer something else, for example USB socket, which will be shown in some detail views while presenting the offer. The shelf placement algorithm is trying to position an offer on as high shelf as possible (provided it has high enough attribute levels).

There is one more alternative design. One can argue that some customers, typically depending on their trip purpose, may care for voluntary changes, solely or primarily, while the others may care about the comfort at check-in and in-flight, forsaking the flexibility to change the ticket altogether. In such a case, the ideal design, to cater for such divergent needs, would be to base shelf definitions on different attributes, and removing the strict increasing requirement. Then the shelves, hypothetically, could be defined as:

- *Shelf O* most basic fare, no attribute level guarantees
- *Shelf R* exchangeable, for free or for fee
- *Shelf Q* at least one piece of checked bag
- *Shelf C* priority check-in, at least recliner seat comfort, free drinks on board

The arbitrary and non-consecutive letters R, Q and C, and not the ordinal numbers, were chosen on purpose to demonstrate that shelves defined in such way are not comparable, one is not better, worse or superset or subset of the other. They serve different purposes. There are two major problems in such structuring. First, if such choices are presented to customers, it would break the very strong cognitive paradigm of comparing the shelves and understanding that the shelves that are higher or more to the right, are better then shelves that are lower or more to the left. Second, the attribute levels in airline offers are correlated, meaning that if you want to include some attribute level, then typically you must have and pay also for the other attributes and their levels. In effect the creation of such offers that would be price efficient (you pay only for what you need) would be very difficult.

One needs to mention one more attempt at defining the shelves based on attributes. For easy comprehension of shelf definition just one attribute is selected, for example seat comfort, and the shelves are defined solely by this attribute. Such a proposal was put forward by Delta Airlines in first half of 2019. Such an approach has one major drawback: the selection of just one attribute is very arbitrary, may be in line with some airline offer structuring assumptions, while it may be completely not in line with other airline, resulting in big problem with presenting possibly whole offer, from all the airlines.

Below you can see the example of shelves defined by the seat comfort, in Sabre's SR360 application:

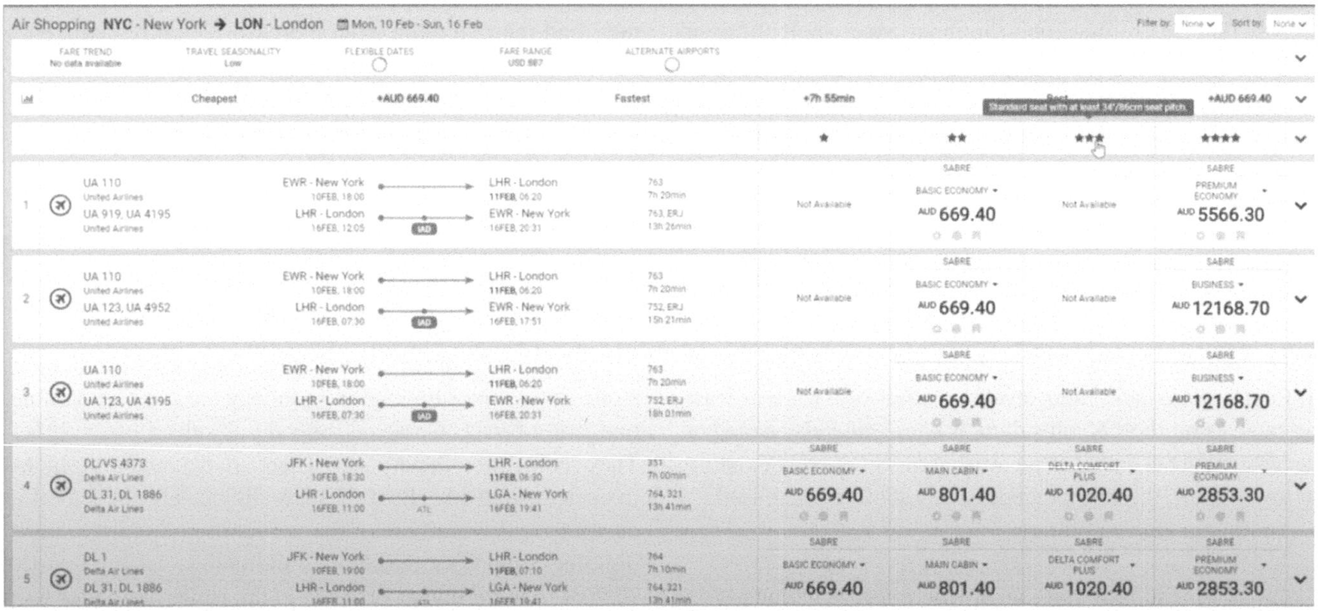

Also, the airline industry fare distribution company ATPCO is also working towards defining an attribute-based proposal—this is a general industry problem with lots of players, and Sabre is right in the middle of it.

Cabin based approach

Another, very entrenched from historical perspective, approach can be basing the shelf definitions just on the airline designated cabin (Economy, Premium Economy, Business, First). This is a very simple and commonly understood definition, but it has three major drawbacks. First, airlines have complete liberty what attributes and levels they include in a cabin. In effect one airline business cabin may have much more goods than the other airline. Surely, in general some services are *typically* included in business class, like priority check-in, better seat, premium food and drinks, but so that it is truly a *guaranteed* by the shelf product, all the attribute levels must be validated, without just relying on the cabin designated by airline. In consequence the cabin would need to be defined based on actual attribute levels guaranteed, which is the essence of the shelves. Second problem of the cabin only approach is the lost revenue from very coarse-grained matching of the demand and supply curves. In practice only up to 3 or 4 cabins are used on particular flight while first we can present on the screen more choices than just 3 or 4 (5 choices seems optimal from choice theory perspective), second the airline offer, typically in form of branded fares, is more broad than just 3–4 brands (5–7 is the typical size).

Solution details

Considering all the options described in the previous section we did market validation and selected the attribute-based approach with attributes strictly increasing.

Now the major question comes what attributes and their levels the shelves should be defined by? What attributes matter for the customers? Do they have same utility? Are they common enough across the airline offers that we can structure possibly all the offers into say 4 shelves?

The first approach to answer these questions is to analyze offers attributes and make the proposal based on general market knowledge of customer preferences. Based on offer analysis the first proposal of Sabre defined shelves was created. Shelves, for US Domestic markets and markets originating from US, the Transatlantic markets in particular, were defined in the following way:

1. *Shelf 0* most basic fare, no attribute level guarantees.
2. *Shelf 1* exchangeable for free or for fee.
3. *Shelf 2* exchangeable and refundable for free or for fee.
4. *Shelf 3* exchangeable and refundable for free or for fee, one piece of free checked bag.

Below an example of attribute-based display in Sabre's SR360 application:

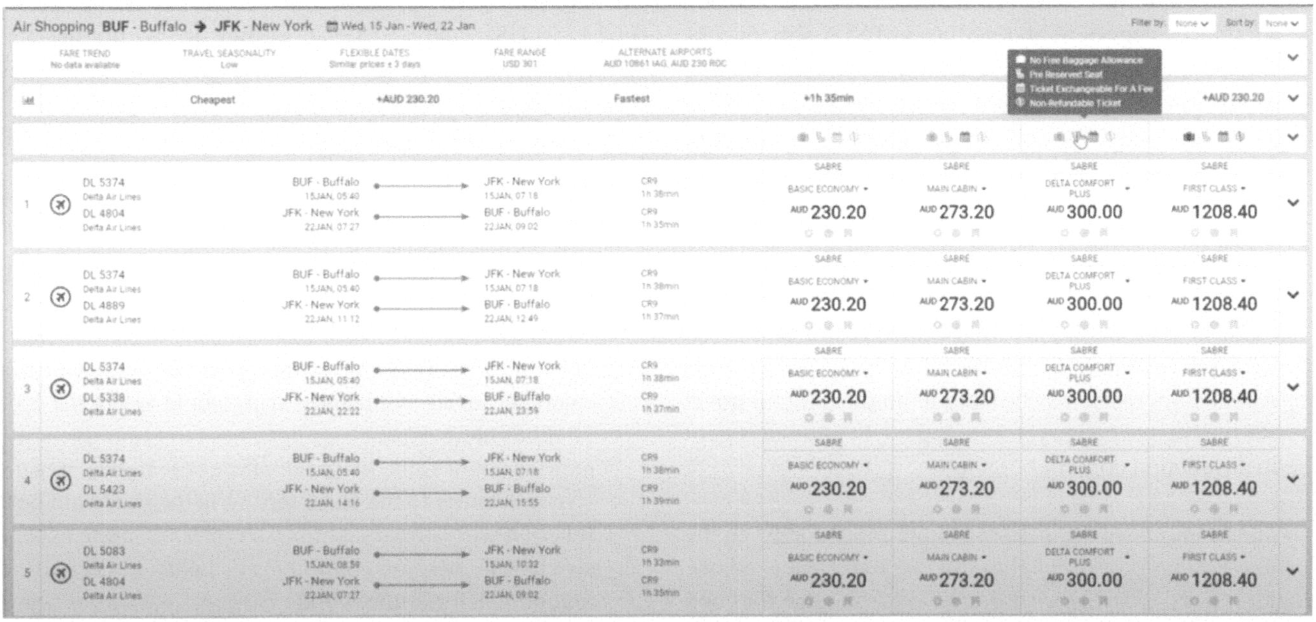

Now, as you can see the end user is able to compare clearly what is included (guaranteed) on a shelf and is able to compare the prices across the offers from different airlines.

This is some progress, but there are two difficulties:

1. The ancillaries typically included differ a lot across different markets. For example, while free checked bags are very uncommon in US Domestic travel, it is typical on Transatlantic travel. Then should the shelf definitions for Transatlantic be different than for US Domestic? Should all the shelves for Transatlantic contain one checked bag? Or maybe we should have second piece of checked bag as the factor differentiating the shelves? Or are there other useful differentiating attributes on that geography?
2. Even within one geography, like US Domestic only, as the competitive situation between different city pairs is different, then the offer variety is sometimes greater and sometimes smaller. So, the same challenge as mentioned for Transatlantic exists also within one geography.

In short you can see then the single uniform shelf definitions are not practical as we end up with lot of empty shelves—offers are not available according to higher shelf definitions or all offers are fulfilling some lower shelf definition.

What can we do about that? Several approaches may be considered:

1. As offers differ between geographic regions and within the geographic regions (differ from market to market), then let's make the shelf definition follow the current market offer and adjust itself according to what is currently offered. We will give example of such approach below.
2. Let shelf definitions differ across geographic regions but keep it same within one region.
3. Let's keep the basic shelf definition scaffolding the same, but try to differentiate shelves further by market by some additional attribute, like pre reserved seats on one market and number of free checked bags on the other.

All these *dynamic* shelves approaches described above were prototyped, as separate shelves *modes* and are now being offered for testing by beta customers.

As you can see, as you adjust the shelf definitions to airline offer on particular markets, there is a spectrum of shelf definitions variability, starting from one uniform shelf definitions, applied globally for all markets, ending at totally dynamic shelf definitions, when shelves can vary from market to market, even within same geographic region (useful example can be different definitions for DFW-ORD market and LAX-NYC market, let alone LAX-HNL market). There is plenty of possibilities in between, for example:

1. Definitions are different across IATA zones or subzones, but same within one IATA zone or subzone.
2. Definitions are same for trips originating at one IATA subzone and arriving at other IATA subzone but different for other IATA subzones departure-arrival pairs.

For example, all trips originating in IATA subzone 11 (North America) and ending in IATA subzone 21 (Europe) would have one and uniform definitions, and it will be typically different than definitions for IATA subzones pair of for example 11 (North America) into 12 (Caribbean Islands).
3. Definitions are same within some origin and destination country of the trip, and different for other pairs of origin and destination country. Then all the flights from United States into Great Britain would have one and same definitions, while the flights from United States to Mexico would have other definitions.
4. Another approach may be same definitions for all travel starting in one country, for example US, without regard to destination country.

All options above are different compromises between the desired shelf properties and the customer experience: on one hand it is good to keep the shelves always the same for good user experience, on the other hand we must adjust the shelf definitions to the market specifics to be able to offer good shelves at all.

Methods for defining optimal shelf structure

Optimization metric

Assuming some level of shelf definitions geography adjustment was selected (options described in previous section) what should be measure of the shelf structure being optimal, that is the best for given geography? As the product goal is to generate incremental revenue the natural metric to optimize is this incremental revenue. This metric or similar is selected and optimized in the works in the field of shelf placement optimization, for example (A Beginner's guide to Shelf Space Optimization using Linear Programming 2016; Corstjens and Doyle 1981; Lim et al. 2004). In our case though, as we were still defining a product proposal, without market adoption yet (so without any feedback signal on revenue) we had to assume some proxy measure for the future incremental revenues. Also, the literature does not have the constraint on shelf structure design that we have, that the next shelf must include all the attributes of the previous shelf. And, much of the literature assumes physical shelves.

We decided that what we will optimize will be the number of *cheap upsells* per itinerary. By *cheap upsell* we understand the offer that is within 100% of the price of the most basic fare of the particular airline within the same flights used. So for example if for an airline A, the cheapest fare costs 200USD (and this offer is placed on shelf 0), then if we have 3 upsell shelves, 1, 2 and 3, with the offers for 250 USD, 350 USD and 600 USD respectively, then, for this itinerary (routing, flights), the number of cheap upsells is 2 (250 USD and 350 USD, as they are within the 400 USD limit). Then the arithmetic average for all itineraries is computed.

Other alternative formulations can be considered as well:

1. Optimize number of upsells in general without respect to the price. Then for three upsell shelves offered, the number of upsell may be up to three. The problem with this metric is that while it is relatively easy to have the three upsell shelves filled, with typically expensive Business and First class fares, then it is very doubtful if a customer, even on business trip, would select the 5 or 10 times more expensive First class fare. Intuitively the metric should value more the offers that are less expensive in relation to the basic fare. Whether it should be a function with a hard cutoff, like our 100% more expensive cutoff, or some continuous or laddered decreasing function, is for future work.
2. Optimize number of upsells, with some 100% or other percent cutoff, or other decreasing function, but not in relation to same airline cheapest offer, but in relation to whole market offer (all airlines) summary statistics. For example, in relation to the cheapest fare altogether or in relation to cheapest fare with same routing or to the cheapest fare with same number of stops

Optimization technique

Assuming we took the number of cheap upsells (price increase up to 100% within same airline itinerary) as the main optimization metric, how do we find the optimal shelf definitions that will optimize that metric? So, having very many possible combinations of attributes and their levels, and considering all the offers available on given geography, for which we are searching for the optimal shelves, and having some business constraints on the shelves structure (like the strictly increasing), how do we find that best shelves structure? The problem seems to lend well for the constraint programming technique (Lustig and Puget 2001): for all the attributes and their levels, given some constraints on that levels, find the combination that maximizes some goal function (in our case the number of cheap upsells).

Another go to solution may be treating this as linear programming or mixed linear programming problem. This is ongoing research at Sabre.

As there are a variety of constraint programming environments available and directly applicable to the problem, please notice the problem specifics—the number of input combinations to evaluate is not trivial. Assuming four shelves, and the attributes to choose from being exchangeability, refundability, seat comfort, pre reserved seats, priority check-in and number of checked bags, the number of combinations for 4 shelves, upon application of all the constraints becomes approx. 30,000. This is not a problem per se, but now for every valid attribute combination creating the shelf definitions you need to evaluate the goal function (the number of cheap upsells) on the market offer available for given geography. Now, sampling only several hundred market pairs for given geography (for example United States—Europe), and only several travel dates, and assuming only 100 itineraries in offer for particular market and travel dates, you need to evaluate the cost function on the hundred thousand of itineraries, for every shelf structure proposal (for every valid attribute combinations). So, the problem is not the number of variants to consider, but the size of the input to execute the goal function for every variant. Taking into consideration the generally known high processing times of the constraint programming engines we decided to implement simple nested loops (attribute values combinations) with constraint filtering in a general compiled programing language. As the time to evaluate the cost function on the hundred thousand of itineraries, for every 30,000 shelf definitions proposal is still considerable we took the advantage of the fact that our best solution does not need to be guaranteed to be globally optimal, and we are still good with an approximate solution that is still for example 99% as good as the globally best solution.

What are the methods available to find such approximate solution? As the solution space (shelf definitions) is multi-dimensional, with typically 50–500 dimensions, and contains multiple deep valleys with regard to the cost function, the gradient based minimum search methods seemed not directly applicable. Instead the problem was treated as a black-box search problem (random optimization). Multiple methods for search for approximate solutions exists: simulated annealing, iterated search, repeated search, hill climbing, Tabu search, to name a few. Upon literature review the *Greedy randomized adaptive search procedure* (GRASP) (Feo and Resende 1995) was selected. For our problem the GRASP method was finding better solutions in limited number of steps than other popular random search methods.

Auxiliary metrics

As we were optimizing the cheap upsells metric, we also monitored the other metrics:

- Number of upsells per itinerary, without respect to the price
- Number of upsells per itinerary within 25%, 50%, 75% of the cheapest fare price
- Number of non-empty shelves (shelves that had at least one offer, from any airline, from any itinerary or routing)
- Diversity measures for seat comfort, brands used and the diversity of other ancillaries that were not part of the shelf definitions. Diversity was measured in terms of differential entropy from the uniform distribution (Kullback–Leibler divergence).
- Kullback–Leibler divergence from the uniform distribution of offers across the shelves—at best all the shelves should be equally full.

All the other metrics were usually strongly correlated to the main optimization metric. If there were any outstanding cases, there were investigated often leading to more market insight and identifying better global solution or finding solutions for markets specific problems.

Future work

Experimentation on shelf definitions

We intend to offer, for every market geography, not just the one seemingly best shelf definitions, but several sets of shelf definitions and let the end user decide which one is the best. As the product must bring revenue and we do not want to lose much revenue for testing non optimal shelves designs, we plan to use Multi Armed Bandits to find and continuously optimize the best shelf definitions. Application of Multi Armed Bandits (White 2013) enables to dynamically allocate the share for the test variants, as opposed to the A/B or multivariate testing in which the share of the variants is fixed, leading to smaller revenue loss from the weaker variants while still having same portion of traffic for weaker variants. On the other hand, the Multi Armed Bandits, like *reinforcement learning*, while optimizing the strong variants (exploiting the strong variants) are still giving some chance to the weak variants (they still explore the so far weak or unknown variants).

When selecting the variants, we plan to generate probably good shelf definitions proposals with the same optimization method but incorporate some diversity measure to optimize as well. We do not want the algorithm to select (for example) the five best shelf definitions proposals that have very close optimization metric values but are very similar to each other. We want proposals with possibly high metric value, but which are possibly much different from each other.

Trip purpose segmentation

We know from market analysis, and from the analysis of business travelers segment in particular, that some customers value some attributes much more than others. For example, for a business traveler the exchangeability of ticket is important, while two free checked bags are probably irrelevant. This may be completely different for a family with children. So, it seems to make sense to differentiate the shelf definitions based on the trip purpose (Vinod 2018). This can be done for example by designing shelf structure (adding more or other constraints in our constraint programming formulation) based on market knowledge and customer surveys, or by incorporating the trip purpose context into the experimentation method itself. In such a case the *contextual bandits* (Langford and Zhang 2008) can be used.

References

A Beginner's guide to Shelf Space Optimization using Linear Programming. 2016. https://www.analyticsvidhya.com/blog/2016/09/a-beginners-guide-to-shelf-space-optimization-using-linear-programming/. Sept 2016.

Corstjens, M., and P. Doyle. 1981. A model for optimizing retail space allocations. *Management Science* 27 (7): 822–833.

Feo, Thomas A., and Mauricio G. C. Resende. 1995. Greedy randomized adaptive search procedures. *Journal of Global Optimization.* 6 (2): 109–133.

Gilbertson, Dawn. 2019. No, you can't move into that empty extra legroom seat on your flight. They cost extra, *USA Today* newspaper. https://www.usatoday.com/story/travel/airline-news/2019/09/18/american-delta-united-extra-legroom-main-cabin-extra-comfort-plus-economy-plus/2312489001/. Accessed 18 Sept 2019.

John Myles White. 2013. *Bandit algorithms for website optimization.* Newton: O'Reilly.

Langford, John, and Ton Zhang. 2008. The Epoch-Greedy Algorithm for contextual multi-armed bandits. In *Advances in Neural Information Processing Systems* 20, Curran Associates, Inc., 817–824.

Lim, Andrew, Brian Rodrigues, and Xingwen Zhang. 2004. Metaheuristics with local search techniques for retail shelf-space optimization. *Management Science.* 50: 117–131. https://doi.org/10.1287/mnsc.1030.0165.

Lustig, Irvin J., and Jean-Francois Puget. 2001. Program does not equal program: Constraint programming and its relationship to mathematical programming, Interfaces, 29–53.

Vinod, Ben, et al. 2018. An approach to offer management: maximizing sales with fare products and ancillaries. *Journal of Revenue and Pricing Management* 17: 91–101.

Waldek, Stefanie. 2019. What's the difference between Premium Economy and Economy Plus?, June 12, 2019, *Conde Nast Traveler.* https://www.cntraveler.com/story/whats-the-difference-between-premium-economy-and-economy-plus. Accessed 12 June 2019.

Tomasz Szymanski was Principal Scientist at Sabre Research, responsible for development of advanced decision support applications in Java and Scala leveraging data science and machine learning. He is passionate about machine learning, artificial intelligence, and applications in travel. After 16 years at Sabre, Tomasz currently works for Nordea bank.

Ross Darrow was Director at Sabre Research, responsible for advanced statistical modeling and decision support. After three decades at American Airlines and Sabre, he recently joined Charter and Go, an innovation start-up for charter airlines, as Chief Scientist.

Journal of Revenue and Pricing Management (2021) 20:330–343
https://doi.org/10.1057/s41272-021-00316-z

PRACTICE ARTICLE

Applying reinforcement learning to estimating apartment reference rents

Jian Wang[1] · Murtaza Das[1] · Stephen Tappert[1]

Received: 20 April 2020 / Accepted: 30 June 2020 / Published online: 16 March 2021

Abstract
In apartment revenue management, rental rates for new and renewal leases are often optimized around a reference rent, which is defined as the "economic value" of an apartment unit. In practice, reference rents are usually estimated using some rules-based approaches. These rules are mostly intuitive to understand and easy to implement, but they suffer from the problems of being subjective, static, and lacking self-learning capability. In this study, we propose a reinforcement learning (RL) approach to estimating reference rents. Our intent is to find the optimal reference rent estimates via maximizing the average of RevPAUs over an infinite time horizon, where RevPAU (Revenue per Available Unit) is one of leading indicators that many apartments adapt. The proposed RL model is trained and tested against real-world datasets of reference rents that are estimated with the use of one rules-based approach by two leading apartment management companies. Empirical results show that this RL-based approach outperforms the rules-based approach with a 19% increase in RevPAU on average.

Keywords Reinforcement learning · Apartment · Reference rent · Revenue management · Dynamic pricing · Artificial intelligence

Introduction

An important decision that apartment operators need to make every day is to set rental rates for new and renewal leases. New leases are signed by prospective tenants who want to lease a unit to reside for a given term, say, 12 months. Renewal leases are signed by existing tenants whose current leases are about to expire. There are three approaches to setting rental rates. The first is to set the rents manually, typically a local owner or manager who uses their own judgment and knowledge to set the rents. This method has proven inferior to a variety of automated approaches, and will not be discussed here. The second approach is to utilize machine learning methodologies, including statistical regression models. In such models, the dependent variables represent the variants of rental rates and independent variables the physical characteristics of property, occupancy, market dynamics, and so on. There is a variety of literature addressing such an approach. For instance, Sirmans and Benjamin (1991) provide a rich literature review. Pagliari and Webb (1996) propose a regression model to set rental rates based on rent concessions and occupancy.

The third approach is to utilize a revenue management (RM) methodology. This approach determines rental rates by balancing demand and supply in such a way that total revenue will be maximized. There are very few publications relevant to this approach. Wang (2008) describes an apartment dynamic pricing system with a focus on setting rental rates for new leases. Such a system uses supply and demand forecasts in conjunction with an estimated price elasticity of demand, centered around a "reference rent". In apartment RM, reference rent is defined as the "perceived economic value" of an apartment unit. It is an economic notion whose value cannot be observed directly. It is believed, however, that reference rents vary over a number of factors including property characteristics, market condition, move-in date, and so forth. How reference rents are estimated will directly determine the optimality of rental rates, which, in turn, will impact the total revenue gain from the leases to be signed.

✉ Jian Wang
Jian.Wang@Realpage.com

Murtaza Das
Murtaza.Das@RealPage.com

Stephen Tappert
Stephen.Tappert@RealPage.com

1 4550 North Point Parkway, Suite 410, Alpharetta, GA 30022, USA

Chapter 10 was originally published as Wang, J., Das, M. & Tappert, S. Journal of Revenue and Pricing Management (2021) 20:330–343. https://doi.org/10.1057/s41272-021-00316-z.

Reference rents are estimated daily or weekly. In practice, the prevailing estimation methods are to use rules-based approaches. A set of rules are specified for a collection of metrics that are understood to encapsulate the market scenarios an apartment operator might face. These rules are simple to understand and easy to implement, but they suffer from three main issues: (1) they are subjective in the sense that they are specified based on the experience and knowledge of individual apartment managers; (2) they are static because they are rarely updated over time even after market conditions may have already changed; and (3) they lack a self-learning capability in terms of detecting and correcting any mistakes in some business cases for which reference rents might have not been estimated appropriately.

In this study, we attempt to overcome these issues by putting forth a reinforcement learning (RL) approach to estimating reference rents. Our intent is to find the optimal reference rent estimates via maximizing the average of revenue per available unit (RevPAU) over an infinite time horizon. Our RL model is trained and tested against real-world datasets of reference rents that are estimated with a rule-based approach by two leading apartment management companies. Empirical results show that this RL-based approach outperforms the rule-based approach with an average of 18% increase in RevPAU.

The remainder of this paper is organized as follows: Sect. 2 reviews literature related to the methods of average-reward RL and its applications in RM. Section 3 proposes a model-based RL approach to computing actions that maximize the expected reward per step. Section 4 describes the datasets and some descriptive statistics. Section 5 presents an empirical study comparing the performance of this RL-based solution with that of the rules-based approach. Finally, Sect. 6 concludes the paper with suggestions for further research and improvements.

Related literature

In practice, the implementation of RM systems always adapts the "divide-and-conquer" strategy. A RM system is usually divided into a sequence of interdependent processes. There are mainly two reasons by doing so. The first reason is because RM problems are so complicated that it is extremely difficult, if possible, to complete the optimization of actionable prices with a single process. The second reason is the end users (e.g., RM managers) usually demand to understand how the prices are optimized before they can put any trusts in the solutions. In other words, they prefer a RM system that is "transparent" in the integral processes for solving optimal prices. This will allow them to better understand and evaluate the roles that the processes play.

The estimation of reference rents is one of fundamental processes in an apartment RM system. It provides reference rent estimates as input to a subsequent process of Optimizer which optimizes rental rates in the apartment RM system. In this research, we separate the process of estimating reference rents as a RL model and the process of Optimizer as part of environment. Notice that these two processes share the same goal: maximizing the total revenue to be gained from the units currently available, or equivalently, maximizing the average of RevPAUs. To our knowledge, there are no publications specifically addressing the issue of reference rent estimation.

From the RL perspective, the process of estimating reference rents can be regarded as a learning agent under the formal framework of a Markov Decision Process (MDP). The task of a learning agent can be classified as either episodic or continuing; episodic tasks end in a finite number of steps, while a continuing tasks repeat indefinitely. In this regard, this process is continuing because it finds and takes actions at each time step over an infinite horizon.

There are a variety of RL publications relevant to discounted optimality of RM for episodic tasks. For example, Raju et al. (2003) investigate the use of RL to determine dynamic prices in an electronic retail market. Schwind and Wendt (2002) propose an RL approach to pricing information services. Their method provides encouraging results for efficient adaptive pricing of resource attribution related to the multidimensional RM problem. Shihab et al. (2019) research the application of deep RL to airline seat inventory control and overbooking based on the data generated by an air-travel market simulator. Sutton and Barto (2018) give an excellent introduction to RL and some applications of episodic tasks. For our purpose, we focus our literature review on the average-reward RL (ARL) that is relevant to the undiscounted optimality framework for continuing tasks.

In ARL, gain-optimal control policies are defined as those sequences of actions that help maximize the expected payoff per step. There are many relevant publications. For instance, Singh (1994) proposes four gain-optimal algorithms. In reality, however, gain-optimality has some intrinsic limitations as an optimality criteria. For example, it cannot distinguish between different policies that all reach gain-optimality but incur varying costs. A more selective criterion of bias-optimality is thus considered. This criterion can filter gain-optimal policies such that the one that reaches gain-optimality with the minimum cost will be selected. Mahadevan (1996a) proposes a model-based ARL algorithm for computing bias-optimal policies for an admission control queuing system. The simulated results show that this algorithm is more efficient than gain-optimal methods by learning bias-optimal policies. Mahadevan (1996b) describes a wide spectrum of ARL algorithms including examples for computing bias-optimal policies.

There is limited literature in the area of RM relevant to using ARL to find an optimal pricing policy. For example, Gosavi et al. (2002) use ARL to develop a strategy for seating allocation and overbooking in order to maximize the average revenue gained by an airline. Raju et al. (2006) apply Q-learning to price products dynamically with customer segmentation. They consider an infinite horizon learning problem where there is no deadline for the sale of stock, and price changes according to queue length and time. Rana and Oliveria (2014) propose a model-free RL approach to establishing a pricing policy that maximizes the revenue for selling a given inventory by a fixed deadline. Bondoux et al. (2017) apply a Q-learning algorithm to find the right fare for a single leg of flight. They argue that the results are promising based on numerical simulations.

Methodology

RL framework

A typical RL framework contains five interacting components: Agent, Environment, State, Action and Reward (Sutton and Barto 2018). Figure 1 illustrates an RL framework for estimating reference rents. Agent is a core process trying to find percentages of reference rent changes as actions. Environment abstracts the world surrounding Agent. It consists of three primary processes: (1) Optimizer takes the reference rent estimates derived from the actions as input to optimize the rental rates for new and renewal leases; (2) PMS (Property Management System) is a mechanism for publishing rental rates, storing leasing activity and managing inventory of units; and (3) Extractor detects and extracts the status of the Environment. It particularly extracts and derives some features of the Environment as States and estimates the immediate Rewards of actions. Both States and Rewards are inputs to the Agent.

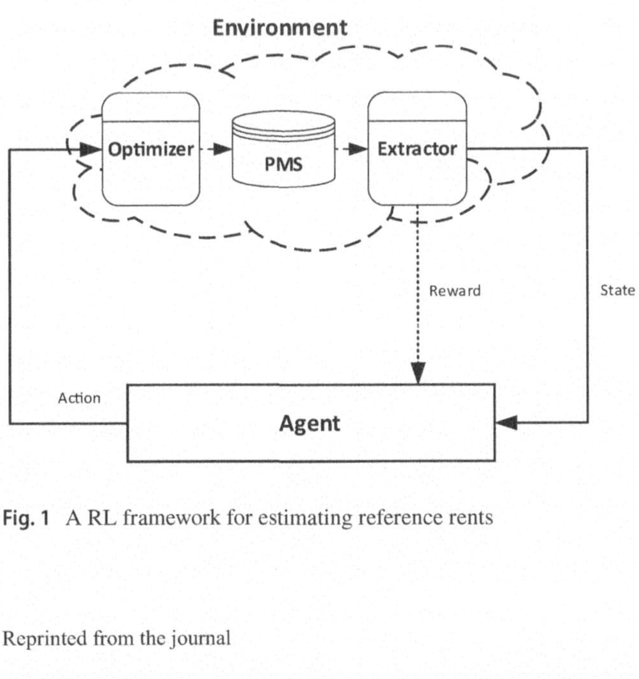

Fig. 1 A RL framework for estimating reference rents

Denote S as a set of finite states. Any state $s_t \in S$ at a time step t can be expressed as a vector-valued tuple$s_t = (s_{1,t}, s_{2,t}, s_{3,t}, s_{4,t}, s_{5,t}, s_{6,t})$, whose features are defined as follows:

- $s_{1,t}$: exposure as the number of current available units divided by capacity
- $s_{2,t}$: leasing velocity as the speed at which recently available units have been leased. During a period of $[t-h, t)$ prior to the current time step t, where h is a hyper-parameter specifying the length of the period, $s_{2,t} = \frac{L_{[t-h,t)}}{U_{[t-h,t)}}$ in which $L_{[t-h,t)}$ represents the number of leases having been signed and $U_{[t-h,t)}$ the number of units available during the period
- $s_{3,t}$: ratio of last market composite over last reference rent. That is, $s_{3,t} = \frac{MC_{t-1}}{RR_{t-1}}$ where MC_{t-1} and RR_{t-1} represent the market composite and reference rent estimated at previous time step $(t-1)$
- $s_{4,t}$: ratio of current market composite over last reference rent. Namely, $s_{4,t} = \frac{MC_t}{RR_{t-1}}$ where MC_t represents the market composite estimated at current time step t
- $s_{5,t}$: market to which a property belongs
- $s_{6,t}$: season that current time step t belongs to

Note that in the calculation of $s_{3,t}$ and $s_{4,t}$, the metrics of market composites MC_{t-1} and MC_t are used. Market composite MC_t at time t is estimated as the weighted average of adjusted competitor rents. The weights represent the pricing influences of competitors to the apartment. Their values can be either specified by apartment managers or they can be estimated based on the data related to searching and leasing activity in the market to which the apartment belongs. The adjustments of competitor rents reflect the relative rent differences between the apartment and the competitors. It is beyond the scope of this study to discuss how to estimate market composites. Interested readers can refer to a study by Noone et al. (2012). For the sake of simplicity, we directly adopt the market composites that are estimated by the apartment.

Denote A as a set of admissible actions. For any state $s_t \in S$ at time t, there corresponds one or more actions $a_t \in A_{s_t}$ that Agent can choose from, where A_{s_t} is a subset of A. Our objective is to find optimal actions for all states such that the average of rewards will be maximized. It needs to be pointed out that if reference rents of RR_t are used directly as actions, there will be a dilemma in model training. On the one hand, given a group of apartments in a market, for example, if all reference rents from the group are used as actions to train one single RL model for all of the apartments in the group, then, most probably, this learned RL model cannot be applicable to them. This is because these apartments may have different values of reference rent estimates for the same state, which cannot be recommended by the learned RL model simultaneously. On the other hand, if reference rents from an individual apartment are used as actions to train one RL model specifically for this apartment only, then there might not be enough data to train the model properly.

To overcome this dilemma, we define actions a_t as a percentage change of current reference rent over the prior one, i.e., $a_t = \left(\frac{RR_t}{RR_{t-1}} - 1\right) \times 100\%$. Once a_t is determined, RR_t can be easily derived as $RR_t = (1 + a_t)RR_{t-1}$ because RR_{t-1} is known. Unlike the actions which directly use reference rents, the actions thus defined focus on the relative changes of reference rents. It will not only enable us to use all reference rents from the group to train an RL model, but also it will help minimize the effect of reference rent differences among the apartments. To simplify the model training, the actions are limited to a finite set $A = \{\delta_L, (\delta_L + 1), \ldots, -1\%, 0\%, 1\%, \ldots, (\delta_U - 1), \delta_U\}$, where δ_L and δ_U are two parameters specifying the lower and upper bounds to be considered.

Associated with any action $a_t \in A_{s_t}$, there exists a transition probability $P_{s_t,s_{t+1}}(a_t)$ describing the likelihood of moving from state s_t to state s_{t+1} after action a_t is taken. It satisfies the relationship $\sum_{s_{t+1} \in S} P_{s_t,s_{t+1}}(a_t) \equiv 1$ for any $s_t \in S$ and $a_t \in A_{s_t}$. Under the assumption of MDP property, $P_{s_t,s_{t+1}}(a_t)$ is fully determined by s_t, a_t and s_{t+1}.

In addition, denote $r(s_t, a_t)$ as the immediate reward received at time $(t + 1)$ for the action a_t taken on state s_t at time t. The reward $r(s_t, a_t)$ is a mapping: $S \times A \rightarrow R^+$, where $R^+ = [0, \infty)$ is an interval of non-negative real values. Specifically, it is estimated as

$$\hat{r}(s_t, a_t) = \frac{1}{\|\bar{U}_t\|}\left(\frac{\sum_{u \in U_t \subseteq \bar{U}_t} lt_u R_u(a_t)}{\sum_{u \in U_t \subseteq \bar{U}_t} lt_u}\right)$$

where $\bar{U}_t$ denotes the set of units available to be rented out during the time period of $[t, t + 1)$, $\|\bar{U}_t\|$ the cardinality of $\bar{U}_t$, and U_t a subset of $\bar{U}_t$ that have been assigned to a lease. lt_u and $R_u(a_t)$ are the lease term in month, and monthly rent optimized based on action a_t, respectively, for the lease tied to a unit $u \in U_t$. It is noted that $\hat{r}(s_t, a_t)$ can be estimated on a weekly level also.

Reference rent estimation algorithm

Optimal actions are found by maximizing the average of rewards over an infinite time horizon. Denote $V^*(s)$ as an optimal value function for any $s \in S$ satisfying Bellman Optimality Equation

$$V^*(s) = \max_{a' \in A_s}\left(r(s, a') - \rho^* + \sum_{s' \in S} P_{s,s'}(a')V^*(s')\right)$$

where ρ^* is some scalar representing the average reward for the optimal policy (Singh 1994). A greedy policy, consisting of optimal actions for all states, can be formed by selecting actions that maximize the right-hand side of the above equation (Gosavi 2004).

Note that in the equation, the difference of $(r(s, a') - \rho^*)$ represents the incremental reward over ρ^* resulting from an action $a' \in A_s$. Over the stages of the infinite horizon, the actions that can increase the average reward (and thus the incremental reward) are to be more preferable. From the perspective of finding the optimal actions only, any value can be chosen for ρ^*. When the value of ρ^* is set as zero, for example, the approaches to optimizing the actions are called value-iteration (Gosavi 2004). The use of any non-zero ρ^* is to assure the numerical stability. When the value of ρ^* is set as some positive value, the approaches to optimizing the actions are called relative value-iteration (Singh 1994).

We attempt to find an optimal policy using sample data that are generated by an existing rule-based approach, which can be regarded as another policy. To avoid numerical instability, we choose a model-based RL approach, one of relative value-iteration methods. The value of ρ^* is set as the average of observed rewards from the samples. Specifically, for any $s \in S$, the optimal action $a^*(s) \in A_s$ can be found using the following off-policy algorithm.

(1) Initialization

(a) Set $V^0(s) = 0, \forall s \in S$.
(b) Set ϵ as error tolerance parameter (e.g. $\epsilon = 0.0000001$)
(c) Set N as maximum number of iterations (e.g., $N = 10,000$)
(d) Denote $\hat{\rho}$ as an estimate for ρ^*. It is approximated as the global average of observed rewards $\hat{r}(s, a)$ over all $s \in S$ and $a \in A_s$

(2) For $n = 0,1,2,\ldots,N$

(a) For each $s \in S$, compute $V^{n+1}(s)$ by

$$V^{n+1}(s) = \max_{a \in A_s}\left(\hat{r}(s, a) - \hat{\rho} + \sum_{s' \in S} P_{s,s'}(a)V^n(s')\right)$$

(b) If $sp(V^{n+1} - V^n) < \epsilon$ or $n > N$, go to step (3); otherwise increment n by 1 and return to step (a). Here, $sp(b)$ is a span function defined as $sp(b) = \max_{s \in S} b(s) - \min_{s \in S} b(s)$

(3) Denote n^* as the n at which the above procedure stops. For each $s \in S$, choose the optimal action $a^*(s)$ as

$$a^*(s) \in \arg\max_{a \in A_s} \left(\hat{r}(s,a) + \sum_{s' \in S} P_{s,s'}(a) V^{n^*}(s') \right)$$

Performance measurement

No processes can be improved without measurement; the process of estimating reference rents is no exception. In practice, occupancy, average rent and RevPAU are three key performance indicators (KPI) that are commonly used to measure the performance of a pricing process. Occupancy is the percentage of the total units that have been leased. Average rent is the sum of the rents from the leased units divided by the number of leased units. RevPAU is the sum of the rents from the leased units divided by the total number of units. Any of these three metrics can be derived from the other twos. Because RevPAU measures the revenue gain per unit directly, it encompasses the influence of both rent and occupancy, and is frequently used to compare the revenue gain between two pricing processes.

Strictly speaking, RevPAU is always tied to a period of time. Traditionally, it is estimated for a period of past, say, a particular historical month. Similar to the estimation of immediate reward $\hat{r}(s_t, a_t)$, monthly RevPAU can be estimated for any historical month of M as follows

$$\text{RevPAU(M)} = \frac{\sum_{\substack{\forall t \le m \\ m \in M}} \left(\frac{\sum_{u \in U_t \subseteq \bar{U}_t} \delta_m(u) lt_u R_u(a_t)}{\sum_{u \in U_t \subseteq \bar{U}_t} \delta_m(u) lt_u} \right)}{\sum_{\substack{\forall t \le m \\ m \in M}} \sum_{u \in U_t \subseteq \bar{U}_t} \delta_m(u) \| \bar{U}_t \|}$$

where $\delta_m(u)$ is a binary variable indicating whether the lease signed at t for unit u is occupying the unit over a date $m \in$ M or not. That is, if the lease is still occupying the unit u over m, $\delta_m(u) = 1$; otherwise, $\delta_m(u) = 0$.

It can be seen that RevPAU(M) results from many prior actions a_t taken at prior times $t \le m \in M$. In practice, the actions a_t that are taken at time t could be optimal or non-optimal. By simply looking at the value of RevPAU(M) alone, we are unable to differentiate the contributions from individual actions. To measure the performance of optimal and non-optimal actions, we bifurcate RevPAU into two perspectives: forward and backward, measuring the revenue impact on the future and past periods, respectively.

Denote λ as action type of optimal or non-optimal actions, and T as a period during which actions are taken. A metric of monthly forward-RevPAU(T, λ) is defined to measure RevPAU that is contributed by the actions a_t of type λ taken during T. Specifically, it is estimated as follows

$$\text{forward-RevPAU}(T, \lambda) = \frac{\sum_{\substack{a_t \text{ is type } \lambda \\ t \in \mathrm{T}}} \left(\frac{\sum_{u \in U_t \subseteq \bar{U}_t} lt_u R_u(a_t)}{\sum_{u \in U_t \subseteq \bar{U}_t} lt_u} \right)}{\sum_{\substack{a_t \text{ is type } \lambda \\ t \in \mathrm{T}}} \| \bar{U}_t \|}$$

The numerator in forward-RevPAU(T, λ) is the sum of averages of monthly revenue from the leases signed at times $t \in$ T whose rents $R_u(a_t)$ are optimized based on actions a_t of type λ. The denominator is the sum of available units at the times when actions a_t are of type λ.

In addition, another metric of monthly backward-RevPAU(M, λ) is defined to measure RevPAU by adding the dimension of action type λ. That is,

$$\text{backward-RevPAU}(M, \lambda) = \frac{\sum_{\substack{a_t \text{ is type } \lambda \\ \forall t \le m \\ m \in M}} \left(\frac{\sum_{u \in U_t \subseteq \bar{U}_t} \delta_m(u) lt_u R_u(a_t)}{\sum_{u \in U_t \subseteq \bar{U}_t} \delta_m(u) lt_u} \right)}{\sum_{\substack{a_t \text{ is type } \lambda \\ \forall t \le m \\ m \in M}} \sum_{u \in U_t \subseteq \bar{U}_t} \delta_m(u) \| \bar{U}_t \|}$$

The numerator in backward-RevPAU(M, λ) is the sum of average monthly revenues from those leases that stay over dates $m \in$ M and whose rents $R_u(a_t)$ are optimized based on actions a_t of type λ. The denominator is the sum of available units at the times when those leases are signed.

Backward-RevPAU(M, λ) can be further extended to backward-RevPAU(M, λ, w) by considering an additional dimension of w, called weeks remaining or weeks left. The weeks remaining of w is the number of weeks between the dates when a lease is signed and when the lease starts. For the sake of brevity, we omit the specification for backward-RevPAU(M, λ, w).

The use of both forward-RevPAU(T, λ) and backward-RevPAU(M, λ) allows us to compare the relative performance of RL- and rule-based approaches. For example, according to the optimality of actions from the RL-based approach, the actual actions, that are derived from the reference rent estimates by the rules-based approach, can be categorized as "optimal" or "non-optimal". In terms of revenue gain, it is expected that forward-RevPAU(T, λ = "optimal") ≥ forward-RevPAU(T, λ = "non-optimal"). If this condition does not hold, then the optimal actions suggested by the RL-based approach will become worse than those by the rule-based approach.

Datasets

The datasets consist of historical reference rent estimates from two leading apartment management companies A and B. There are about 5 years of history for both companies. Of the raw datasets, we have removed about 6% of data, which are related to missing states, misspecification of process parameters, or abnormal actions outside of the interval of $[-5\%, +5\%]$. Based on the cleaned datasets, Table 1 summarizes some descriptive statistics including the numbers of markets, properties, units and leases, minimum, median and maximum of reference rents and their changes. Company A has more properties than company B thus having more leases.

It can be seen that reference rents for company A are also larger. For example, the minimum, median and maximum of reference rent estimates for company A (and B) are \$800 (\$800), \$2513 (\$1678) and \$13896 (\$9909), respectively. In addition, the data cleaning results in the percentage changes of current reference rents over last reference rents for both companies A and B falling into the interval of $[-5\%, 5\%]$. The action space is thus limited to the set of $\{-5\%, -4\%, \ldots, 0\%, \ldots, 4\%, 5\%\}$, in which the lower and upper bounds of -5% and 5% are set for the parameters of δ_L and δ_U, respectively.

Furthermore, there are total 27 markets in which 7 of them are common to both companies. Specifically, company A has 261 properties of 79532 units belonging to 17 markets, while company B has 65 properties of 19961 units belonging to 17 markets. Figure 2 illustrates the market distributions in which the x-axis denotes the market codes, and y-axis the percentage of observations by markets.

For the state space, the features for the state $s_t = (s_{1,t}, s_{2,t}, s_{3,t}, s_{4,t}, s_{5,t}, s_{6,t})$ at each time step t are defined as follows. The feature of $s_{1,t}$ is set as high, medium or low if the value of exposure at t falls in the interval of $[0\%, 5\%)$, $[5\%, 9\%)$ or $[9\%, 100\%]$, respectively. The feature of $s_{2,t}$ is set as high, medium or low if the value of leasing velocity at t falls in the interval of $[0\%, 20\%)$, $[20\%, 35\%)$ or $[35\%, 100\%]$, separately. The hyper-parameter h specifying the historical period used to estimate leasing velocity is chosen as 5 weeks. The feature of $s_{3,t}$ (and $s_{4,t}$) is set as above, equal or below when the ratio of last market composite over last reference rent (and ratio of current market composite over last reference rent) at t belongs to the interval of $[1.001,3)$, $[0.999,1.001)$ or $[0.25,0.999]$, respectively. The feature of $s_{5,t}$ is the market code that the property belongs to. Finally, the feature of $s_{6,t}$ is set as the season for the month that time step t falls in. Specifically, $s_{6,t}$ is high if the month is between May and August; low if the month is between November to February; and medium otherwise.

The above definitions for the state features result in $4131(= 3 \times 3 \times 3 \times 3 \times 17 \times 3)$ possible states, from which only 2117 and 2196 distinct states are actually observed for companies A and B, respectively, which represent around 50% of all possible states. Figure 3 shows state distributions for both companies, each of which is sorted by the frequency of occurrence of its own states. It can be seen that the states for both companies are not distributed uniformly implying that numbers of observations across the states are not balanced. This may result in better estimators for some states and worse estimators for others.

Figure 4 displays the distributions of reference rent estimates. The distributions are unimodal for both companies, The reference rent estimates are clustered around the medians of \$2513 and \$1678 for companies A and B, respectively. However, company A has a longer tail indicating that the changes of its reference rent estimates might be more volatile. This is confirmed by Fig. 5 in which the action distributions are displayed. In Fig. 5, actions for both companies are centered around the median of 0% but with company A having longer tails on both sides. This means that the changes of reference rent estimates for company A are larger.

Actions for all of the states are not always distributed the same. Figure 6 displays action distributions for the top 10 states that occur most frequently by company. The x-axis represents actions, and the right y-axis indicate the state codes for the companies. In the action distribution of any given state, there are two solid vertical lines and one dotted vertical line. The solid lines indicate the lower and upper bounds of a protection interval with its center at the median action of the state. The dotted line represents an optimal action which is found from the candidate actions inside the protection interval. This will be explained further in next section. In addition, it can be seen that action distributions for company A look more uniform than company B. This again shows that the selection of actions for company A is more volatile because all of the actions have a similar chance of getting selected.

Table 1 Some descriptive statistics for the cleaned datasets

Company	Properties (markets)	Leases (units)	Min and max lease month	Min ref rent (change)	Med ref rent (change)	Max ref rent (change)
A	261 (17)	138,054 (79,532)	Jun 2013–Apr 2019	\$800 (−5%)	\$2513 (0%)	\$13,896 (+5%)
B	65 (17)	38,887 (19,961)	Jul 2013–Feb 2019	\$800 (−5%)	\$1678 (0%)	\$9909 (+5%)

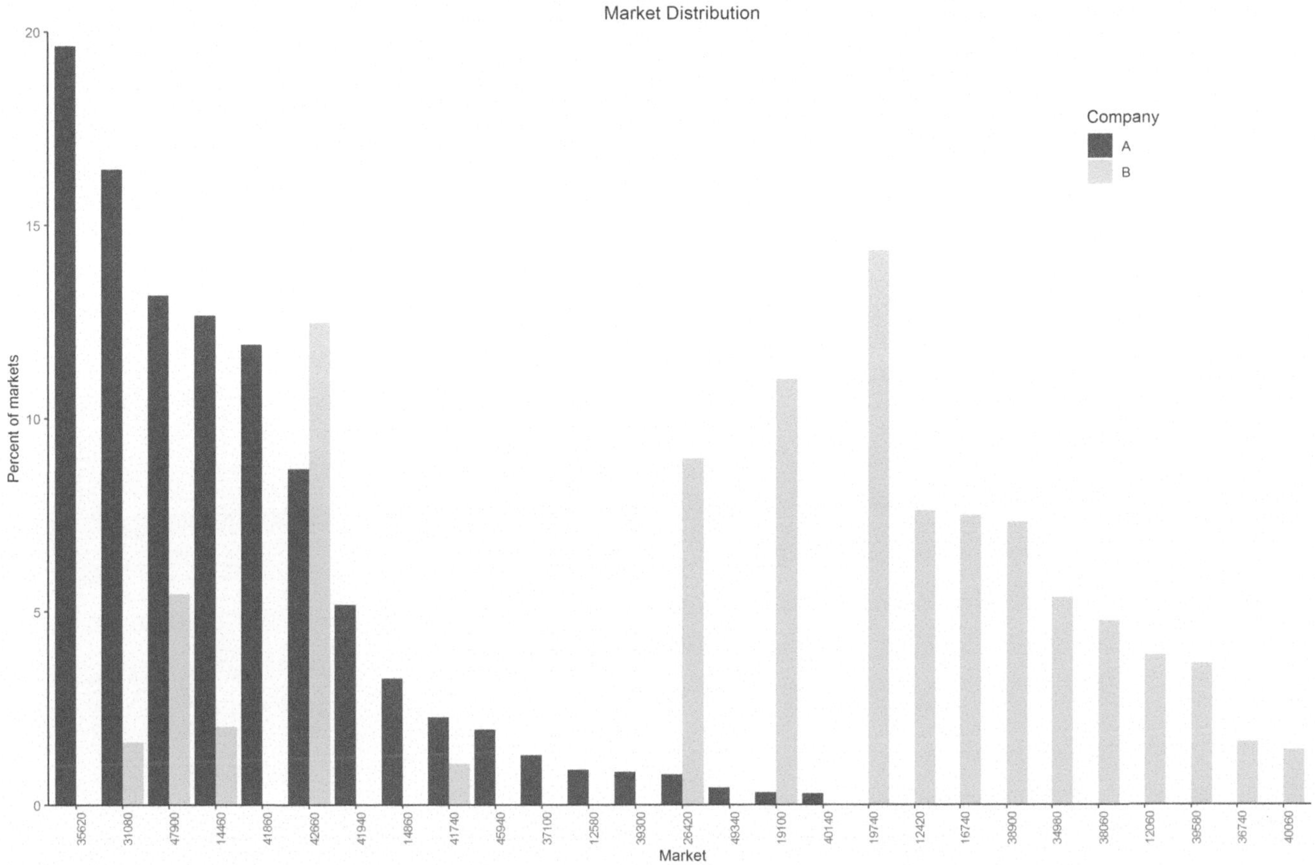

Fig. 2 Market distribution for companies A and B

Reward distributions are summarized in quantiles in Table 2. The mean and median of rewards for company A (and B) are \$62 (\$26) and \$0 (\$0), respectively. Notice that 75% of quantiles are \$18 and \$0 for companies A and B. This implies that company B has a larger fraction of actions with \$0 of rewards which infer the different pricing tactics that the two companies might have adapted when they estimate reference rents. As a matter of fact, over the past 5 years, company B estimated reference rents everyday but company A only did once a week. The daily estimation of reference rents for company B has thus resulted in many actions with \$0 of rewards because apartments do not have leases signed every day.

Finally, Fig. 7 shows the heatmaps of transition probabilities for the top 10 states by median actions. The rows denote the top 10 states over which median actions are taken, and columns the resulting states including the same 10 top states and a virtual state of "other" which buckets all of other states. The scale of greyness measures the magnitudes of probabilities. It can be seen that the diagonal entries are darker for both companies indicating that more than 50% of chances these top 10 states would transit back to themselves after a median action is taken. In addition, if one of the top 10 states does not transit back to itself after a median action is taken, it would more likely move to a state other than the top 10 states.

Empirical results

As mentioned early, Fig. 6 not only shows the action distributions for the top 10 states, but also it shows the optimal actions along with the lower and upper bounds of a protection interval. Of the total 20 optimal actions for both companies, only 7 of them stay the same as median actions (centers of protection intervals), and the other 13 optimal actions are at either lower or upper bounds. This implies that taking median actions does not always help approach the maximal of average of rewards.

Table 3 summarizes the performance in terms of weekly forward-RevPAU(T, λ) where T is chosen about 1050 days and λ indicates the type of actions. For company A, there are 367323 optimal actions representing 36% of all actions. Forward-RevPAU s are \$55 and \$48 for optimal and non-optimal actions, respectively. The increase of forward-RevPAUsof optimal actions over non-optimal

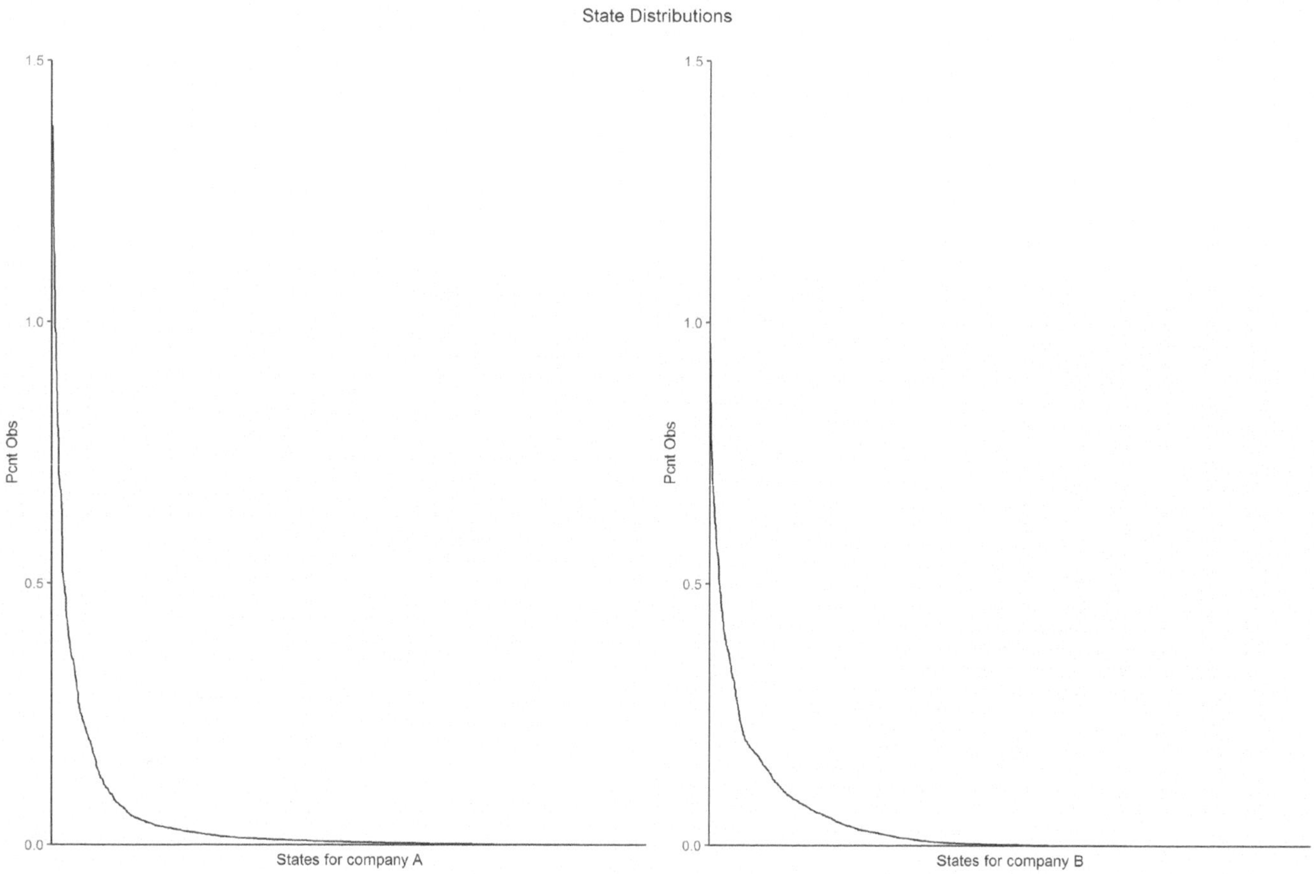

Fig. 3 State distributions for companies A and B

actions is $15\%\left(=\frac{\$55}{\$48}-1\right)\times 100\%$. In addition, the value of ROI is 9% as estimated by $\left(\frac{\$55}{36\%*\$55+64\%*\$48}-1\right)\times 100\%$. This value assumes that if the non-optimal actions were replaced by optimal actions, there would be 9% improvement over the current forward-RevPAU. Similarly, increases of forward-RevPAU and ROI for company B are 24% and 18%, both of which are larger than company A. For company B, one reason that its ROI is larger than property A might be because there are more non-optimal actions that can be improved.

Figure 8 displays the averages of weekly backward-RevPAU(M, λ, w), in which M is selected as 11 months, λ denotes action type and w as 65 weeks-left. For company B, backward-RevPAU s of optimal are always larger than those of non-optimal actions at every weeks-left, which is expected. The overall averages of backward-RevPAU(M) over the 11 months for optimal and non-optimal actions are \$100 and \$92, respectively. That implies that backward-RevPAU for optimal actions is about $9\%\left(\approx\left(\frac{\$100}{\$92}-1\right)\times 100\%\right)$ higher than that for non-optimal actions.

On the other hand, for company A, backward-RevPAU s of optimal actions are larger than those of non-optimal actions over all except for a few of weeks left. We have not figured out why this happens yet. One plausible explanation might be due to the weak convergence of optimal solutions. However, the overall average of backward-RevPAU(M) over the 11 months for optimal actions is \$128, which is slightly (about 1%) higher than \$127 for non-optimal actions.

Conclusion

In this research, we propose an average-reward based reinforcement learning to estimate reference rents. We extend the traditional metric of RevPAU to the new metrics of forward-RevPAU and backward-RevPAU. An empirical study is done using the real datasets from two leading apartment management companies. It is shown that this RL-based approach has outperformed a rule-based approach in terms

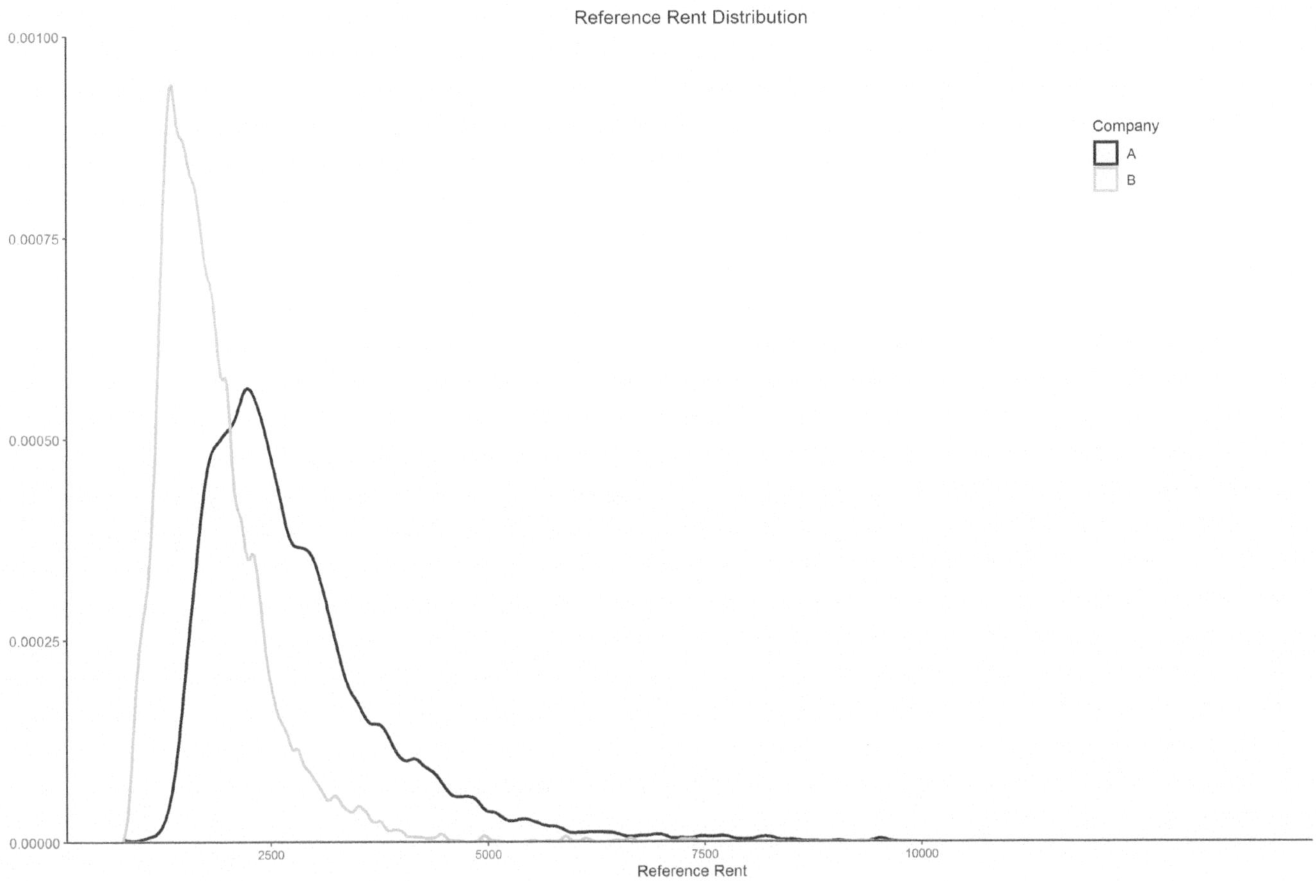

Fig. 4 Distributions of reference rent estimates for companies A and B

of both forward-RevPAU and ROI increases. It has also outcompeted the rule-based approach for majority of weeks left in terms of backward-RevPAU. However, there are at least two aspects that need to be further researched so that this RL-based approach can be deployed in production with confidence.

The first aspect is the handling of unseen states. As stated earlier, only 50% of possible states have been observed from the datasets. As a result, the solutions that are recommended by this RL-based approach might be optimal only for the observed states but not for the unseen sates. There are two options to address this issue. The first option is to invoke ϵ -greedy approach by randomly selecting an action for any unseen state. This approach is risky in the context apartment RM, as the revenue of any single transaction is very high. As the reference rent will affect all transactions conducted while it is in place, an improper estimate of reference rent will impact a large fraction of total revenue (Wang 2008). The second possibility is to invoke the rules-based approach when previously unseen states are encountered. This option is relatively safe, as it will be no worse than the current application. The caveat is that any actions thus proposed are not necessarily optimal. To resolve this dilemma, we need to determine which states should be encompassed by the RL approach, and which states should be handled by the existing rules-based method.

The second aspect is the lack of visibility of cause and effect in an RL system. Although the rules-based approach does not perform as well as its counterpart RL-based approach, it has attracted many users because it connects the actions to the states intuitively. On the contrary, the RL-based approach has a difficulty in explaining how the actions are determined; this "black box" nature is common to many real-world applications of reinforcement learning methods (Arnold et al. 2019). As a future research topic, we plan to find a more linear or self-evident RL-based approach which can explain how optimal actions are determined, at least for those states that occur frequently.

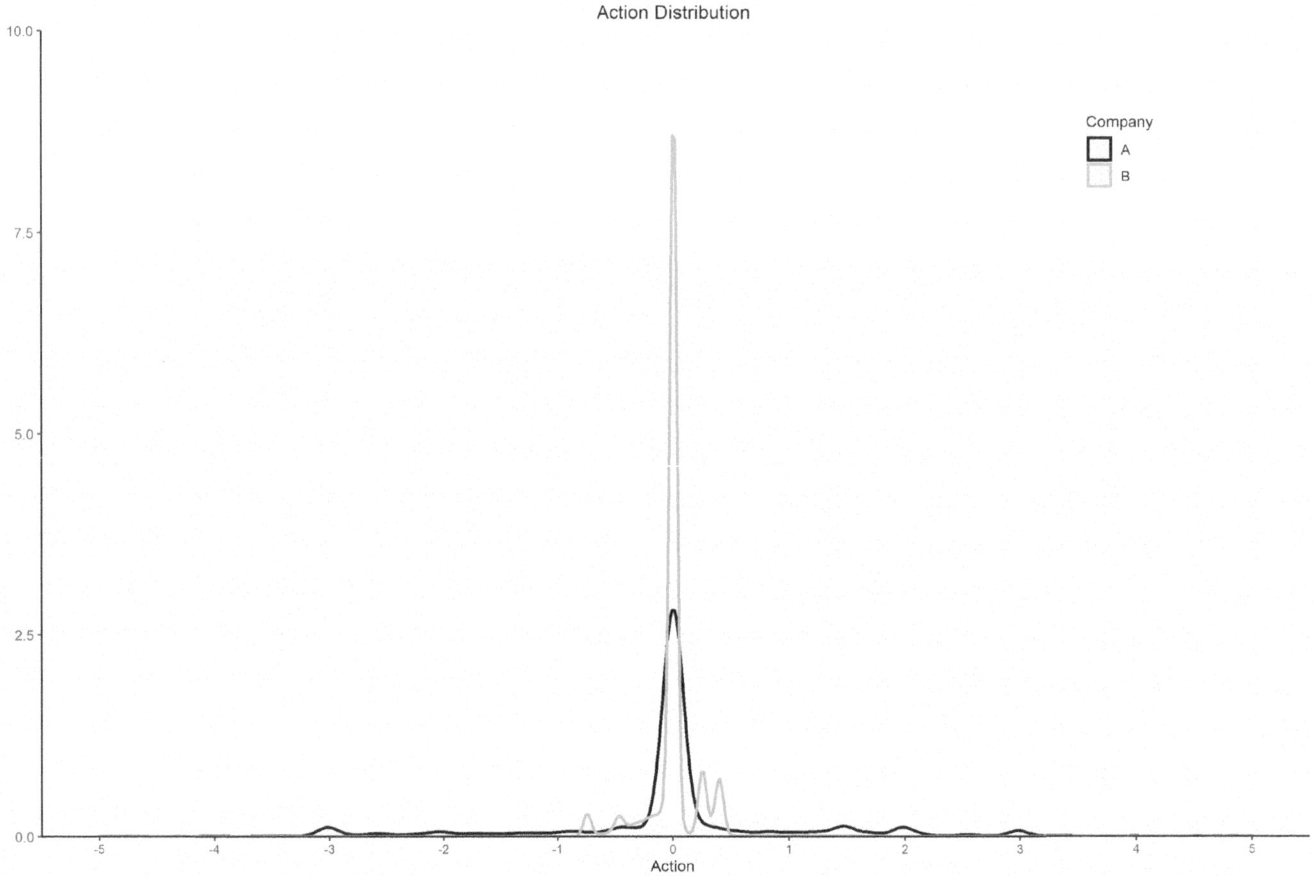

Fig. 5 Action distributions for companies A and B

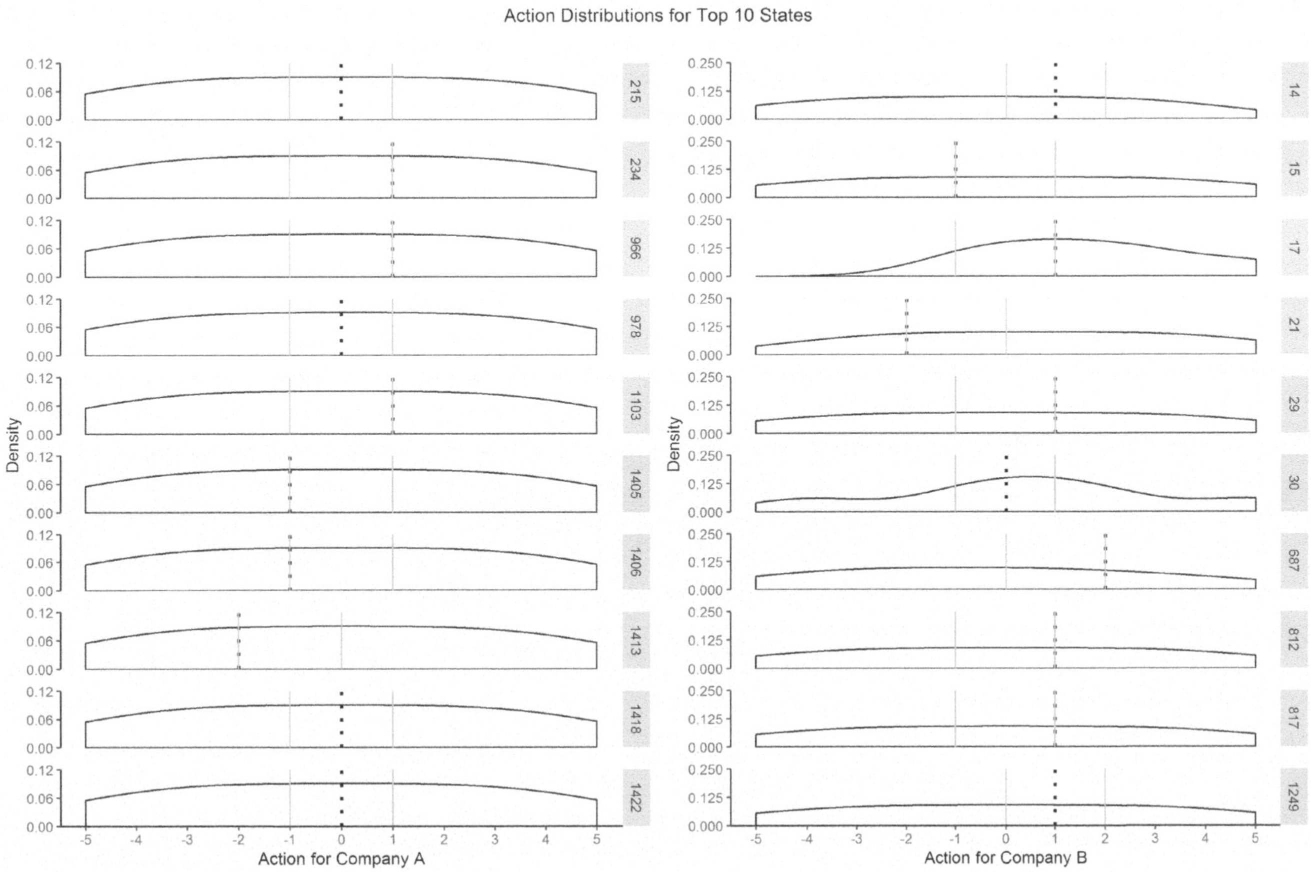

Fig. 6 Action distributions for top 10 states for companies A and B

Table 2 Reward quantiles for companies A and B

Company	Min	25% quantile	Mean	Median	75% quantile	Max
A	\$0	\$0	\$62	\$0	\$18	\$1351
B	\$0	\$0	\$26	\$0	\$0	\$7490

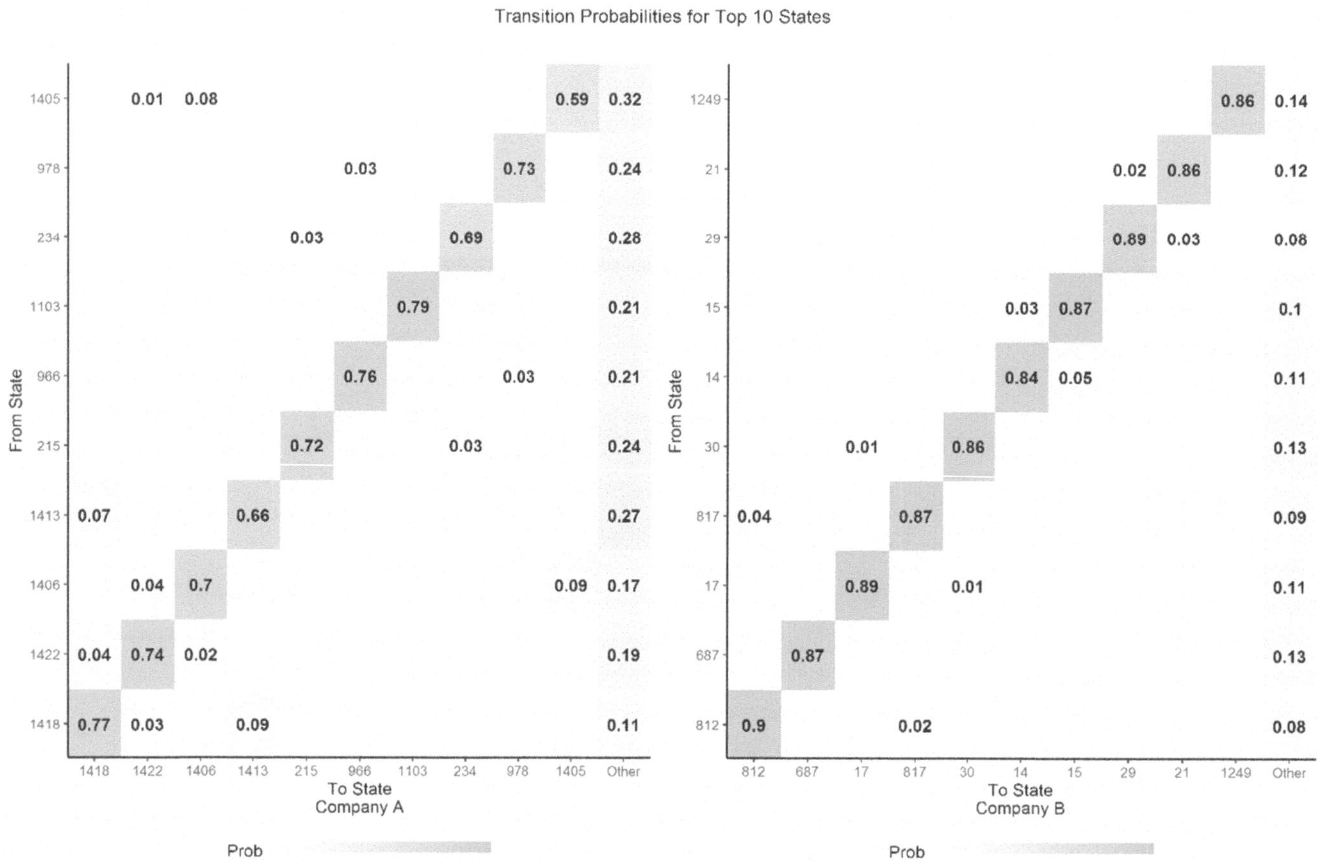

Fig. 7 Heatmaps of transition probabilities for top 10 state for companies A and B

Table 3 Performance Comparison by forward-RevPAU

Company	Action type	Actions (percentage)	foreward-RevPAU	foreward-RevPAU Increase (ROI)
A	Optimal	67,323 (36%)	$55	15% (9%)
	Non-optimal	119,469 (64%)	$48	
B	Optimal	141,310 (22%)	$31	24% (18%)
	Non-optimal	368,503 (78%)	$25	

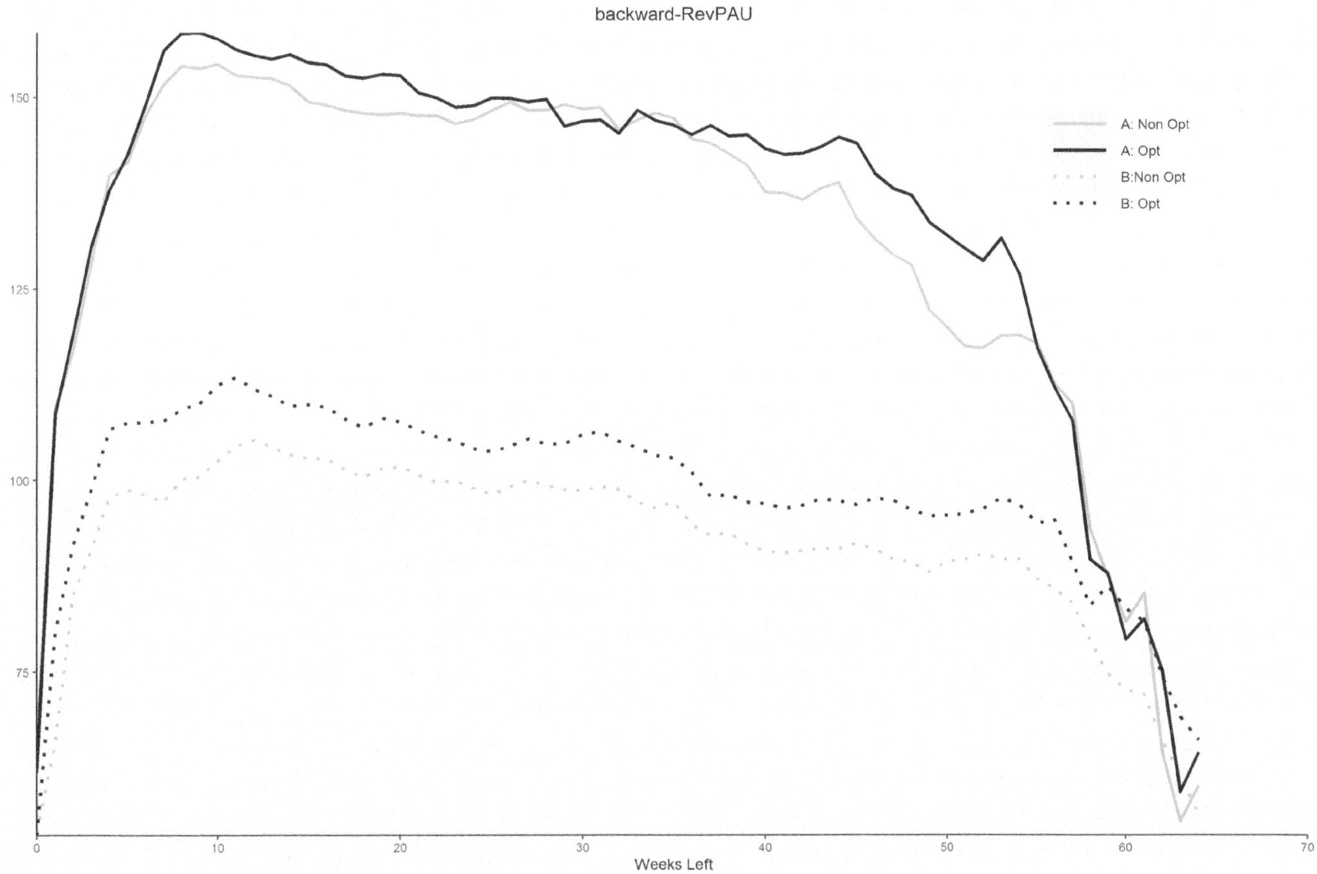

Fig. 8 Comparison of backward-RevPAU for companies A and B

References

Arnold, G.D., D. Mankowitz, and T. Hester. 2019. Challenges of Real-World Reinforcement Learning. arXiv:1904.12901 [cs.LG].

Bondoux, N., Q. Nguyen, T. Fiig, and R. Acuna-Agost. 2017. The End of Airline Revenue Management as We Know It? (Deep) Reinforcement Learning for Revenue Management. 57th Annual AGIFORS Symposium, 2017, Volume 2017—Conference Proceeding.

Gosavi, A. 2004. Reinforcement Learning for Long-Run Average Cost. *European Journal of Operational Research* 155: 654–674.

Gosavi, A., N. Bandla, and T.K. Das. 2002. A Reinforcement Learning Approach to a Single Leg Airline Revenue Management Problem with Multiple Fare Classes and Overbooking. *IIE Transactions* 34: 729–742.

Mahadevan, S. 1996(a). An Average-Reward Reinforcement Learning Algorithm for Computing Bias-Optimal Policies. AAAI-96 Processings.

Mahadevan, S. 1996(b). Average Reward Reinforcement Learning: Foundations, Algorithms, and Empirical Results. *Machine Learning—Special Issue on Reinforcement Learning* 22:159–195.

Noone, B.M., L. Canin, and C.A. Enz. 2012. Strategic Price Positioning for Revenue Management: The Effects of Relative Price Position and Fluctuation on Performance. *Journal of Revenue and Pricing Management* 12 (3): 207–220.

Pagliari, J.L., and J.R. Webb. 1996. On Setting Apartment Rental Rates: A Regression-Based Approach. *Journal of Real Estate Research* 12 (1): 37–61.

Raju, C.V.L., Y. Narahari, and K. Ravikumar. 2003. Reinforcement Learning Applications in Dynamic Pricing of Retail Markets. Proceedings of the IEEE International Conference on E-Commence.

Raju, C.V.L., Y. Narahari, and K. Ravikumar. 2006. Learning Dynamic Prices in Electronic Retail Markets with Company Segmentation. *Annals of Operations Research* 143: 59–75.

Rana, R., and F.S. Oliveria. 2014. Real-Time Dynamic Pricing in a Non-stationary Environment Using Model-Free Reinforcement Learning. *Omega* 47: 116–126.

Schwind, M., and O. Wendt. 2002. Dynamic Pricing of Information Products Based on Reinforcement Learning: A Yield-Management Approach. In *KI 2002: Advances in Artificial Intelligence*, ed. M. Jarke, G. Lakemeyer, J. Koehler. KI 2002. Lecture Notes in Computer Science, vol. 2479. Berlin: Springer.

Shihab, S.A.M., C. Logemann, D.G. Thomas, and P. Wei. 2019. Autonomous Airline Revenue Management: A Deep Reinforcement Learning Approach to Seat Inventory Control and Overbooking. arXiv:1902.06824v2 [cs.AI]. Accessed 13 June 2019.

Singh, S.P. 1994. Reinforcement Learning Algorithm for Average-Payoff Markovian Decision Processes. AAAI'94 Proceedings of the 12th National Conference on Artificial Intelligence, vol. 1, 700–705

Sirmans, G.S., and J.D. Benjamin. 1991. Determinants of market rent. *Journal of Real Estate Research* 6 (3): 357–379.

Sutton, R.S., and A.G. Barto. 2018. *Reinforcement Learning: An Introduction*, 2nd ed. Cambridge: The MIT Press.

Wang, J. 2008. A Realization of an Apartment Dynamic Pricing System. *Journal of Revenue and Pricing Management* 7 (3): 256–265.

Jian Wang is Vice President of Data Analytics at RealPage. He has more than two decades of experience in developing pricing and revenue management solutions for diverse industries.

Murtaza Das is a Senior Data Scientist at RealPage where he researches and develops Machine Learning and Artificial Intelligence models geared towards multifamily housing solutions.

Stephen Tappert is Vice President of Data Science at RealPage. He has worked with applied mathematics and predictive modeling in the real estate industry for more than 20 years.

Journal of Revenue and Pricing Management (2021) 20:344–350
https://doi.org/10.1057/s41272-021-00317-y

PRACTICE ARTICLE

Machine learning approach to market behavior estimation with applications in revenue management

Nitin Gautam[1] · Shriguru Nayak[1] · Sergey Shebalov[1]

Received: 12 April 2020 / Accepted: 15 September 2020 / Published online: 15 March 2021

Abstract
Demand forecasting models used in airline revenue management are primarily based on airline's own sales data. These models have limited visibility into overall market conditions and competitive landscape. Market factors significantly influence customer behavior and hence should be considered for determining optimal control policy. We discuss data sources available to airlines that provide visibility into the future competitive schedule, market size forecast, and market share estimation. We also describe methodologies based on Machine Learning algorithms that can use to forecast these quantities and explain how they can be leveraged to improve pricing and revenue management practices.

Keywords Demand forecasting · Competitive-aware revenue management · Integrated commercial planning

Introduction

An important component of revenue management practice is demand forecasting. Over the years, researchers and practitioners have been developing various approaches to address this challenge. For an overview of the forecasting techniques adopted by the airline industry over the years see, for example, Azadeh et al. (2015) and Strauss et al. (2018). While these techniques might significantly vary in assumptions and methodologies, most of them are based on the airline's data and hence have a limited view of overall market trends and competitive behavior. Meanwhile, the transparency provided by the internet and aggregation capabilities of Online Travel Agencies created an environment where offers from all airlines serving a market are easily available to consumers. Therefore, demand patterns are significantly affected by these factors, and modeling them in revenue management systems would improve forecast accuracy and hence generate better pricing and availability policies.

Revenue management demand forecasting can be viewed as a combination of several steps and each of them can benefit from additional information describing the state of the market:

1. *Untruncation*: This process is used to reconstruct historic demand from availability data and bookings observed subject to availability. If a specific product was not available for some time, demand for that product likely shifted to other products. This might be a higher booking class on the same itinerary, a different itinerary of the same airline, a different airline, a different mode of transportation, or a decision not to travel at all. Demand for the closed products cannot be observed directly and hence it is estimated based on how demand for the open products changed during that interval. If an increase in demand for open products is not significant in a market where the airline dominates, it would indicate that the closed products are not very popular. However, if an airline only provides a small fraction of all possible travel options, then demand for close products might have been significant, but was lost to the competition. Therefore, accurate information about the airline's market share is important.
2. *Forecasting*: Once historical demand is estimated, it needs to be mapped into the future departures. At this step, trend and seasonality of demand are modeled. An airline's own data might not be sufficient to recognize the behavior of the whole market. It is beneficial to understand the total market size and be able to predict how it will change in the future.
3. *Willingness to pay estimation*: Finally, to compute the optimal control policy for future departures, it is impor-

✉ Sergey Shebalov
sergey.shebalov@sabre.com

[1] Sabre Airline Solutions, Southlake, TX, USA

Chapter 11 was originally published as Gautam, N., Nayak, S. & Shebalov, S. Journal of Revenue and Pricing Management (2021) 20:344–350. https://doi.org/10.1057/s41272-021-00317-y.

tant to understand what travel options are available to consumers and how they select between them. Since travelers consider options offered by all airlines in a market, revenue management systems should recognize those options and adjust the control policy it produces accordingly.

In this paper, we will consider several data elements that can be used by a revenue management system to account for market conditions and competition—future schedules for all airlines present in a market, total market size forecast, and market share estimation. We will describe data sources that can provide this information, methodologies that are used to analyze it, and potential improvements in revenue management practices these data can lead to.

Table 1 SVO-DBX market in June 2019

Flight	Equipment	Expected local traffic
June schedule as of March		
EK-132	A380	192
EK-134	B777	129
SU-524	B777	118
SU-520	B777	113
QF-8132	Codeshare	98
June schedule as of April		
EK-132	B777	175
EK-134	B777	148
SU-520	B777	156
SU-524	B777	128

Future competitive schedule

Airlines typically publish the initial schedule for distribution 330–360 days before operations and it becomes available from companies like OAG and Cirium. If the competitor's schedule has changed from the previous year, it would impact the demand and hence the airline's own forecast should be adjusted. For example, if an airline were the only carrier providing direct service in a market last year, but this year a competitor introduced a direct service as well that would likely lead to a reduction of demand. This adjustment is not straightforward, however, as the introduction of a service from a new airline also often leads to market stimulation, i.e., generation of additional demand for travel. Airlines' schedules do not stay fixed after they are published. They are continuously adjusted to better fit new demand patterns, react to changing requirements from regulators, accommodate airline's own initiatives, and satisfy many operational constraints.

These changes should be monitored, so an airline can react in a timely matter. For example, Table 1 shows how service options provided by Aeroflot (SU), Emirates (EK), and Qantas (QF) on Moscow to Dubai market for June 2019 departures changed as between March 2019 and April 2019:

In April, Emirates downgauged EK-132 from A380 to B777 and Qantas removed codeshare from that flight. This change raised expected traffic for Aeroflot flights in the same market. A revenue management system based on the airline's own information only would take several weeks to recognize the uptake in bookings, correct the forecast, and hence adjust the optimal control policy. By that time, most of the booking window would already have passed and hence the opportunity would be lost.

If an analyst is aware of such change when it is published, the analyst can adjust availability and obtain incremental revenue. However, manual monitoring of such changes is realistic only for major cases like entering a new market or an increase in the frequency of service. Subtle schedule adjustments like downgauge or departure time change are impossible to spot and analyze in a timely manner. Meanwhile, a 15 min change in departure time of a competitor's flight can create a connection in a market that airline did not serve before and hence impact demand distribution in that market. The good news is that competitors' schedule is typically already analyzed by the airline's scheduling department. The revenue management system only needs to enable additional data feed from a scheduling system and highlight the changes that might significantly impact performance for further analysis by an analyst.

Future market size

An important step of demand forecasting is modeling year-over-year trend and seasonality. Revenue Management systems currently adopted in the industry normally use a variation of a time-series forecasting technique to estimate these two factors either for each service or for the portion of a market they serve. However, if an airline only serves a small portion of the market, this approach based only on airline's own bookings will still miss general tendencies present in the market. With so much transparency provided by the internet, consumers can easily compare offers from different airlines and hence airlines need to model all potential travelers and understand their behavior. This understanding can help airlines to design optimal pricing and availability policies.

Information that helps to get a better picture of how each market behaves is available. Many consulting companies, industry analysts, and airline own strategy and network planning departments use this information to understand a market and estimate its future potential. First, historical

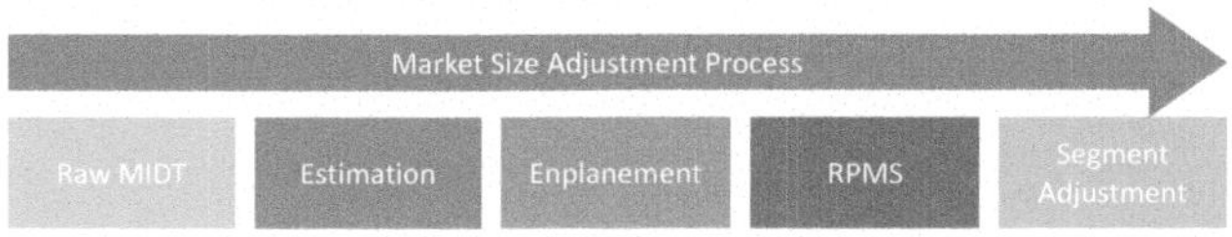

Fig. 1 Market data adjustment process

bookings from major GDS systems like Sabre, Amadeus, and Travelport are collected into a dataset called MIDT. Then various external data sources are referenced to estimate GDS presence in each market and pseudo-MIDT bookings are created to estimate the total market size. For LCC and other carriers that do not distribute through GDS, itineraries are built based on their schedules. The quality service index (QSI) of a schedule is used to assign passengers on various itineraries. Once the base OD itinerary database is created, it is adjusted with external data sources to create the final estimate of market size.

There are more than 50 external data sources that publish data at various levels that can be used to calibrate the OD passenger traffic. The data can be reported at the airline level, segment level, airport level, OD level, country level, regional level, or industry level. These data sources can also be supplemented with web-scraped data to fill in gaps. The adjustment process is a stepwise process as shown in Fig. 1.

For example, enplanement data from airports are calibrated against airline reported data to provide a realistic view of available capacity and load factors. Flight data reported by aviation authorities are matched with itinerary data contained in enhanced MIDT to make sure they are in synch. Each of these adjustments has multiple sanity checks built into the process. Any data used for calibration are checked for consistency and completeness before it is consumed. At each level of adjustment, appropriate constraints are defined and applied. For example, the sum of traffic on all itineraries going through a specific flight cannot be higher than the capacity of that flight. Floor and ceiling threshold values for adjustment factors have been determined based on analysis on a vast amount of historical data and are applied. These values are subjected to continuous evaluation and changes.

In many cases, historical information is reported monthly. Thus, the outcome of the Market Size Estimation process is a time-series representing monthly market sizes for any pair of cities in the world served by air. This time-series is generally available for a span of several years and hence can be used for predicting the total market size in the future. To illustrate this concept, we generate market size forecasts using several different techniques. All of them use historical observations of monthly traffic $y_1, \ldots, y_{T-1}$ to construct forecasts as y_{T+h}. Typically, we use history for 5 years (60 monthly data points) to predict demand for the next 3 years (36 monthly data points). For this study, we used 3000 major markets in all regions of the world. We selected an equal number of short-haul, medium-haul, and long-haul markets. We considered city-pair data from January 2010 to December 2015 as training data and January 2016 to December 2018 data as test data.

As our baseline model, we used naive forecasts. Here, we set all forecasts to be the value of the previous year's passenger count. For the second approach, we used auto.arima()function in R which combines unit root tests, minimization of the AIC, and MLE to obtain an ARIMA model forecast. Finally, as the third approach, we used time-series regression based on gradient boosting. Table 2 shows the performance of each method. We can clearly see that time-series regression based on GBM works better than ARIMA and significantly better than a naïve forecast.

Another metric that helps visualize forecast accuracy aggregated over multiple entities is the Area Under the Absolute Percentage Error (APE) curve. Figure 2 shows this metric for all three models we experimented with. Again, the GBM-based model works better as the area under the curve is more compared to the other two curves.

While the overall performance of the GBM-based regression is very good, there are still cases where forecasting algorithms perform poorly. For example, in 2017 Doha to Dubai market's passenger count dropped by a big margin due to the diplomatic crisis between UAE and Qatar. As shown in Fig. 3, forecasting the data using 2010 to 2015 for the next 3 years results in unreliable forecasting.

A special case of market size forecasting occurs when the market is stimulated with additional frequency by one of the airlines. We observed about 49,000 instances in the same 3000 markets sample of an airline starting a new direct service in a market and operating it continuously for at least 12 months. We included more features into the GBM approach, such as passenger counts in before and behind airports, flight count, number of seats, and equipment used on a new service. This allowed us to achieve a 12.5% error rate in predicting market size for these markets. For example, when Ryanair started a new service from Brussels to Budapest, they operated it 5 times a week using 194 seat aircraft and were able to significantly increase market size. Figure 4 illustrates the change and forecast produced by our model with each point corresponding to a month of operations.

Table 2 Error metrics and evaluation

Metrics	Naive forecast	ARIMA	GBM-based regression
MAPE	21.68	18.76	12.71
WMAPE	15.72	13.6	10.14
RMSPE	42.87	33.07	23.28
Slope of APE regression line	0.1815	0.172	0.1277
Area under APE curve	0.1774	0.1679	0.1235

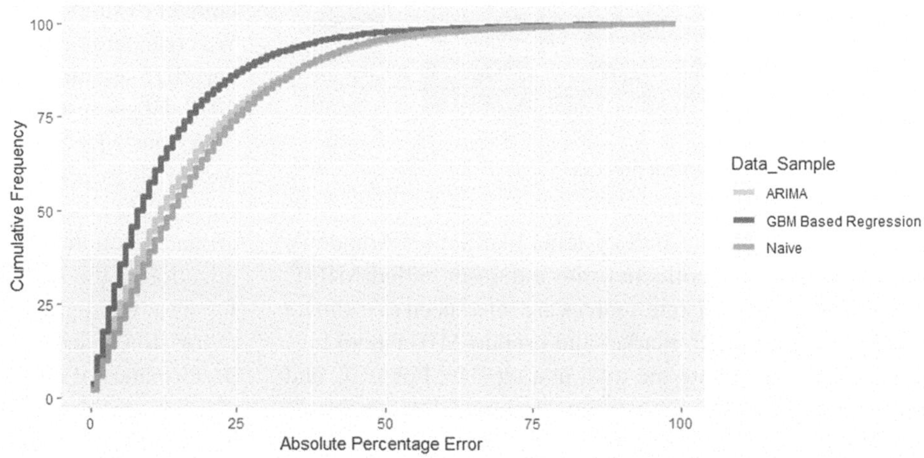

Fig. 2 Area under the absolute percentage error (APE) curve

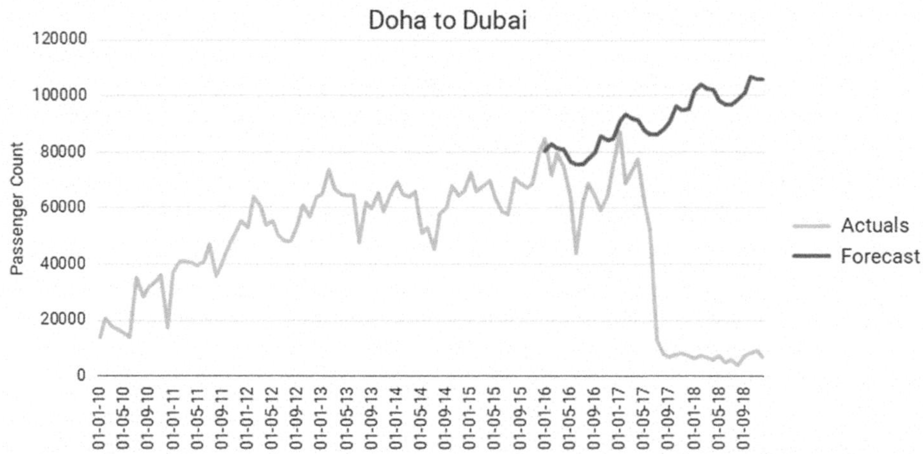

Fig. 3 Doha to Dubai market passenger traffic from 2010 till 2018

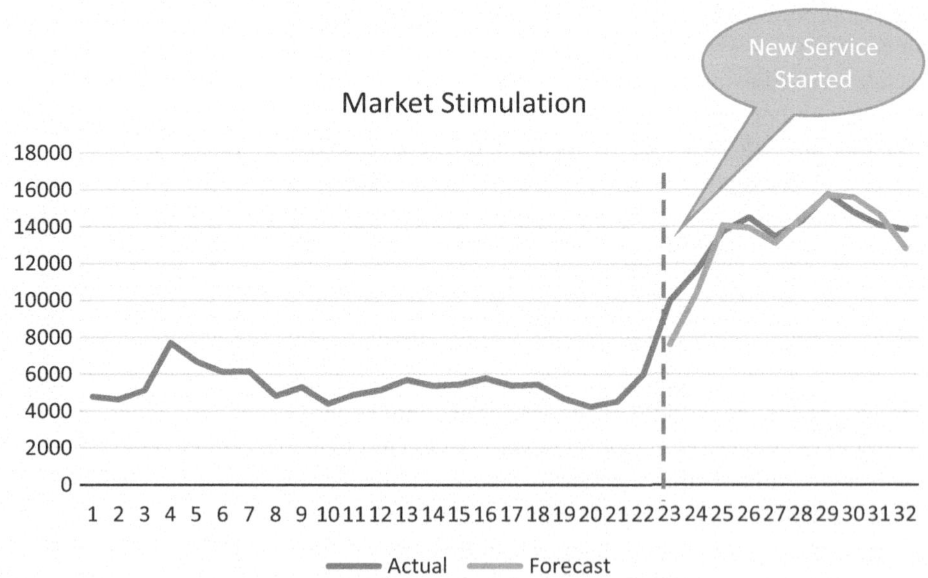

Fig. 4 Brussels to Budapest market stimulation by Ryanair

While the prediction of total market sizes described above already reaches a satisfactory accuracy level to be used in practice, there are several opportunities to further improve its performance:

- *Additional data*: the volume of travel between two cities is affected by a variety of macroeconomics, demographics, social and other factors. Training models on more data from sources like World Bank (Garrow 2016) that includes GDP, Per Capita Income, Population, Literacy rate, and other data elements can help discover hidden patterns and dependencies. Another promising direction is the use of Shopping Data that represents requests for travel submitted to Global Distribution Systems or directly to airlines. These requests indicate the true intentions of customers and hence can help to recognize unserved and underserved markets with little booking volumes.
- *Enhanced algorithms*: enhanced forecasting algorithms based on dynamic regression models and artificial neural networks might produce better results as they are able to extract hidden patterns and incorporate unrelated data sources. Hierarchical forecasting should be applicable to this problem as data often exist on different aggregation levels. Ensemble approaches seem to be promising as well since markets often behave differently and hence a combination of different algorithms would likely work better than one.
- *Better model special conditions*: often a market is influenced by special conditions and hence cannot be accurately modeled by a general approach. In addition to Market Stimulation caused by the entry of a new airline described above, examples of such conditions include the construction of a new runway, weather event with severe impact on infrastructure that requires long recovery like major hurricane or earthquake, global sport or social events like World Cup and many others. The impact of such events should be modeled through customized models that are trained on relevant data and analyze similar historic occurrences.

Revenue Management practice can use forecast for the total market size as a reference and validation point for the demand forecast it produces. For example, assume the sum of demand for all itineraries offered by an airline in a given market for a month of departures is 3000 passengers. Assume also that capacity share for the airline, i.e., number of seats over the total number of seats in the market is 20%. It is reasonable to conclude that the total market size is about 15,000 passengers per month. If that number is significantly different from the estimate computed based on industry statistics, then the additional analysis is required to understand the reasons for such difference and correct assumptions used the estimation methodologies, so they converge. Similarly, market size varies significantly in different seasons; while that pattern is not detected from the airline's own bookings, there might be inefficiencies in control policy that is not flexible enough to explore seasonal behavior and should be corrected. Finally, markets are often changing dynamically due to external reasons that form the demand for travel. Recognizing that trend early provides an opportunity to proactively benefit from these changes rather than react to them. Market size forecast constructed using information from all airlines serving the market can indicate these changes before they become visible in the airline's own bookings.

Future market share

As we mentioned above, scheduling information for most airlines is available from companies like OAG and Cirium (www.oag.com, www.cirium.com). Schedules are typically published for up to 330–360 days forward and hence can be used to predict market share for the whole booking horizon controlled by revenue management. There are two main steps required to make that prediction. First, itineraries for all markets and all airlines serving these markets need to be constructed. This is usually done by specifying the rules for legal connection between two flights. Examples of such rules include minimum and maximum connection times, if flights marketed by two different airlines can be connected, how long can the connecting trip be, compare to a straight line between origin and destination, etc. Then, the market share for each itinerary is determined based on attributes, such as departure time, elapsed time, number of stops, type of aircraft, the position of an airline in the originating city, etc. The most popular model used to calculate market share is the Multinomial Nested Logit model that is calibrated using Maximum Likelihood Estimation (Garrow 2016). Recently, American Airlines reported significant improvement in forecasting accuracy if the market share model is calibrated using Random Forest (Agrawal and Dasgupta 2019). That calibration is done by training the model to match expected market share for each itinerary in the market to actual market share obtained from historical bookings for all airlines.

Estimating market share only from the airline's own sales data is complicated (Talluri and Tekin 2018). Additional knowledge of market share and total market size discussed in the previous section are valuable to the revenue management system as they can be used to estimate the demand for an individual itinerary. That demand can be compared with the forecast produced by the revenue management system and adjustments can be made either manually or automatically. For example, if competitor airline added a new flight in a market, then demand for all existing flights would be reduced. Revenue management system demand forecast

Table 3 Share distribution on Moscow to Istanbul market

SVO-IST		VKO-IST		MOW-IST	
Airline	IST (%)	Airline	Share (%)	Airline	Share (%)
SU	92	TK	96	SU	35
LO	3	J2	3	TK	47
SV	5	LO	1	KK	10

based on historical observations will require several weeks to recognize that change and adjust, while forecast based on information for the whole market will provide that insight as soon as the competitor's schedule is published.

Another important factor a revenue management analyst should consider in the analysis of market share is if the airline operates out of a multi-airport city. An airline's own bookings might only paint a partial picture and not reveal all factors influencing customer behavior. For example, Table 3 describes how market share estimation compares on the Moscow to Istanbul market:

Both Aeroflot and Turkish Airlines revenue management and pricing departments might assume a dominant position in Moscow, based on their view of operations in respective base airports. This assumption can drive an aggressive upsell strategy resulting in a risk of losing customers to strong competition present in the same catchment area.

Conclusion

The airline industry captures a large amount of information describing the demand for travel. Recent advances in Data Management capabilities and Machine Learning algorithms allowed us to use that information for further enhancement of existing decision support systems. One such opportunity is leveraging information traditionally used Network Planning and Scheduling to improve the accuracy of demand forecasting produced by revenue management systems. Competitive schedule, market size, and market share estimations can be used to adjust demand forecast in case market behavior is changing due to competitive actions or external factors (Fig. 5).

As an additional benefit, this practice would better align business processes in Pricing and Revenue Management departments with Network Planning and Scheduling. All these departments have revenue maximization as the main objective and all of them require accurate demand forecasting to achieve it. Providing them with all available data for training models and ensuring consistency of the results they use for making decisions would ultimately lead to improved performance of the whole Commercial Planning organization.

In this paper, we discussed how information about market conditions available to the Network Planning department can help improve the Revenue Management practice. The converse is also very promising. Revenue Management has good visibility into bookings for future departures and demand elasticity with respect to price and time dimension of the booking curve. These data can be helpful in making better decisions on network planning and capacity allocation.

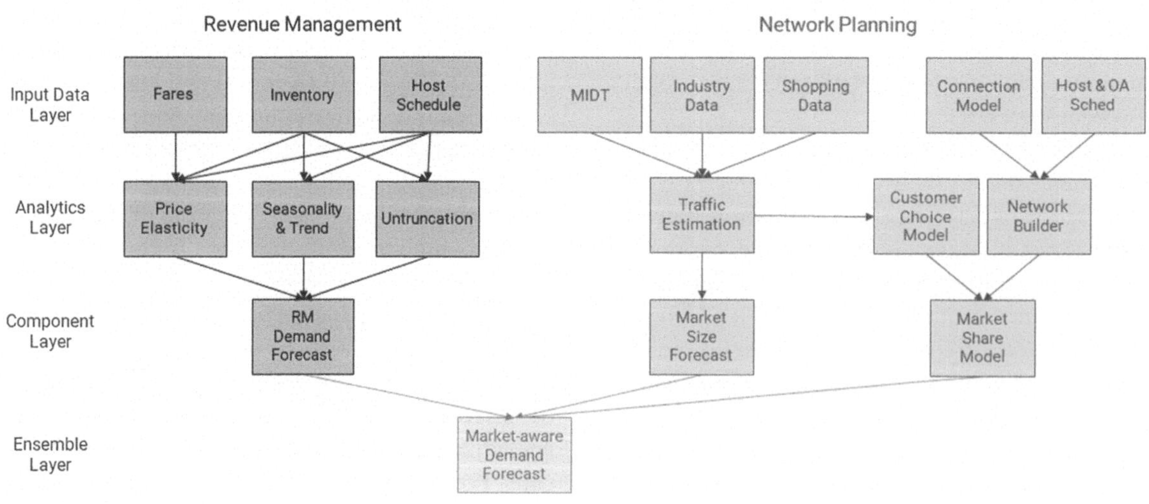

Fig. 5 Market-aware demand forecasting

Reference

Agrawal D., and J. Dasgupta. 2019. *Efficient network planning using heuristics and machine learning*. AGIFORS SSP Study Group.

Azadeh, S.S., P. Marcotte, and G. Savard. 2015. A taxonomy of demand uncensoring methods. *Journal of Revenue and Pricing Management* 13: 4402015.

Garrow, L. 2016. *Discrete choice modelling and air travel demand: Theory and applications*, 1 edition. London: Routledge; (May 23, 2016), ISBN-13: 978-0754670513, ISBN-10: 0754670511.

https://databank.worldbank.org/home.aspx

https://www.oag.com/

https://www.cirium.com/

Strauss, A.K., R. Klein, and C. Steinhardt. 2018. A review of choice-based revenue management: Theory and methods. *European Journal of Operational Research*. https://doi.org/10.1016/j.ejor.2018.01.011.

Talluri, K., and M. Tekin. 2018. *Estimating market size from sales data*. AGIFORS RM Study Group 2018

Nitin Gautam serves as Manager Operations Research at Sabre Airline Solutions. Before joining Sabre, he has worked at startups heading the data sciences team. He possesses extensive experience in Machine Learning in the domains of Airline, Text-Analytics, Retail Optimization, Advertisement, Computer Vision, etc. Earlier in his career, he headed product development teams working for Hitachi, Microsoft, etc. He earned a Masters in Artificial Intelligence with a specialization in Machine Learning from the University of Edinburgh in 2011.

Shriguru Nayak is a senior analyst in Sabre Airline Solutions Operations Research and Data Science team. Prior to joining Sabre, he worked in Living Analytics Research Centre, IBM, IIITB, and Viralmo. He holds a master's degree in Information Technology from IIIT- Bangalore.

Sergey Shebalov is Vice President of Research at Sabre. He leads a team responsible for designing and developing decision support systems for retailing, distribution, fulfillment, and operations in the travel industry. Prior to joining Sabre, he worked at Carmen Systems and United Airlines. He earned a PhD in Mathematics from University of Illinois at Urbana-Champaign in 2004.

Journal of Revenue and Pricing Management (2021) 20:351–367
https://doi.org/10.1057/s41272-021-00318-x

PRACTICE ARTICLE

Multi-layered market forecast framework for hotel revenue management by continuously learning market dynamics

Rimo Das[1] · Harshinder Chadha[1] · Somnath Banerjee[1]

Received: 10 February 2020 / Accepted: 18 August 2020 / Published online: 30 March 2021

Abstract
With the rising wave of travelers and changing market landscape, understanding marketplace dynamics in commoditized accommodations in the hotel industry has never been more important. In this research, a machine learning approach is applied to build a framework that can forecast the unconstrained and constrained market demand (aggregated and segmented) by leveraging data from disparate sources. Several machine learning algorithms are explored to learn traveler's booking patterns and the latent progression of the booking curve. This solution can be leveraged by independent hoteliers in their revenue management strategy by comparing their behavior to the market.

Keywords Revenue management · Forecast · Unconstrained demand · Constrained demand · Market · Machine learning

Introduction

The revenue management process has three core principles: pricing, revenue management, and product distribution (Vinod 2004). All three principles should be in synergy and work together in a unified manner to maximize revenue and profits. In order to harmonize these three principles and a hotel's financial goals, accurate forecasting is critical for decision making.

In the last decade, with the maturity of the internet, economic growth and shift in consumer behavior have introduced millions of new travelers. The TripAdvisor and Oxford global travel market study stated that in 2017, the travel industry is valued at 5.29 trillion USD, and the global travel market has grown 41.9% in the last 10 years (Oxford 2018).

It is essential for hoteliers to broaden their revenue strategy to respond to these dynamic changes in the hospitality industry. In economics, a market is a place where buyers and sellers come together to exchange some products or goods (Blackmar 1912). In the hotel industry, the idea of a market can be conceptualized by defining a geographical area, where sellers are an arbitrary set of hotels, and buyers are travelers looking for accommodation. Not only can a market be defined by the geographical area, but it can also be defined by a combination of hotels in the same star category having similar pricing or reservation trends.

A vast majority of legacy hospitality systems are dated and limited in their interaction capabilities, making the transfer of clean data challenging from the onset. Beyond system intelligence issues, revenue managers face the challenge of seemingly endless new streams of data being accessible ranging from market fluctuations, competitor pricing to impactful events, varying air traffic, and even minute factors like changing weather that impact demand and revenue. Managing these technical complications and balancing a near-limitless array of data points are not feasible to do manually in a timely and cost-effective way. Traditionally, revenue management forecasts are rigid which have tended to operate largely in broad strokes (business rules) or are otherwise based on conventional modeling techniques which often lack optimal predictive efficiency. These methods are primitive and inaccurate because they do not capture the exogenous factors affecting the hotel, nor are they capable of handling very large datasets that are exploding in volume and velocity. Here, a machine learning architecture is

✉ Rimo Das
rimo@lodgiq.com; rimo@outlook.com

Harshinder Chadha
harshinder@lodgiq.com

Somnath Banerjee
somnath@lodgiq.com

[1] LodgIQ, 440 N Wolfe Rd, Sunnyvale, CA 94085, USA

Chapter 12 was originally published as Das, R., Chadha, H. & Banerjee, S. Journal of Revenue and Pricing Management (2021) 20:351–367. https://doi.org/10.1057/s41272-021-00318-x.

introduced that more accurately forecasts by capturing the hidden patterns in the massive data.

Literature review

There has been a tremendous amount of research in demand forecasting in the field of revenue management. Rajopadhye et al. (2001) built an algorithm to forecast the unconstrained hotel room demand. The forecasting algorithm in the article was constructed by using two inputs: short-term and long-term forecasts. The long-term forecast uses a Holt–Winters model to forecast the room demand, and the short-term forecast uses the advanced reservations, cancelation rate, the net turndowns, and net demand booking profile. In Weatherford and Kimes (2003), the authors describe three types of forecasting methods: historical, advanced, and combined booking models (a combination of historical and advanced) for revenue management. Furthermore, the authors explain the issues in revenue management forecasting citing level of aggregation, forecast accuracy, breadth of data, and outliers. In the paper (Zakhary et al. 2009), the authors proposed a Monte Carlo simulation for forecasting hotel arrivals and occupancy. In their research, they address the revenue management forecasting issues (Weatherford and Kimes 2003) detailed by modeling the effects that seasonality, reservations, cancelations, length of stay, group reservations, and trends have on hotel arrivals and occupancy. The authors use hotel data to forecast arrivals and occupancy at an aggregated daily and weekly level (Zakhary et al. 2009). All of the research written above is focusing on demand forecasting at the property level. For this specific topic in market forecasting, there has been some research on how a market affects revenue management and tourism. In the paper (Sigala 2015), the author puts forward a "learning with the market" framework in revenue management and conducts several focus groups with revenue management professionals to ascertain the viability of a market-driven framework. Several concepts such as market dynamics, market engagement, and social media are explored to determine the impact on tourism markets (Sigala 2015).

In the paper (Cooper et al. 2006), the authors present an analysis of the underlying errors in revenue management models. The authors state that in practice, assumptions may be incorrect and model parameters are unknown. Moreover, in their research, they detail how controls based on erroneous assumptions can lead to poor forecasting results because the historical data depend on past protection levels, which inevitably leads to modeling errors. Cooper et al. (2006) proposed an iterative approach where assumptions and model parameters are updated as the observed data are collected and reflected in the forecasting-optimization process. Group forecasting accuracy has been an important area of research within demand forecasting. Kimes (1999) researched group forecasting accuracy for hotels and wrote about the challenges in group forecasting. The author found that inaccurate group forecasting has more impact during high occupancy days and noted that the transient segment is severely impacted because the willingness to pay is higher for the transient segment. Through the analysis, Kimes found accuracy can be affected by (not limited to lead time), type and size of the hotel, dependence on group business, frequency of forecasting updates, or occupancy of the hotel. This paper aims to combine demand forecasting concepts, a market-driven approach to revenue management leveraging machine learning to build a model that forecasts market demand.

Research methodology

In this section, the overall objective and approach are introduced along with the factors that impact the machine learning framework. First, multi-textured data such as market reservations (powered by Amadeus Business Intelligence), market pricing records, and events are collected from different sources. Thereafter, these data are explored to understand the characteristics of the data. Then, various factors are considered which impact a market and would be crucial for the machine learning algorithms to be trained on them. After that, a set of machine learning algorithms are used to develop the market forecast model.

Data categories

In the previous sections, it was stated that a market is defined as a collection of hotels in a geographic region. For the purpose of this research, a total of 97 hotels are chosen to illustrate the New York City market with a total capacity of 38,745 rooms. The most critical metrics considered at the market level that impact daily demand are given in Table 1:

Table 1 Market KPIs

Metric	Definition	Expression
Capacity	Total available rooms in the market	$\sum$ capacity
Occupancy	Percentage of occupied rooms in the market	$\frac{\sum \text{Rooms sold}}{\text{Capacity}}$
ADR	The average price paid per room in the market	$\frac{\text{Revenue}}{\text{Rooms sold}}$
RevPAR	The average daily rooms revenue generated per available rooms in the market	$\frac{\text{Revenue}}{\text{Capacity}}$
Revenue	Sum total revenue from the individual hotels comprising the market	$\sum$ revenue

The purpose of the market forecast model is to forecast the unconstrained and constrained demand on a daily basis, in a defined market (aggregated and segmented).

Seasonality

Butler (1998) defines seasonality as "the temporal imbalance in the phenomenon of tourism" and that it can be determined by various factors such as the number of visitors, expenditure of visitors, transportation disturbances, and attractions. In "The Measurement of Seasonality and Its Economic Impacts'' (Bar-On 1999), the author describes seasonality as a phenomenon that can be defined as effects occurring each year at a similar time span and magnitude, and these effects can be but are not limited to climate patterns, festivals, events, and local holidays. Figure 1 displays the occupancy (number of visitors who visited the New York City Market/capacity of the market) during the years, 2017 and 2018. From the graph, the cyclical nature of the occupancy is an example of seasonality.

For a deeper view of the seasonality, Figs. 2 and 3 display the average occupancy by month and day of the week in the New York City market during 2017 and 2018. From Fig. 2, January and February are the softest periods, and the occupancy progressively climbs in the following months. Similarly, in Fig. 3, the occupancy on Sunday and Monday is significantly lower compared to the rest of the week where higher demand can be exhibited. Even though these visitor trends are specific to the New York City market, the seasonality effects can be observed for any defined markets.

Market segments

Segmentation is defined as "the process of dividing consumers into distinct groups with similar sets of needs and wants to require the development of marketing mixes by the hotel" (Ivanov 2014). Identifying the right customer is a very important aspect of revenue management. To that end, hotel revenue managers must identify the booking patterns of a potential segment and then determine if the desired pricing strategy can achieve the hotel's revenue goals.

At the market segmentation level, market data (powered by Amadeus Business Intelligence) are broadly classified into the following market segments:

1. Discount—customers from advanced purchase, Online Travel Agencies (OTAs), promotional/package rates, and frequent guest program redemptions.
2. Retail—customers from rack rates, best available rates, non-discounted, non-affiliated, and non-contracted customer demand.
3. Negotiated—corporate negotiated and consortia rates.
4. Qualified—customers from qualifying affiliation discounts, such as AAA, senior discounts, and employees.

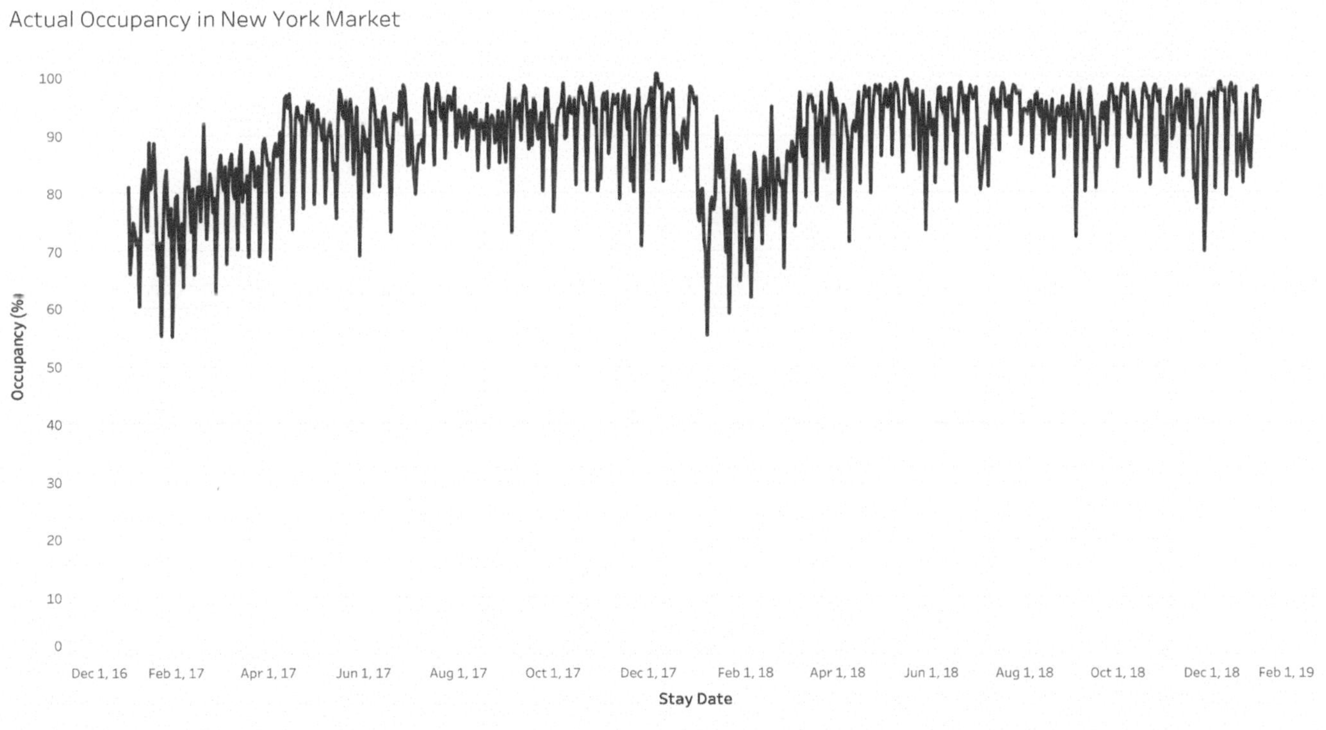

Fig. 1 Actual occupancy from 2017 and 2018

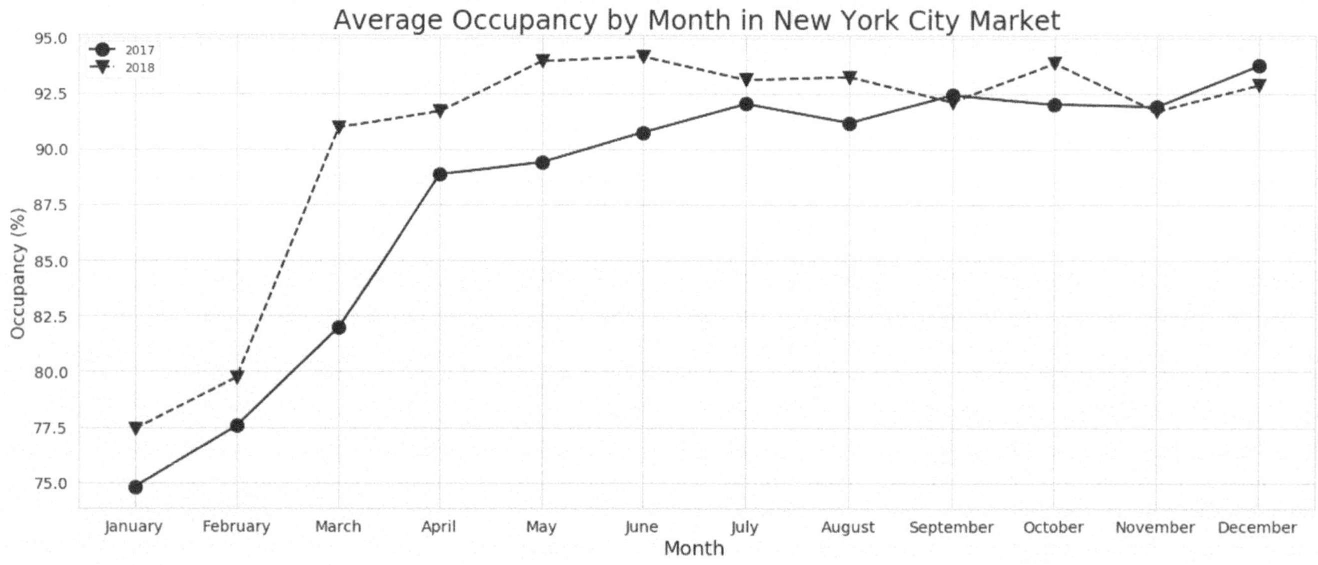

Fig. 2 Monthly seasonality

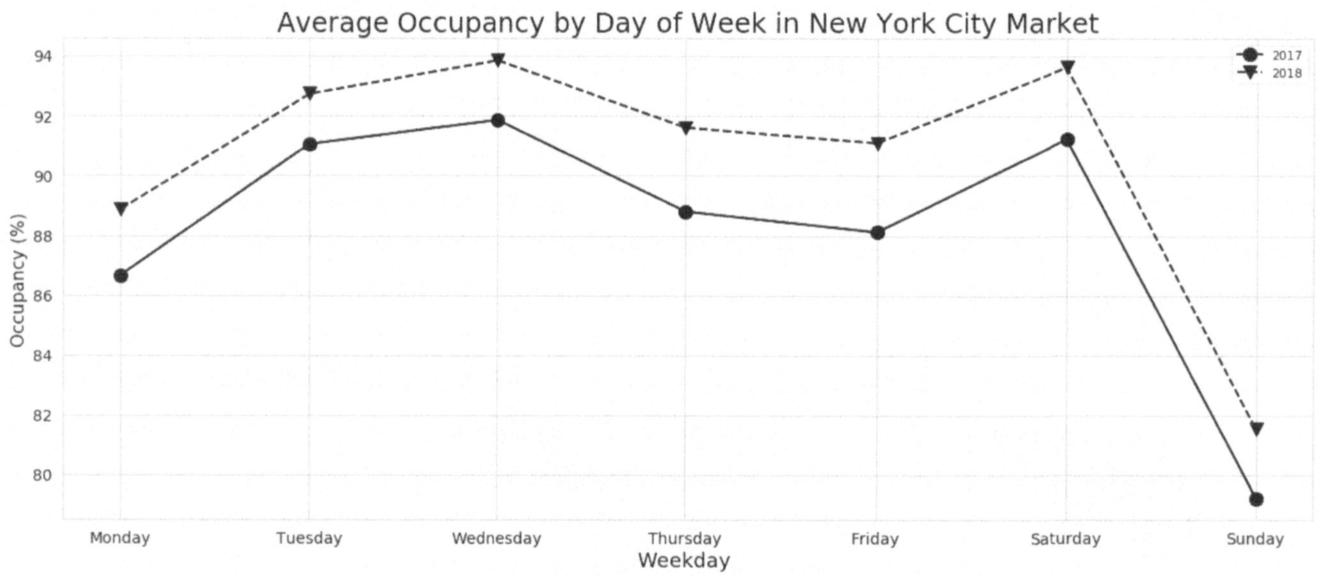

Fig. 3 Weekday seasonality

5. Wholesale—customers from wholesalers, consolidators, and tour operators.
6. Group—customers from group tours, domestic and international groups, association, convention, and corporate groups.

In Fig. 4, a plot of the booking pace can be seen of all these segments.

From Fig. 4, the booking trends for the various market segments can be seen. The Discount and Retail segments have an inflection point where the Discount segment overtakes the Retail segment.

Group segment

Among the several market segments, the Group segment is a highly differentiated segment as the trends are varied from the regular demand. The Group market segment can be split into two distinct categories: **Group-Sold** and **Group-Committed** segments. The Group-Committed segment

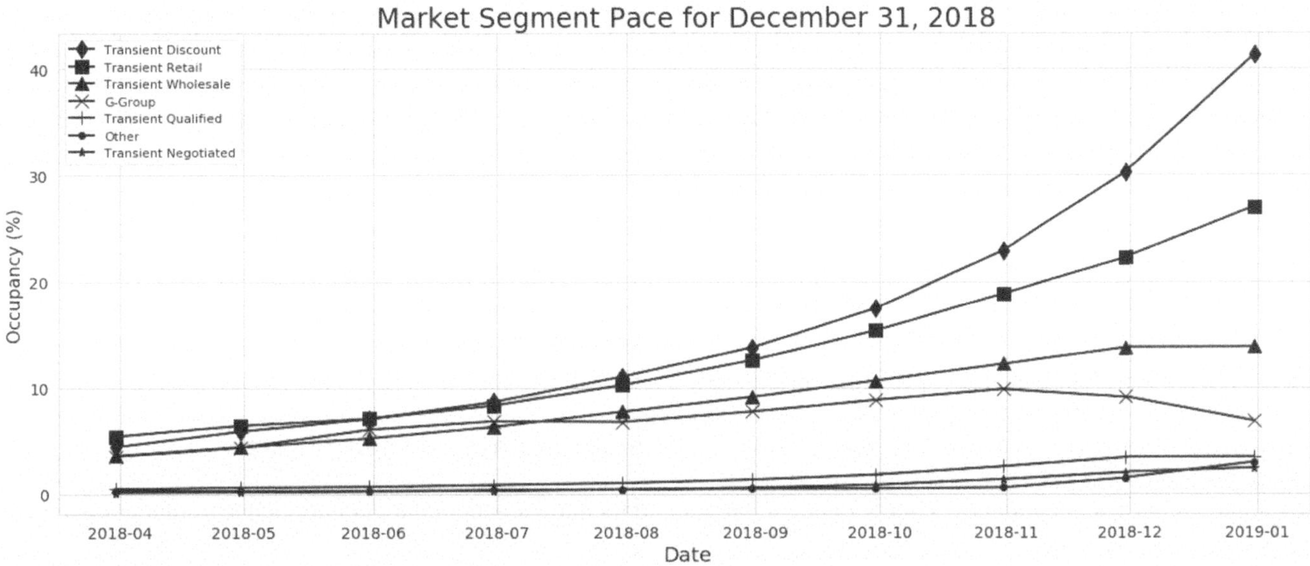

Fig. 4 Market segmented pace for 12/31/18

reflects customers who have signed a contract with the intent to commit to a block of rooms and have officially paid the terms of the contract. The Group-Sold segment reflects the customers who have reserved rooms but have not signed a contract or paid for the rooms. As the segments get closer to the lead time, the demand from the Group-Sold segment transfers over to the Group-Committed segment.

The difference between the final occupancy from Group-Sold and the Group-Committed blocks is known as the group wash. From Fig. 5, it can be seen that as we move closer to the stay date, the demand block moves from sold segment to committed, while there remains an unused demand block or the group wash. These characteristics make that the segment behave a little more erratically in comparison to the other market segments. Moreover, the effect of seasonality causes a multiplicative effect.

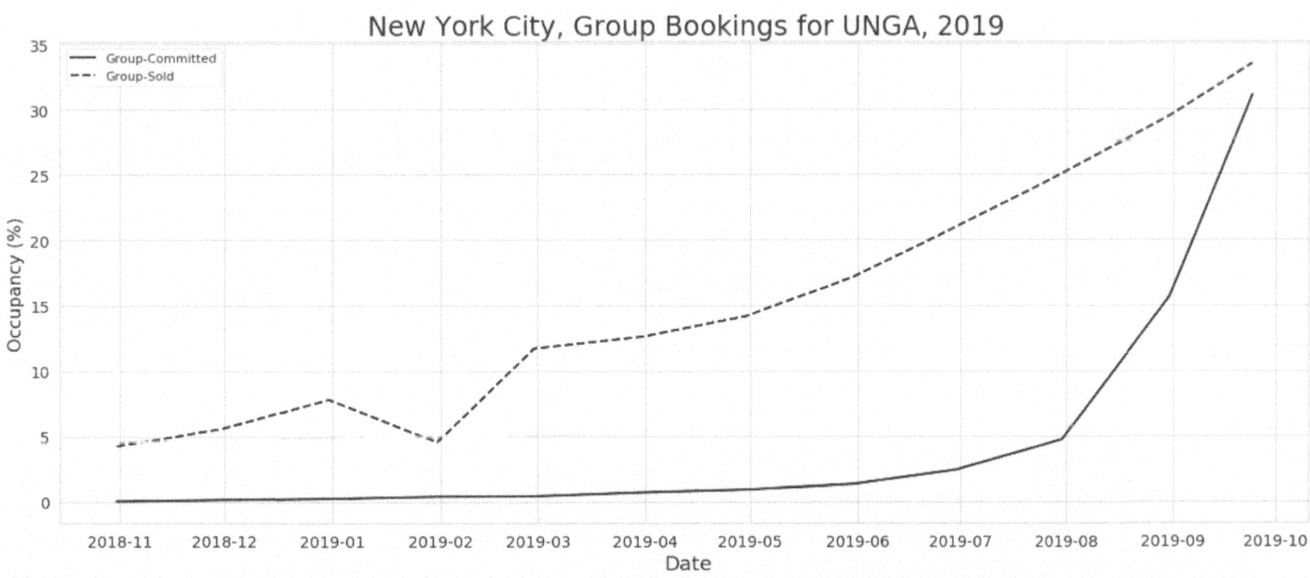

Fig. 5 Group booking pace for 24/09/19

Holidays and events

In tourism and the lodging industry, in particular, seasonality has been the focus of several studies (Butler 1998; Bar-On 1999). In fact, the term "holiday," within this context of seasonality in tourism, has typically assumed a period spanning several days, called "Shoulder Days", (even weeks or months) compared to the shorter (single day) application in finance and economics (Demicco et al. 2006).

Holidays introduce a year-to-year shift in arrivals when more than a month is involved. This occurs when travel related to the event or holiday takes place during the days immediately before and/or after the holiday and these days belong to the month prior to or just after the month of the holiday itself (Bureau of Transportation Statistics). Holidays in the US such as Labors Day, Easter, Memorial Day, Thanksgiving, and Christmas also introduce year-to-year variation when they affect more than one month (Bureau of Transportation Statistics).

From Fig. 6, the number of travelers shows a decrease from the Tuesday (November 20, 2018) before Thanksgiving Day (November 22, 2018) and remains high through Thanksgiving Day and Black Friday (November 23, 2018), and falls on Sunday (November 25, 2018). The days before Thanksgiving Day often fall in November while days after that occur in December. Thanksgiving Day, therefore, has the potential to impact both November and December (Fig. 7).

Events also potentially impact more than one month, because it may occur in one month this year but in another month the next year (DreamForce at San Francisco, for instance). Moreover, there is also a change of locations of events, (SuperBowl, for instance) because it may occur in one location this year but in another location the next year. These kinds of fluctuations should be included in the model to control for the year-to-year variation which local holidays and events induce in seasonal movements.

During these days, the market demand has a higher deviation than the other normal days. This is critical as they are typically interested from a business perspective, to tailor strategies structured around the emerging opportunities (or lack thereof).

Span of data and sample size

The multi-textured data are divided into two groups. Figure 8 shows data separation for training and testing of the machine learning models (Fig. 9).

Machine learning

In this research, several machine learning models are utilized that combine historical market-level data and build features that deal with seasonality and events to more precisely forecast demand. Bootstrap Aggregate methods and Gradient Boosting machines are utilized to achieve improved forecasting performance and model interpretability. This work paved the way to interpret how the machine learning models work and determining what features would impact the forecast. A complete survey of the

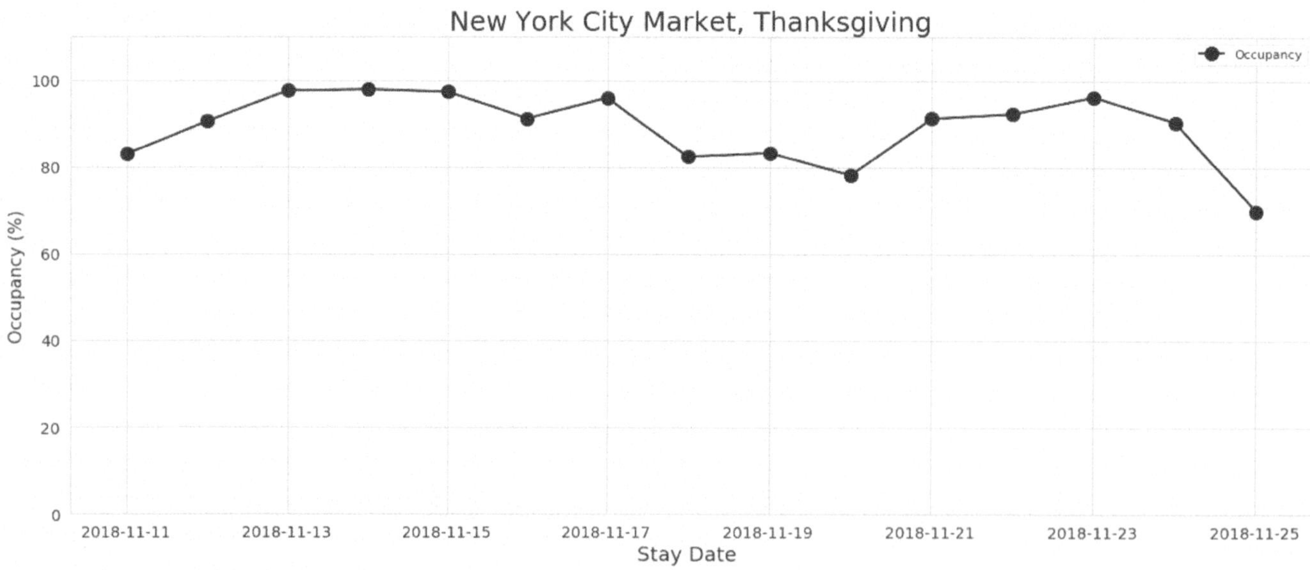

Fig. 6 Thanksgiving 2018

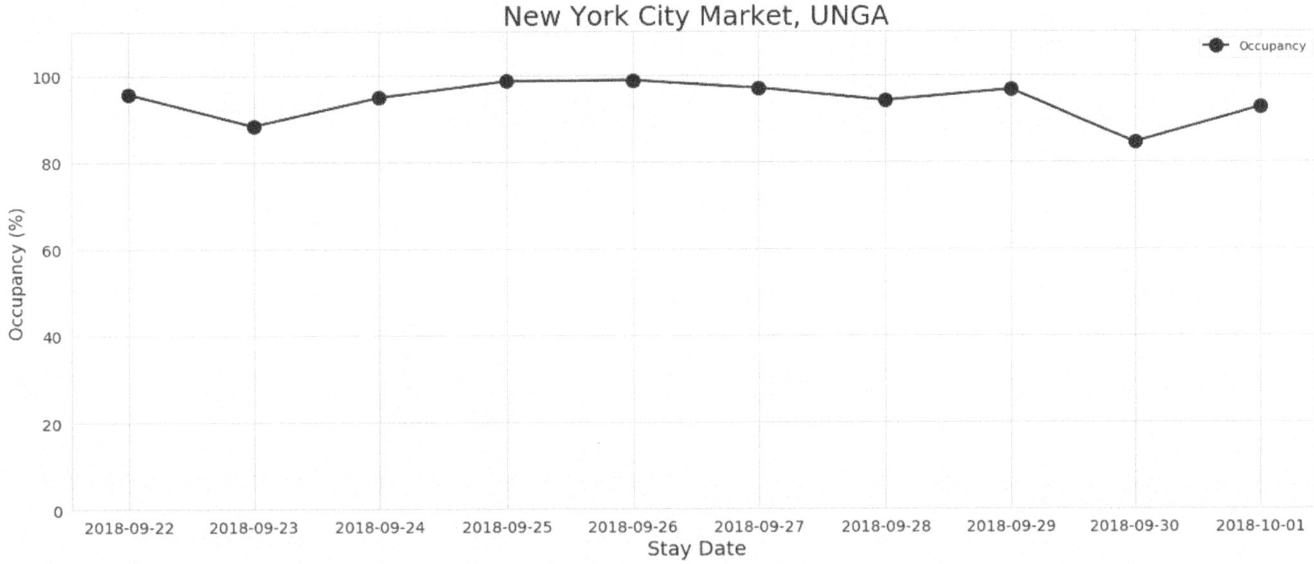

Fig. 7 United nations general assembly 2018

Fig. 8 Training and test set

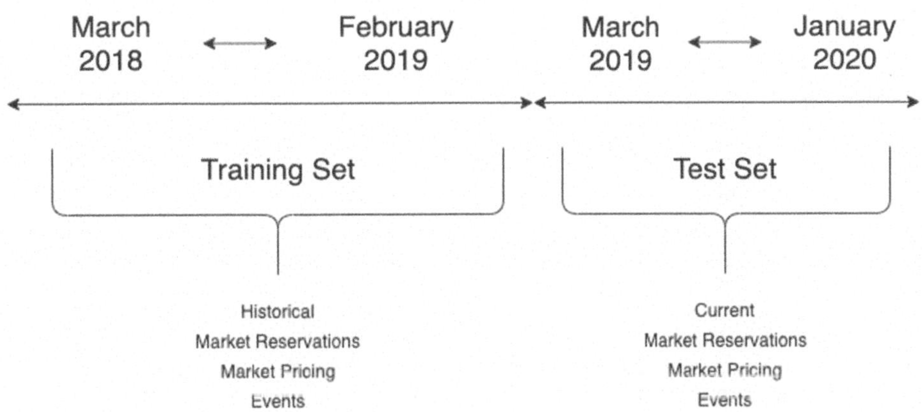

topic "model interpretability" is beyond the scope of this paper (please refer to SHAP (Lundberg and Lee 2017) for an overview).

Bootstrap aggregating

Bootstrap Aggregating (Breiman 1997) involves building each model in the ensemble using a randomly drawn sample of the data. The predictions are made by averaging the predictions of each decision tree. This method of random sampling elements of the predictor space adds more diversity which reduces the variance of the trees at the cost of equal or higher bias. Random Forest is the main representative of bagging (Amit et al. 1997) in which prediction by an ensemble of trees is more accurate than that of any individual tree.

GradientBoost

The idea behind 'boosting' in machine learning is to train weak learners in a gradual, additive, and sequential manner, each trying to improve over its prior learner. This method fits the new predictor of the residual errors made by the previous predictor for every incorrect observation at every iteration. Gradient Boosting Decision Tree (Friedman 2001) uses Gradient Descent to find the shortcomings in the previous learner's predictions. The boosting algorithms used in this research are as follows:

1. XGBoost (Chen et al. 2016).
2. LightGBM (Ke et al. 2020).
3. CatBoost (Dorogush et al. 2018).

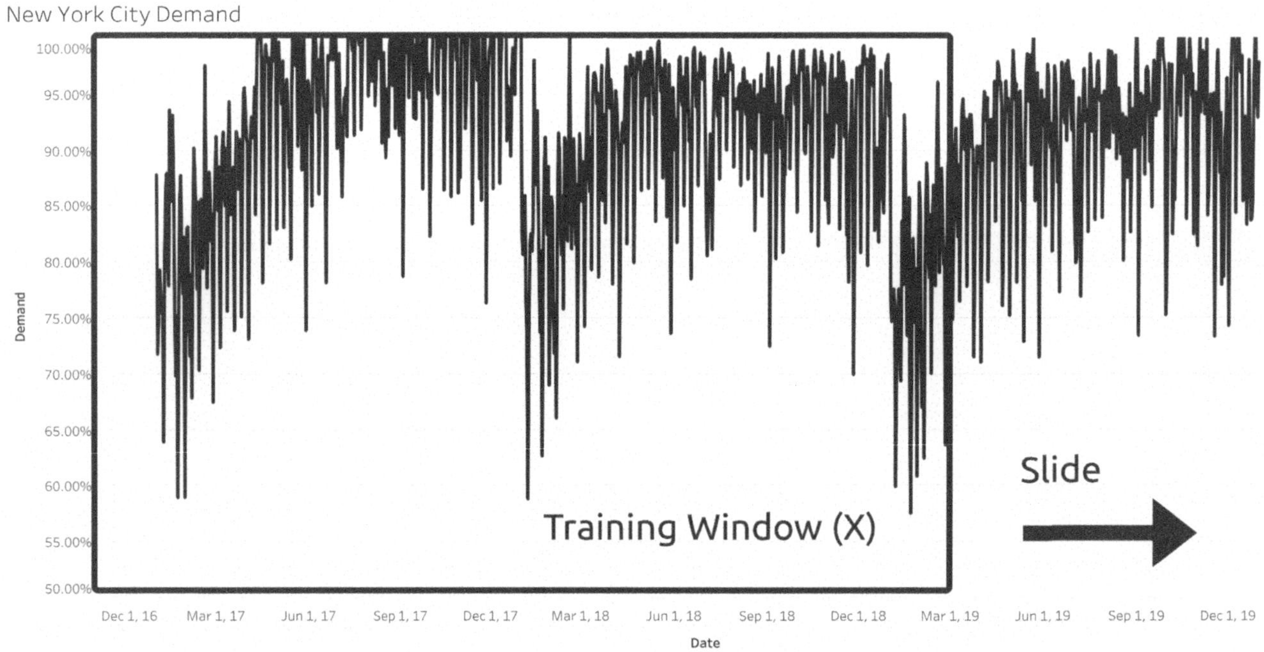

Fig. 9 Training data with sliding window

Stacking

Stacking (Breiman 1996a, b) is an ensemble machine learning modeling technique where multiple unique models are created, and it uses their output as the input for the meta-classifier/regressor to make a prediction. Unlike bagging and boosting, stacking may be (and normally is) used to combine models of different types. One of the advantages of stacking is that it leverages the strength of each individual estimator by using their output as the input of a final estimator. This reduces the bias any individual model might have.

Feature engineering

As mentioned in previous sections, there are several factors that affect market demand: market reservations, seasonality, events, market pricing. In this section, raw data are transformed to extract meaningful features which directly influence our model.

The lead time or days before arrival (DBA) is the time between the date of engagement/reservation and the reservation check-in date. In this research, DBA is divided into groups (DBA buckets) of fixed width. The DBA buckets are formed according to the assumption that the booking pace in these buckets follows the same trend (DBA Bucket: 0–3, 4–7, 8–14, 15–30, 31–45, 46–60, 61–90, 91–365).

A binary feature is added which handles events data. If there is an event, then the variable indicates 1, otherwise it indicates 0. Also, as the shoulder dates are adjacent to event dates, these dates are dynamically estimated based on the event type. These dates are then designated as 1.

Several features are used to account for seasonality patterns. Due to periodic nature, a series of sine and cosine functions are employed in which frequencies are integer multiples of the fundamental seasonal frequency.

$$\begin{aligned} \mathrm{WS}_{\cos} &= \cos\frac{nD\pi}{7} \\ \mathrm{WS}_{\sin} &= \sin\frac{nD\pi}{7} \\ \mathrm{MS}_{\cos} &= \cos\frac{nM\pi}{12} \\ \mathrm{WS}_{\sin} &= \sin\frac{nM\pi}{12} \end{aligned},$$

where *WS* and *MS* are defined as the weekday and monthly seasonality, respectively. n is an integer multiple that can either be 2 or 4. D is the numerical representation of the day of the week and is defined by the interval [0,6] where 0 and 6 correspond to Monday and Sunday, respectively. M is the numerical representation of the month of the year and is defined by the interval [1,12] where 1 and 12 correspond to January and December, respectively.

Moreover, demand categories are created by sorting data values into equal bins of the market segmented and overall actual market occupancy over historical data time periods.

$$\mathrm{demand}_{\mathrm{buckets}} = \begin{cases} 1, & \min(X_t) \le x < \min(X_t) + w \\ 2, & \min(X_t) + w \le x < \min(X_t) + 2w \\ \dots & \\ N, & x \ge \min(X_t) + (N-1)w \end{cases}$$

where w is the width of an interval. X_t is an actual occupancy array over a specified time period. For example, this could correspond but not limited to the actual occupancy weekly, monthly, or quarterly level. N is the number ins. x represents an element in the array X_t. The piecewise function represents what category the element x belongs to with 1 representing the weakest demand category and N representing the strongest demand category.

Market prices are determined by taking the median of the lowest OTA prices for a single-day LOS grouped by hotels star rating in a market. Length of stay (LOS) is defined as the number of room nights a guest stays in a hotel. Online travel agencies (OTA) are travel websites selling travel products to consumers.

$$price(star) = [\min(OTA_{star,1}, \ldots, \min(OTA_{star,n})]$$
$$MP = median(price(star)),$$

where *star* represents the hotel star rating of interest. OTA represents the listed price offerings from a hotel in the channel. The variable n represents the number of hotels that comprise the market. MP represents the median of the lowest OTA prices for a given hotel star rating. Features such as moving averages and pickup (which is a subtraction based on a rolling window) are created. Pickup in the hospitality industry is the net change of a quantity during a defined time span. The moving average is calculated as below:

$$\mathrm{MA} = \frac{\mathrm{MP}_1 + \mathrm{MP}_2 + \mathrm{MP}_3 + \cdots + \mathrm{MP}_n}{n},$$

where MP is the market price, n is the number of days, and MA is the moving average.

Another technical indicator used is plus–minus which gives the number of days the market price has increased or decreased over a rolling window. Mathematically,

$$\Delta\mathrm{MP} = \frac{\mathrm{MP}_i - \mathrm{MP}_{i-1}}{\mathrm{MP}_{i-1}},$$

$$sign(i) = \begin{cases} -1, & \Delta\mathrm{MP} < 0 \\ 1, & \Delta\mathrm{MP} \geq 0 \end{cases},$$

$$plus\text{–}minus(i, n) = \sum_{t=i}^{n} sign(i),$$

where i and n are the specified date range. ΔMP is the percentage change in the market price. "plus–minus" is the sum of sign changes over a span of dates. With these transformed pricing features, the underlying trends of the market pricing can be uncovered.

Table 2 has information on the possible feature set used in the market forecast model.

The entire feature set in Table 2 is not necessarily needed for a market. Hundreds and thousands of markets can be defined with unique hotels—each with its own defining characteristics and features. With that in mind, each market can have a different set of features. This can also be described as global and local feature sets. Local features belonging to a distinct market can be determined by a multi-step process. The first step involves using a Recursive Feature Elimination algorithm. To begin with, the target variable (demand) is first trained on the whole feature set, and the importance of each feature is obtained. Then, the least important features are removed from the current set of features, and the regression metric is checked again. This gives the first set of features that can be removed with cross-validation. The next step is implementing a permutation importance algorithm to rank the features. This helps in removing the ones with negative or small importance.

Table 2 The feature set for modeling

Category	Sub-category	Frequency
Reservations	Rooms Unsold group rooms Cancelations Pickup Moving average	Daily
Market segmentation	Group Retail Discount etc.	N/A
Day of the week	Monday Tuesday etc.	N/A
Month of the year	Jan Feb March etc.	N/A
Day of the week seasonality	$\cos\frac{2D\pi}{7}$ $\sin\frac{2D\pi}{7}$	N/A
Month of the year seasonality	$\cos\frac{2M\pi}{12}$ $\sin\frac{2M\pi}{12}$	N/A
Lead time	DBA DBA bucket	Daily
Demand	High Medium Low	Daily
Average daily rate	Market segment rate Market rate Moving average Pickup	Daily
Market pricing	Market rate Moving average Plus–minus	Daily
Events	Yes No	Daily
Flights	Flight arrivals Flight departures	Daily

Unconstrained and constrained demand

In the lodging industry, unconstrained and constrained demand are significant pieces of information for a hotel. Unconstrained demand in a market is the total demand interested in lodging in a particular market and is irrespective of the capacity in that market. In practicality, the capacity cannot be exceeded which makes it necessary to constrain the total demand to the capacity of the property. The market forecaster forecasts the unconstrained demand by the market segment regardless of the supply of rooms in the market. In addition, the forecasts are price agnostic and do not include day rooms or oversold rooms. The forecasted unconstrained demand is the demand based on the arrival time and the current lead time. Thereafter, the forecasted unconstrained demand is constrained by the constraining equation given below to compute the demand based on the supply of the market. Thereafter, the demand is aggregated by segment which is total unconstrained and constrained forecasts.

$$\begin{aligned} d_{uncons} &= \text{model output} \\ D_{uncons} &= \sum d_{uncons} \\ k &= \frac{1}{D_{uncons}} \\ d_{cons} &= kd_{uncons} \\ D_{cons} &= \sum d_{cons} \end{aligned},$$

where d_{uncons} represents the forecasted unconstrained segment demand. D_{uncons} represents the aggregate of d_{uncons}. The variable k from represents the constraint constant which is the inverse of D_{uncons}. The variable d_{cons} represents the segmented constrained demand which is the product of the constraint constant and unconstrained segment demand. D_{cons} is the total constrained demand and the aggregation of the d_{cons}.

Modeling

The goal of the market forecasting framework is to reduce latency, scale both horizontally (markets), and vertically (dynamic feature selection). The arrival process (in-flow of demand) of each lead date for each market segment is stochastic in nature. Therefore, the distribution of bookings over the lead time of each date for each segment is considered in the training model. Moreover, data streams change over time leading to the change of underlying (static) model relationships and parameters in the data between the features and the demand variable. For example, there are perpetual changes of event dates, conference locations, and traveler preferences are always evolving. Accordingly, such changes need to be taken into account to update the model parameters and maintain high performance. To address this issue, sliding window X (training data) is drifted by a single increment to generate training data on receiving the latest data stream. This helps in preserving dynamic training instances positioned within the window and discarding the past instances that minimize the generalization error.

Using these defined time windows, the machine learning model is trained by minimizing the loss function such as mean absolute percentage error (MAPE). This loss function is further segregated by lead-time buckets (0–3, 4–7, 8–14, 15–30, 31–45, 46–60, 61–90, 91–365 days).

Results and discussions

In this section, the results of the various machine learning algorithms are compared and analyzed to measure the performance of each forecast model. Each model was trained on two years of historical segmented and aggregate market reservations and actual data, starting from March 1, 2017, to February 28, 2019. For model evaluation, a test set with segmented and aggregate market reservations and actual occupancy from March 1, 2019, to September 30, 2019, was used. Each algorithm's performance is evaluated by using MAPE based on one day, monthly, and aggregate error reports. The MAPE of each model is based at the aggregate level, which is the total constrained demand, D_{cons}.

As new information pours in daily, the market forecast model runs, and a new set of predictions are made. Consequently, measuring the accuracy of a forecast on one particular day is not an accurate approach, but rather, the whole set of forecasts must be considered for the arrival date.

Five machine learning regression algorithms and a single time series model are used for building the market forecast model: SARIMAX, LightGBM, Catboost, Random Forest, XGBoost, and Stacked (Stacking). SARIMAX is a Seasonal Autoregressive Integrated Moving Average model that supports exogenous variables. SARIMAX is used as the baseline model to provide the required metric of comparison. Three machine learning algorithms are leveraged to build the Stacked Regressor, viz. the LightGBM and Catboost Regressors were used as the inputs to the XGBoost Regressor. The MAPE was calculated by comparing the one-year forecast based on the date of the forecast and the respective actual occupancy. For example, a forecast made on Jan 1st, 2019 will have predictions of the occupancy for Jan 1st, 2019 to Dec 31, 2019. Subsequently, the actual occupancy for each date is compared to the set of predictions and the MAPE is computed. In the following table, a forecast was made as of March 1, 2019 for the 365 days. Subsequently, the actual occupancy of the market was used to determine the MAPE. In Table 3, a summary of each algorithm's error can be seen.

After calculating the MAPE for each algorithm, the top three performing machine learning models were the LightGBM, Random Forest, and Stacked models. Each of them shows an improvement of at least 2% from the baseline time series model, SARIMAX. The top three performing models are evaluated further in this section.

In a market-level forecast context, to better understand the performance of the models, the error is tracked yearly, monthly as well as at daily levels, especially, for event days. The error uses the aggregate actual and constrained forecasted demand, D_{cons}. For these three models, one-day error reports were generated for specific event days in the New York City market. The one-day error report can be formulated by the grouping the forecast for a particular date in days before arrival (DBA) buckets and calculating the MAPE for each of these DBA buckets. The selected days were Easter, Memorial Day, Independence Day, and Labor Day. Figures 10, 11, 12, and 13 provide the daily accuracy tracker for these days, respectively. The daily accuracy tracker measures the total accuracy of the model. The predictions are tracked at each lead time and then compared against the actual occupancy of the market. Thereafter, the MAPE would be calculated by grouping the lead time into buckets (DBA Bucket: 0–3, 4–7, 8–14, 15–30, 31–45, 46–60, 61–90, 91–365).

$$L = j - i + 1$$

$$MAPE(D,i,j) = \frac{100}{j} \sum_{t=i}^{j} \left| \frac{A(D) - F_t(D)}{A(D)} \right|$$

$$Curve = [(dba_{0,3}, MAPE(Date, 0.3)), \ldots, (dba_{91,365}, MAPE(Date, 91, 365))]$$

D can be specified as a singular date or a range of dates. A and F are the actual and forecasted values, respectively. The variables i and j represent the DBA interval, where $j > i$. L is the span of the DBA bucket. The *curve* is the list of ordered pairs representing the accuracy curve. This provides a view of how the forecast performed as the lead time progressed to the arrival date. For example, to track the accuracy of May 27, 2019, the forecast at each DBA (lead time) leading up to the arrival date would be compared with the actual occupancy (say, 70%). The forecasts from Stacked in the DBA (0–3) bucket are 73.93%, 74.75%, and 75.78%. Using the equations above the MAPE would be 6.73%.

Table 3 Model comparison

Model	MAPE (%)
SARIMAX	5.20
Random forest	3.77
CatBoost	4.20
LightGBM	3.66
Stacked	3.77
XGBoost	3.95

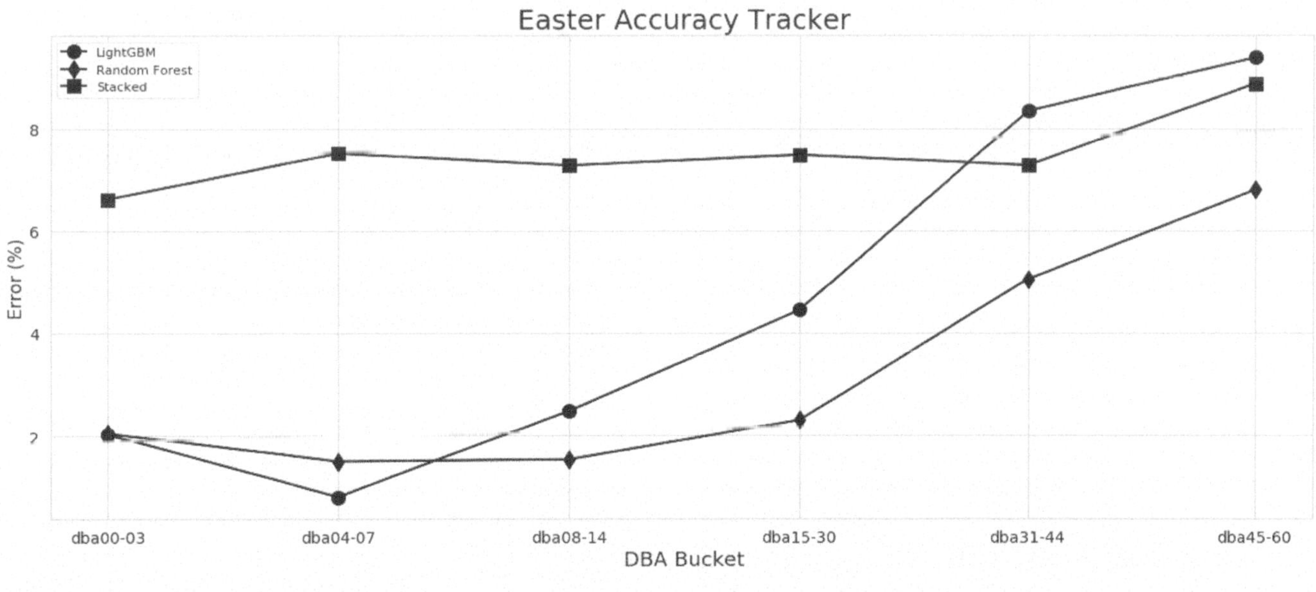

Fig. 10 1-Day accuracy report for Easter (04/21/19) in the New York City market

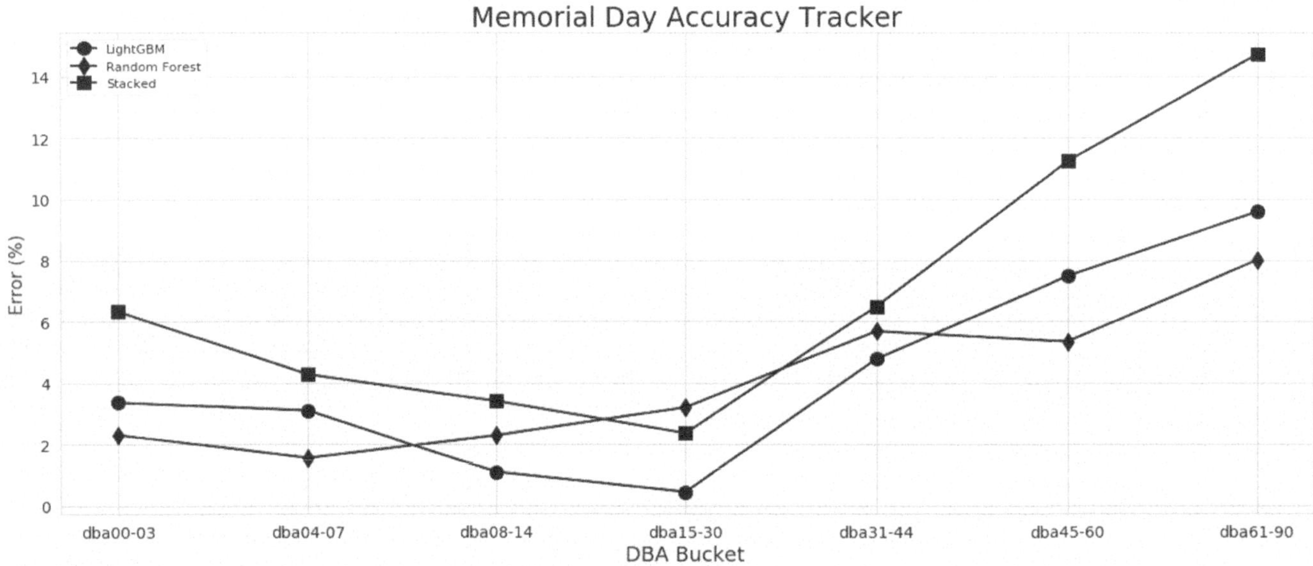

Fig. 11 1-Day accuracy report for Memorial Day (05/27/19) in the New York City market

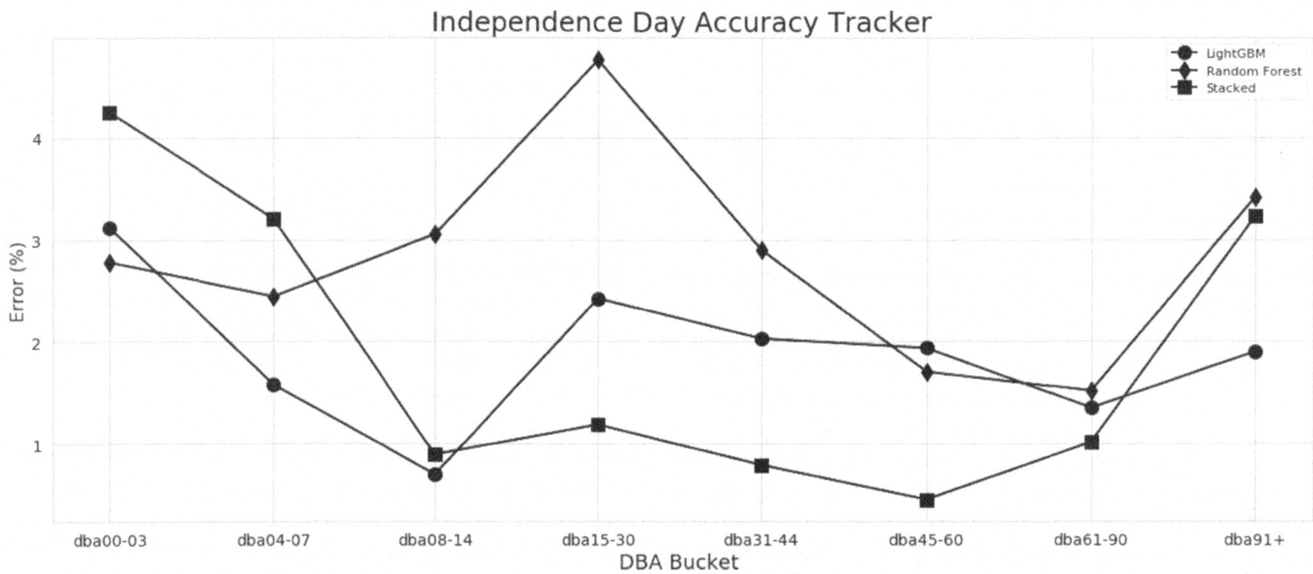

Fig. 12 1-Day accuracy report for Independence Day (07/04/19) in the New York City market

Table 4 presents the MAPE and percentage of Over/Under-forecasted dates. The MAPE is calculated by comparing the actual occupancy with all of the forecasts by lead time. The percentage of Over/Under-forecasted dates are determined by summing the negative and positive errors by the lead time for each model and stay date. From Table 4, it can be observed that the three models under-forecasted for a majority of lead dates in three out of the four event dates. Moreover, the stacked model does not perform well in comparison with the other boosting models in terms of model performance. On the other hand, the performance of the other two models is comparable with one outperforming the other in some dates and vice versa.

Table 5 outlines the comparison of the best and worst forecast errors (absolute) for each model. The worst and best forecast dates are determined by searching for the highest and lowest error for all stay dates grouped by lead time. While LightGBM and Stacked models are able to give the best performance at 65 days before arrival (DBA), Random Forest achieves the same performance at only 17 days before arrival (DBA). Furthermore, the worst forecast performance for the Stacked model is at 3 months before

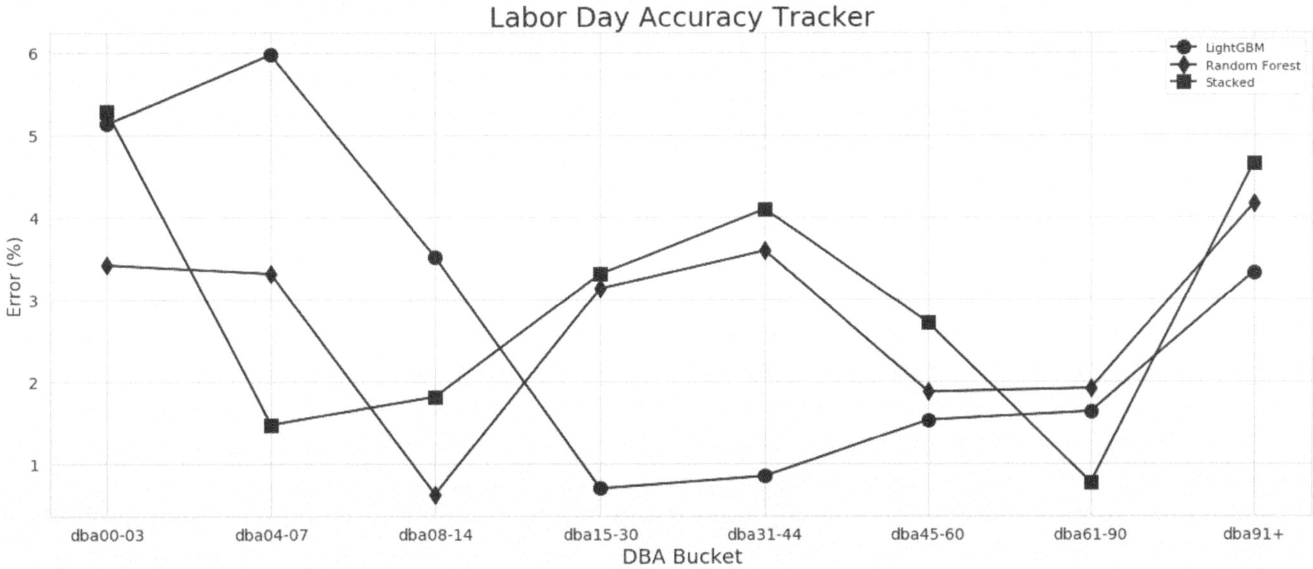

Fig. 13 1-Day accuracy report for Labor Day (09/02/19) in the New York City market

Table 4 Event day error

Stay Date	Model	MAPE (%)	% of dates Over-forecasted	% of dates Under-forecasted
April 21, 2019	LightGBM	5.43	7.69	92.31
	Random forest	3.46	7.69	92.31
	Stacked	7.52	0	100
May 27, 2019	LightGBM	5.52	23.86	76.14
	Random forest	5.27	7.95	92.05
	Stacked	8.78	0	100
July 4, 2019	LightGBM	1.81	60.32	39.68
	Random forest	2.79	95.24	4.76
	Stacked	1.72	75.40	24.60
September 2, 2019	LightGBM	2.59	37.63	62.37
	Random forest	3.30	9.14	90.86
	Stacked	3.54	16.13	83.87

Table 5 Comparison of best and worst forecast error

Model type	Error category	Stay date	Absolute difference (%)	DBA
LightGBM	Best	8/16/2019	0	65
	Worst	7/1/2019	13.72	122
Random forest	Best	9/2/2019	0	17
	Worst	9/1/2019	14.07	184
Stacked	Best	8/16/2019	0	65
	Worst	5/27/2019	14.13	87

arrival, while LightGBM and Random Forest forecasts the worst 4 and 6 months before arrival, respectively.

For a more holistic view of model performance, the monthly accuracy tracker provides a perspective of how the models performed. The monthly MAPE is calculated by comparing the actual occupancy of the entire month with the forecasts by lead time.

Table 6 presents these results from March to September. The Stacked model has the highest under-forecasting percentage for all months while the LightGBM and Random Forest models have a mix of over and under-forecasting.

Table 6 Monthly error comparison

Month	Model	MAPE (%)	% of dates Over-forecasted	% of dates Under-forecasted
March	LightGBM	1.97	78.83	21.17
	Random forest	2.16	85.89	14.11
	Stacked	3.69	13.10	86.90
April	LightGBM	1.99	52.33	47.67
	Random forest	2.13	68.10	31.90
	Stacked	2.99	17.28	82.72
May	LightGBM	2.34	52.03	47.97
	Random Forest	2.56	48.60	51.40
	Stacked	2.81	20.23	79.77
June	LightGBM	1.90	34.20	65.80
	Random forest	2.12	33.09	66.91
	Stacked	2.24	14.45	85.55
July	LightGBM	2.90	37.94	62.06
	Random Forest	2.96	42.57	57.43
	Stacked	3.05	14.70	85.30
August	LightGBM	2.65	43.41	56.59
	Random forest	2.50	45.43	54.57
	Stacked	2.45	31.55	68.45
September	LightGBM	3.22	63.51	36.49
	Random forest	3.49	50.66	49.34
	Stacked	3.05	46.67	53.33

Table 7 Aggregate model comparison

Model	MAPE (%)
LightGBM	2.65
Random forest	2.76
Stacked	2.78

Moreover, the LightGBM model has the lowest MAPE for five out of the seven months ranging from 1.90 to 3.22%.

At last, model performance is presented in an aggregate view. Table 7 presents a model comparison of the aggregate MAPE for all lead dates leading to stay dates.

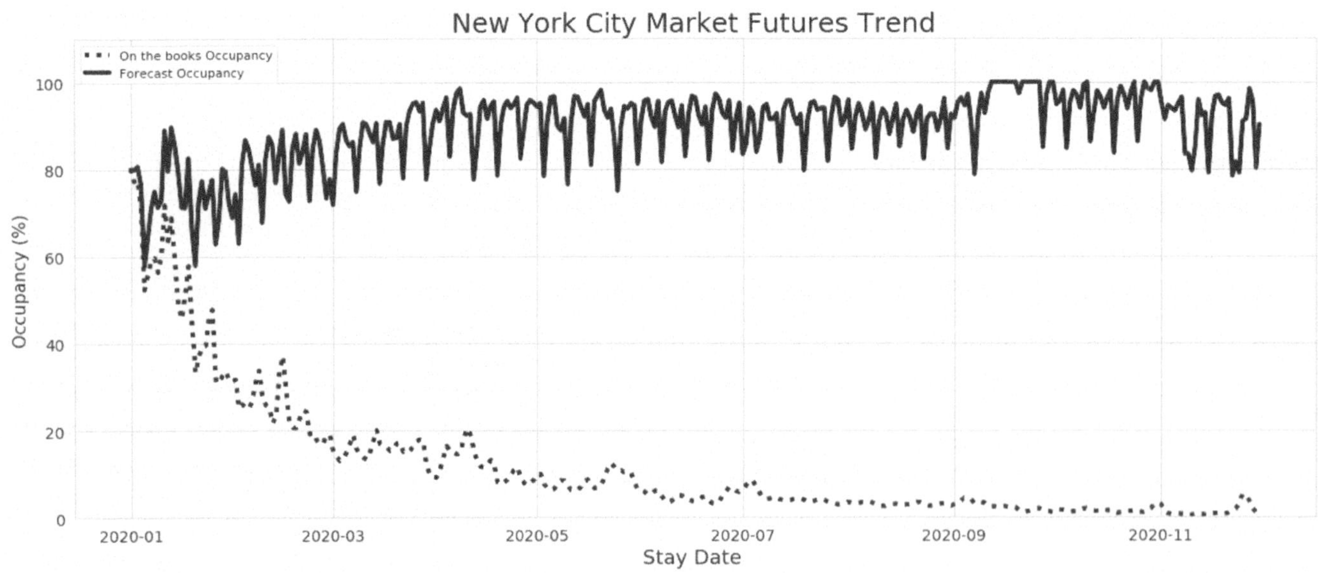

Fig. 14 Forecast as of Jan 1, 2020

To summarize the results and discussions above, the LightGBM tends to perform better when compared to other models. This indicates that the LightGBM-based market forecast model can be useful to a hotel revenue manager and it instills a satisfactory level of trust.

Forecast Interpretation

Market forecast architecture (Fig. 14) provides valuable insights for 365 days into the future that offers visibility into market performance. This information is crucial for hoteliers to anticipate dynamic changes in the market and formulate strategies to tackle those conditions. The forecast measures provide an understanding of the market booking data and pace to compare against their hotel booking movements. They would be able to have knowledge of the market intelligence on future dates for each stay date and accordingly formulate a strategy to complement and/or outpace the market. With booking lead times statistics, hoteliers would be able to make better decisions around the availability of rooms.

All these performance measures are needed both at an aggregate level and also at a segmented level to facilitate prescriptive actions at a particular segment level. Moreover, they would also have the knowledge of market pricing categorized by hotel rating and, thus, strategize hotel pricing movements. Finally, the algorithm also provides information on events' so that the revenue managers can take early action. Therefore, with the forecast market demand for each stay date in the future, hoteliers would be well equipped to keep prices up to date.

As an illustration, when the market demand is high (major conferences, concerts, festivals, etc.), the hotel would want to increase the prices, put stay controls, and also at the same time adjust their distribution strategy to facilitate maximum revenue during those nights. Therefore, in the New York City market, on high-demand nights like United Nations General Assembly Debate (September 22, 2020) (Fig. 15) or Black Friday (November 27, 2020) (Fig. 16), travelers usually book their trips beforehand (i.e., at much higher days before arrival) compared to average market demand nights and hoteliers would want to set their pricing strategy early enough to capture the demand.

On the other hand, during low market demand, the hotel would want to switch over its focus on increasing its occupancy through price and promotions. For example, low demand nights like Mother's Day (10th May 2020) (Fig. 17), in order to drive occupancy, the hotelier would want to have a different pricing strategy.

Conclusion

Unlike previous works, this research focused on having the knowledge of the market, market segments, and what affects consumers' patterns of travel by lead time. To account for accuracy, scalability, and performance, the most fitting way to maximize the market forecasting accuracy is to leverage machine learning methods for forecasting. This research determined the most appropriate machine learning algorithm to forecast market demand (both aggregated and segmented) and dealt with the seasonality and events of a market. The LightGBM model provided the best error rate with

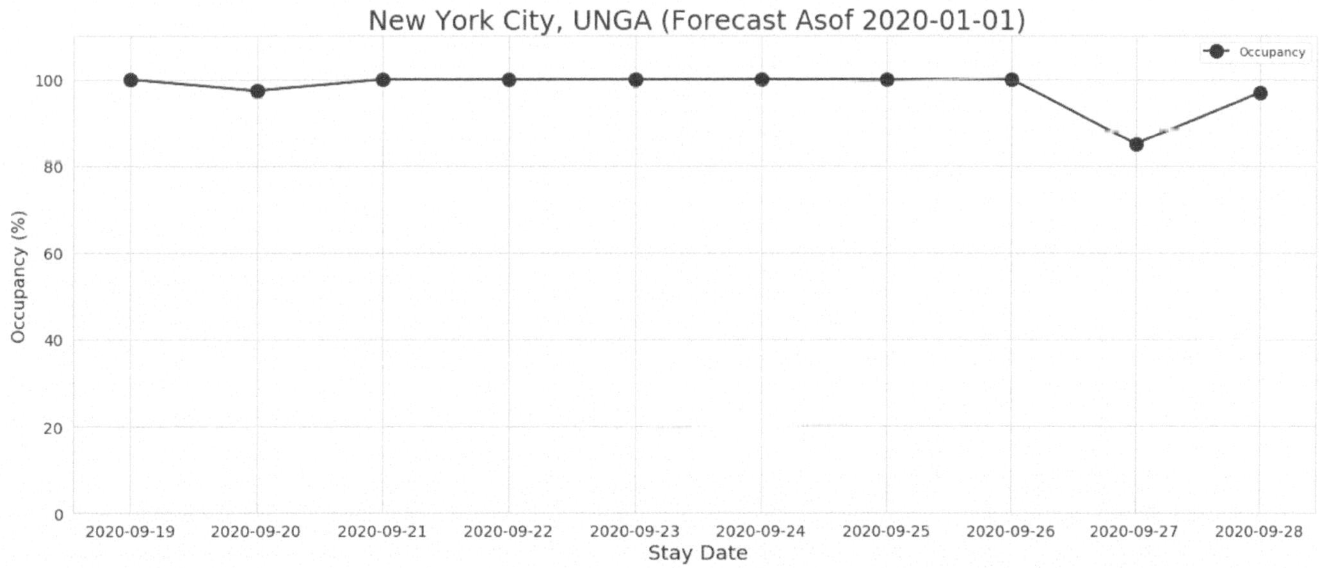

Fig. 15 (Pre-COVID-19) UNGA 2020 forecast

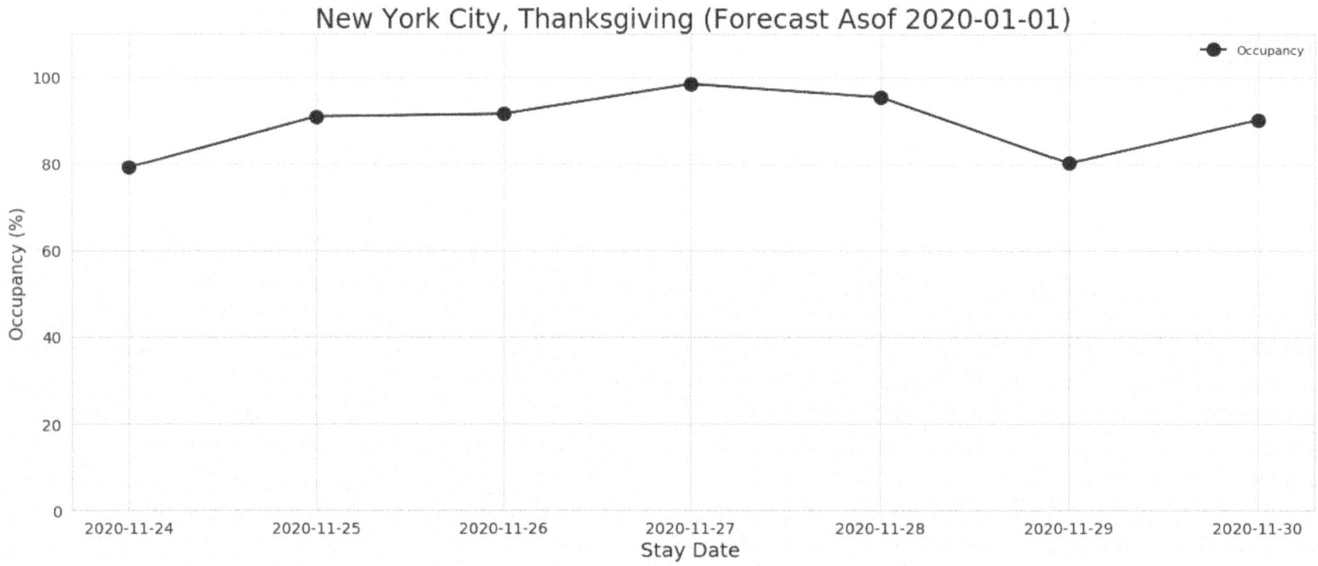

Fig. 16 (Pre-COVID-19) Thanksgiving 2020 forecast

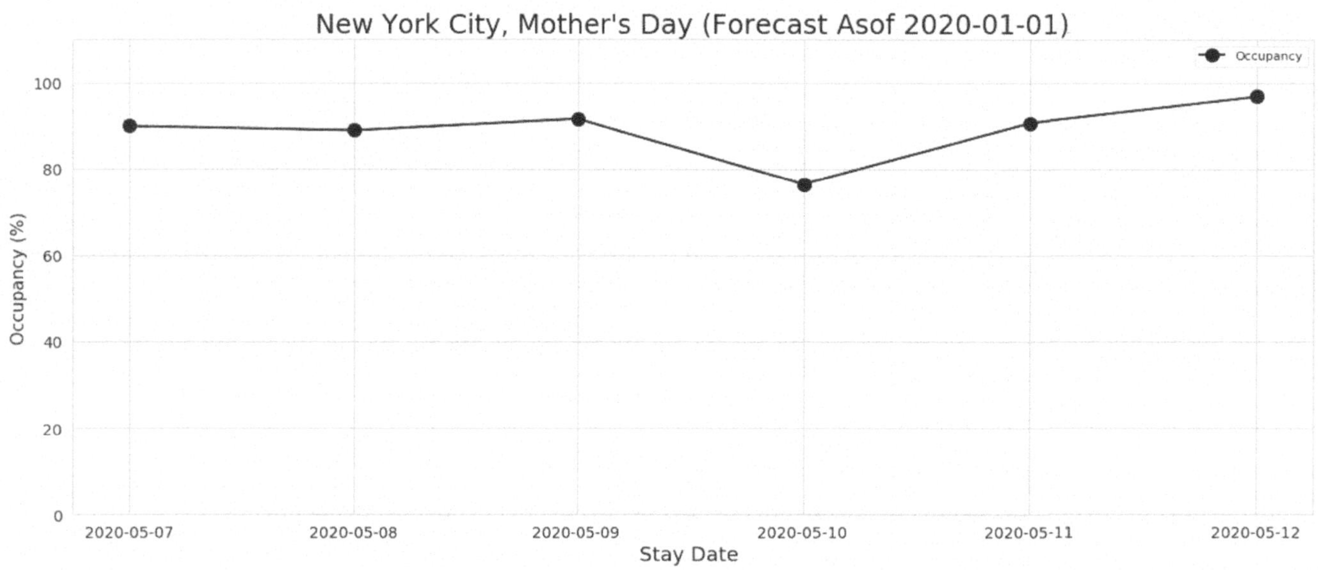

Fig. 17 (pre-COVID-19) Mother's Day 2020 forecast

a MAPE of 2.65% at an aggregated market-level compared to other models and traditional time series model (SARIMAX). Accordingly, a forecasting framework is engineered which reduces latency, can scale both horizontally and vertically, can handle model decay, and supports real-time data processing.

However, the research is subject to a few limitations like the availability of data and, more importantly, data integrity. Proper care has been taken in data pre-processing, and the results are reliable and accurate that the machine learning models perform very well in comparison to traditional time series models, like SARIMAX. However, there might be a few instances where the market might not possess enough data for machine learning models to be deployed. Under such cases, the concept of "progressive enhancement and graceful degradation" is proposed. Depending on the data availability, an applicable number of features are created, and a machine learning model is deployed. For instance, if only market pricing records are available, but no market reservation records, then demand is forecasted based on market pricing (recall, the concept of price demand elasticity).

Future research should also include a study of a combination of machine learning and statistical modeling to check what, if any, additional forecast accuracy can be achieved and also make it more scalable. Moreover, a study of the amount of historical data necessary to train the machine learning forecast model would be helpful. Secondly, a market is also affected by unexpected weather events like hurricanes, a pandemic like COVID-19, or cancelation of a big event like Mobile World Congress, 2020, Facebook Developer Conference, 2020. Therefore, future research may apply the concept of "now-casting" to deal with such uncertainties and adjust the forecasted demand pattern.

References

Amit, Y., S. Bauer, R. Kohavi, L. Breiman, T. K. Ho, J. Kleinberg, Y. Freund, P.L. Bartlett, and J.C. Lee. 1997. Random Forests. Machine Learning. Kluwer Academic Publishers, January 1, 1997. https://link.springer.com/article/10.1023/A:1010933404324.

Bar-On, Raphael Raymond. 1999. The Measurement of Seasonality and Its Economic Impacts. *Tourism Economics* 5 (4): 43758. https://doi.org/10.1177/135481669900500409.

Blackmar, Frank W. 1912. *Economics*. New York: Macmillan Co.

Breiman, Leo. 1996. Bagging Predictors. *Machine Learning* 24 (2): 12340. https://doi.org/10.1007/bf00058655.

Breiman, Leo. 1996. Stacked Regressions. *Machine Learning* 24 (1): 4964. https://doi.org/10.1007/bf00117832.

Butler, Richard. 1998. Seasonality in Tourism: Issues and Implications. *The Tourist Review* 53 (3): 1824. https://doi.org/10.1108/eb058278.

Chen, Tianqi, and Carlos Guestrin. XGBoost. Proceedings of the 22nd ACM SIGKDD International Conference on Knowledge Discovery and Data Mining - KDD 16, 2016. https://doi.org/https://doi.org/10.1145/2939672.2939785.

Cooper, W. L., T. Homem-de-Mello, and A. J. Kleywegt. 2006. Models of the spiral-down effect in revenue management. *Operations Research* 54 (5): 968–987. https://doi.org/10.1287/opre.1060.0304.

Demicco, Fred J., Yan Lin, Ling Liu, Ldia Rejt, Srikanth Beldona, and Diccon Bancroft. 2006. The Effect of Holidays on Hotel Daily Revenue. *Journal of Hospitality & Tourism Research* 30 (1): 11733. https://doi.org/10.1177/1096348005284489.

Dorogush, Anna Veronika, Vasily Ershov, and Gulin Andrey. 2018. CatBoost: Gradient Boosting with Categorical Features Support. arXiv.org, October 24, 2018. https://arxiv.org/abs/1810.11363v1.

Friedman, Jerome H. 2001. Greedy Function Approximation: A Gradient Boosting Machine. *The Annals of Statistics* 29 (5): 11891232. https://doi.org/10.1214/aos/1013203451.

General. Seasonally Adjusting Vehicle Miles Traveled—Bureau of Transportation Statistics. Accessed February 27, 2020. http://www.bts.gov/explore-topics-and-geography/topics/seasonally-adjusting-vehicle-miles-traveled.

Global FocusNorth America. How the Sharing Economy Is Transforming the Short-Term Rental Industry. Knowledge@Wharton. Accessed February 27, 2020. https://knowledge.wharton.upenn.edu/article/short-term-rentals-the-transformation-in-real-estate-and-travel-set-to-check-in/.

Ivanov, Stanislav. 2014. *Hotel Revenue Management: From Theory to Practice*. Zangador Ltd.

Ke, Guolin, Qi Meng, Thomas Finley, Taifeng Wang, Wei Chen, Weidong Ma, Qiwei Ye, and Tie-Yan Liu. LightGBM: A Highly Efficient Gradient Boosting Decision Tree. Accessed February 27, 2020. https://papers.nips.cc/paper/6907-lightgbm-a-highly-efficient- gradient-boosting-decision-tree.pdf.

Kimes, S.E. 1999. Group forecasting accuracy in hotels. *Journal of the Operational Research Society* 50 (11): 1104–1110. https://doi.org/10.1057/palgrave.jors.2600770.

Lundberg, Scott, and Su-In Lee. 2017. A Unified Approach to Interpreting Model Predictions. arXiv.org, November 25, 2017. https://arxiv.org/abs/1705.07874.

Oxford Economics Global Travel Market Study. TripAdvisor Insights, October 25, 2018. http://www.tripadvisor.com/TripAdvisorInsights/w2841.

Rajopadhye, Mihir, Mounir Ben Ghalia, Paul P. Wang, Timothy Baker, and Craig V. Eister. 2001. Forecasting Uncertain Hotel Room Demand. *Information Sciences* 132 (1–4): 111. https://doi.org/10.1016/s0020-0255(00)00082-7.

Sigala, Marianna. 2015. From Demand Elasticity to Market Plasticity: A Market Approach for Developing Revenue Management Strategies in Tourism. *Journal of Travel & Tourism Marketing* 32 (7): 81234. https://doi.org/10.1080/10548408.2015.1063801.

Vinod, Ben. 2004. Unlocking the Value of Revenue Management in the Hotel Industry. *Journal of Revenue and Pricing Management* 3 (2): 17890. https://doi.org/10.1057/palgrave.rpm.5170105.

Weatherford, Larry R., and Sheryl E. Kimes. 2003. A Comparison of Forecasting Methods for Hotel Revenue Management. *International Journal of Forecasting* 19 (3): 40115. https://doi.org/10.1016/s0169-2070(02)00011-0.

Zakhary, Athanasius, Amir F. Atiya, Hisham El-Shishiny, and Neamat E. Gayar. 2009. Forecasting Hotel Arrivals and Occupancy Using Monte Carlo Simulation. *Journal of Revenue and Pricing Management* 10 (4): 34466. https://doi.org/10.1057/rpm.2009.42.

Rimo Das serves as a Data Scientist at LodgIQ, where he is working on dynamic pricing, demand forecasting, and uncertainty estimation. He earned an MS in Business Analytics from University of California, Irvine, and an MBA from Symbiosis International University with a concentration in marketing.

Harshinder Chadha serves as a Data Scientist at LodgIQ, where he is working on demand forecasting and pricing optimization. He earned a BS in Mechanical Engineering from San Jose State University.

Somnath Banerjee serves as the CTO of LodgIQ, an AI/ML-based revenue management platform. He started his career at Texas Instruments Research and later worked at Paul Allen's Interval Research. He founded and operated a multinational technology company for 8 years and led its successful acquisition. He is also very passionate about cryptocurrencies, algorithmic trading, and digital art. He earned a Bachelor of Technology in Electronics and Communication Engineering from IIT, Kharagpur and an MS in Computer Science from Southern Methodist University.

Journal of Revenue and Pricing Management (2021) 20:368–375
https://doi.org/10.1057/s41272-021-00319-w

PRACTICE ARTICLE

Artificial Intelligence in travel

B. Vinod[1]

Received: 20 February 2020 / Accepted: 18 July 2020 / Published online: 17 March 2021

Abstract

Over the past four decades Operations Research (OR) has played a key role in solving complex problems in airline planning and operations. Over the past decade Artificial Intelligence (AI) has seen a rapid growth in adoption across a range of industry verticals such as automotive, telecommunications, aerospace, and health care. It has been acknowledged that while adoption of AI in the travel industry has been slow, the potential incremental value is high. This paper discusses the role of AI and a range of applications in travel to support revenue growth and customer satisfaction.

Keywords Travel · Artificial Intelligence · Machine learning · Robotic process automation · Cognitive insight · Cognitive engagement · Adoption · Experimentation

Overview

Airline deregulation in 1979 resulted in an explosion of scheduled services and air fares. This led to the creation of new value propositions to solve business problems involving millions of decision variables using operations research techniques (Cook 1998; Vinod 2014). Related competencies such as advanced statistical modeling and computer science algorithms also support the creation of these new solutions to generate revenues or reduce costs that lead to operation efficiencies in flight scheduling, reservations, air shopping, retailing, airline pricing, revenue management, crew planning, airline operations, flight planning, staff planning, cargo and customer service.

While Operations Research came into existence to solve logistical problems during the Second World War (1939–1945), the concept of Artificial Intelligence came into existence in 1950 when Alan Turing, wrote his landmark paper "Computing Machinery and Intelligence" (Turing 1950). This paper attempts to answer the question "Can machines think?" This is what led to the Turing Test, which is a test of a machine's ability to exhibit intelligent behavior that was comparable to a human. Some researchers now propose the Winograd Schema Test as an alternative (Levesque 2011). The term "Artificial Intelligence" was coined five years later, in 1955, by John McCarthy, a math professor at Dartmouth College, who developed LISP, the second oldest high-level programming language (after Fortran) and influenced the development of ALGOL.

Artificial Intelligence as a discipline is extremely broad and embraces the concept of machines being able to carry our tasks in a way that we would consider smart. Most applications of AI are within the domains of Natural Language Processing, Machine Learning and Deep Learning, though Reinforcement Learning is increasing in popularity (Hao 2019a). Machine Learning is all about giving machines access to data so that they can learn over time, adapt, and refine predictive capabilities over time. Machine learning also represents a fundamentally different approach to creating software. The machine learns from examples, rather than being explicitly programmed for a specific outcome.

In travel, AI is in the early stages of adoption, unlike industrial verticals such as automotive, life sciences, telecommunications and education where significant advances have been made (Chui et al. 2018). For instance, AI holds much promise in education and China is pioneering AI-enabled personalized tutoring for students as an alternative to human tutoring (Hao 2019b) which could reshape education as we know it today. Nevertheless, the promise of AI in travel is to overcome complexity and simplify the travel experience.

In the early days, adoption of AI was slow, and it is only in the last decade that various industry verticals, including

✉ B. Vinod
benvinod@yahoo.com

[1] Charter and Go, 2804 Crestridge Ct, Grapevine, TX 76051, United States

Chapter 13 was originally published as Vinod, B. Journal of Revenue and Pricing Management (2021) 20:368–375. https://doi.org/10.1057/s41272-021-00319-w.

travel, have been adopting AI. However, McKinsey & Company (Chui et al. 2018) state that while the potential incremental value for AI in travel is high, it lags other industry verticals such as Electronics/Semiconductors, Aerospace and Defense, Automotive, Healthcare Systems, etc.

Technology hype cycle

An interesting model that explains the progress of Artificial Intelligence is the hype cycle by Gartner (Goasduff 2019). Gartner provides a graphic representation of the maturity, adoption, and trends of new technologies in the AI sector.

Trough of Disillusionment: Interest wanes as experiments and implementations fail to deliver.

Slope of Enlightenment: More instances of how the technology can benefit the enterprise start to crystallize and become more widely understood.

Plateau of Productivity: Mainstream adoption starts to take off. Criteria for assessing provider viability are more clearly defined. The technology's broad market applicability and relevance are clearly paying off.

Today we live in an AI-enabled landscape (Musser 2019). Adoption of AI should be done with small incremental initiatives rather than taking on a complex problem (Davenport

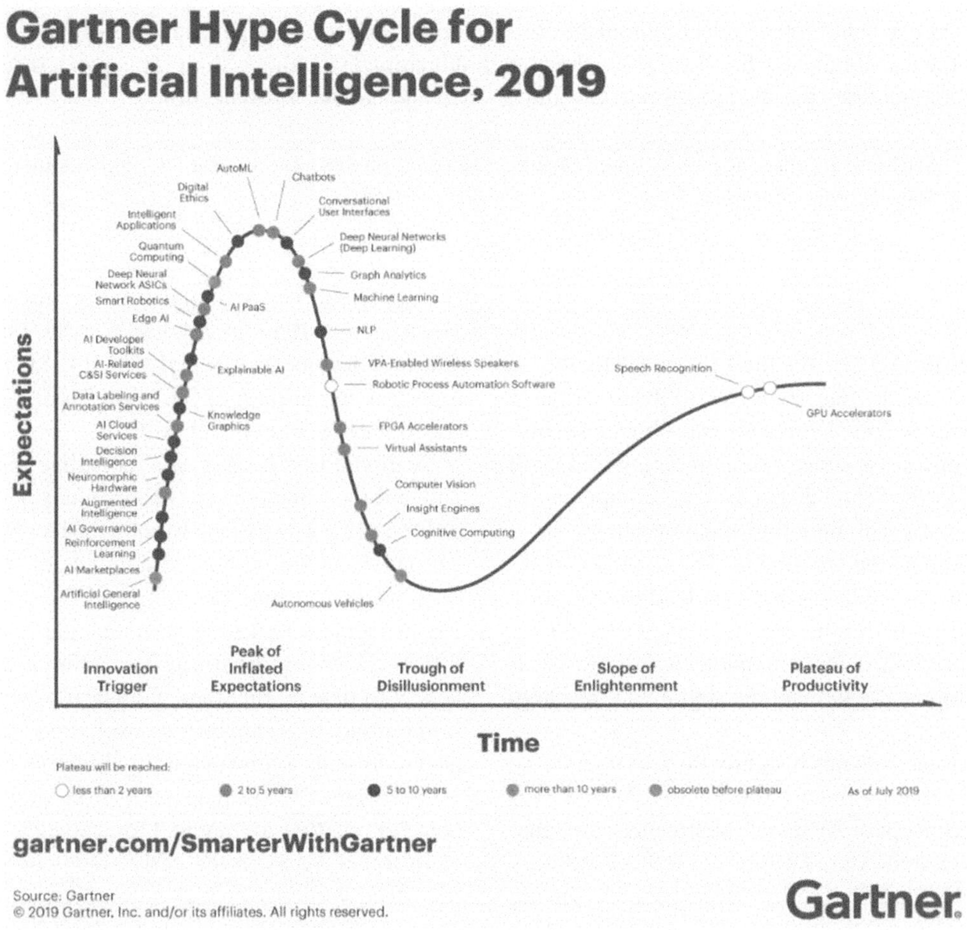

The five phases in the hype cycle are:

Innovation Trigger: A potential technology breakthrough kicks things off and triggers significant publicity. Often no usable products exist, and commercial viability is unproven.

Peak of Inflated Expectations: Early publicity produces a number of success stories—often accompanied by scores of failures. Some companies take action; many do not.

and Ronanki 2018). They further elaborate that robotic process automation projects have seen the most success to date, followed by cognitive insight and cognitive engagement.

Role of AI in travel

Thus, the question is, *what is the role for Artificial Intelligence in travel?* Adoption of AI in the travel industry is in the early stages. Much talked about use cases by airlines are speech recognition, facial recognition for passenger identification at airports, chatbots to respond to basic customer queries, use of biometric data to speed passengers through airports and AI powered autonomous multi-lingual robots (BreakingTravelNews 2019) to guide passengers through the airport terminal. However, the potential for AI lies beyond these use cases.

Many of the core advances in advanced decision support modeling with operations research can be enhanced with AI, which is at the intersection of technologies that reason, interact, and learn. There are opportunities to introduce the concept of continuous learning without human intervention. Traditional optimization techniques such as the branch-and-bound algorithm for solving certain classes of mixed integer programs can potentially benefit from machine learning to explore the branch-and-bound tree (Khalil et al. 2016). Specifically, they address the Generalized Independent Set Problem (GISP) class of problems where finding good solutions is challenging and report improvements in primal integral and time to best incumbent and the benefit lies in the method learning to run heuristics more aggressively early in the search. Bengioet al. (2020) report on using machine learning systematically to apply heuristics to solve combinatorial optimization problems. Besides core optimization, AI can augment and complement a range of use cases that use Operations Research techniques.

Here are examples from revenue management and travel in general.

Revenue opportunity model–pattern recognition to recommend inventory control changes

A recent Harvard Business Review survey with 250 executives identified the primary goal for Artificial Intelligence (AI) is to enhance existing products (Davenport and Ronanki 2018). In revenue management, the revenue opportunity model (ROM) (Ratliff et al. 2013) is a post departure decision support tool that provides insights into incorrect inventory control decisions that were made and also identifies improvement opportunities based on computed ROM metrics. For an airline analyst to make inventory control changes requires an in-depth knowledge of a range of ROM metrics such as unconstrained and constrained demand, spill rates and recapture. Due to the complexity of the decision-making process combined with the non-uniform skill and experience level of the airline analysts, consistency in decision-making is difficult to achieve. Toyoglu (2019) developed an expert system based on information collected from revenue management experts. This is a pattern recognition model that recommends an action based on the scenario and observed ROM metrics. Expert consultants provide their feedback for various scenarios on the action to be taken. This information is then combined with a machine-learning model that constantly learns from the outcomes and refines the recommendations over time to make consistent, repeatable recommendations. Besides the revenue upside of consistent decision making with ROM, this is a prime example of robotic process automation that delivered a 20-fold increase in airline analyst productivity.

Offer management

Global airline ancillary revenues in 2019 exceeded 109.5 billion dollars (IdeaWorks and Cartrawler 2019), which is a five-fold increase in ancillary revenues reported in 2010 of $22.6 billion. Offer Management is the process of selling the right bundle of base airfare and air ancillaries to the right customer at the right price at the right time (Vinod 2017; Vinod et al. 2018). Offer management extends the traditional revenue management process of optimizing allocations for the base fare to include air ancillaries offered by an airline. At its core, offer management begins with customer segmentation based on context for travel. Segmentation is a clustering or unsupervised learning problem. The context for travel is implicitly known based on basic information provided by the customer during the shopping process such as advance purchase, length of stay, Saturday night stay and number in party. This is followed by a recommendation engine that recommends air and ancillary bundles for each customer segment. An offer engine personalizes and prices the bundle for a segment of one (Vinod 2017; Vinod et al. 2018).

Model accuracy of a machine learning based recommendation model, like a statistical model, declines after a model is deployed because the data the model was trained on becomes less valid as consumer preferences change over time. To address the phenomena of *concept drift* (Brownlee 2017), a test and learn experimentation model using reinforcement learning, frequently referred to as a multi-armed bandit, can continuously adapt the product recommendations to changing consumer behaviors, new entrants and new product offerings from existing competitors. Following the lead of non-travel-related companies such as Netflix and Amazon, mass customization is the way of the future in travel to create a unique offer for a specific named customer.

Demand forecasting

Demand forecasting is dominated by time series models and customer choice modeling (CCM) techniques such as the Multinomial Logit (MNL) to forecast airline demand.

To overcome the independence of irrelevant alternatives assumption, nested logit models are more robust since it can represent similarities, measured by the covariance of error terms, from a set of alternatives. Machine learning techniques such as Random Forest (RF), Gradient Boosted Trees (GBT), Support Vector Machines (SVM) and Artificial Neural Networks (ANN) can model non-linear relationships and interaction terms, allow for collinearity and offers flexibility to automatically learn customer behavior implicitly. Various studies indicate that in several scenarios machine learning models outperform traditional methods.

Seat pricing

Determining availability and price by seat when the request is made is considered as the holy grail of revenue management. Today, based on the section of the airplane, seat prices are static by product (e.g. premium economy, economy, main cabin). The opportunity for variable seat pricing depends on how customers value a specific seat on an airplane. There are several factors that need to be considered to determine a dynamic price for a seat: the seat type (aisle, window, exit row, premium economy (wider pitch), basic economy), length of haul (short, medium, long), market, schedule attributes (aircraft type, departure time of day, non-stop versus connection), number in party (seats required together), channel, etc. These factors influence the probability of purchase of a specific seat. Monetizing seats with a dynamic seat price are an active area of research by leading airlines today with various machine learning methods being tested.

Customer segmentation

The age of intelligent retailing has arrived. Airlines want to segment customers that go beyond traditional booking classes. The idea of segmentation is to create unique personas—subsets of the travel population who have similar preferences and characteristics. In its simplest form, personas can be created based on business rules without using any customer data. However, the disadvantage is the absence of knowledge of hidden signatures in the data. More sophisticated clustering algorithms are based on unsupervised learning or supervised learning. Unsupervised learning can use customer attributes and a distance metric to measure similarity among customers, which are used in various clustering techniques such as hierarchical clustering, k-means, sequential k-means, k-mode and k-median. These techniques can be used to segment customers across a range of dimensions such as advance purchase, length of stay, mid-week vs weekend, number in party, length of haul, etc.

Supervised learning methods (Liu et al. 2000; Kothari et al. 2016) can often find simple, easily interpreted segments. Alternately, a latent class model which assumes there are latent (non-observable) variables in the data that can define segments have demonstrated to outperform unsupervised learning techniques.

Test and learn experimentation with the multi-armed bandit

A fundamental building block for intelligent retailing is leveraging a reinforcement learning technique called the multi-armed bandit to manage online controlled experiments to learn customer behavior. The goal of the experiment is to find the best or most profitable action and the randomization distribution can be updated in real time as the experiment progresses. This approach is used in air shopping experiments to control the cost of shopping with cache updates, measure shopping diversity that maximizes conversion rates, determine the best fare with schedule/fare/ancillary attributes to display on multi-month calendars, hotel retailing with in-room and non-room ancillaries are offered to customers, and to recommend air bundles (base fare + air ancillaries) to customers.

The above represents just a sample of leveraging AI in travel. In addition, there are a range of new value propositions such as controlling the cost of air shopping, detecting robotic shopping requests based on pattern recognition, demand forecasting, dynamic packaging of air and non-air products such as hotel and car based on the context for travel, anomaly detection of hotel rates, identification of duplicate properties from multiple suppliers, endpoint security (Columbus 2019), fraud detection, managing corporate travel spend, hotel product and rate normalization across aggregators, design of user interfaces based on click-through behavior, reservations workflow automation and anomaly detection of internal systems to prevent failures that can benefit from machine learning techniques and AI.

These examples in the travel domain rely on AI/ML exclusively.

Name recognition

Identification of first and last names on passports from countries that are not ICAO (International Civil Aviation Organization) compliant causes two issues. First is the inability to use the MRZ (Machine Readable Zone) to delimit first and last names which prevents kiosks to process the name information and for PNR and departure control system (DCS) customer lookup. Second, is the inability to pass accurate APIS (Advanced Passenger Information System) data when the flight is closed.

The name recognition problem begins with matching individual names to ethnicities which can be solved using supervised learning techniques. A range of traditional machine learning techniques such as support vector machines, bagging, boosting, random forest and deep learning techniques can be used. Studies indicate that Recurrent Neural Networks (RNN) with Long Short-Term Memory (LSTM) marginally outperformed the traditional machine learning models. Once the name ethnicity is determined a rules engine can be used to determine the first name and last name of the traveler.

Biometric boarding

Several airlines worldwide have implemented biometric boarding at airports with facial recognition software. SITA working with JetBlue and the US Customs and Border Protection (CBP) was one of the first deployments of facial recognition at airports to verify customers at the gate at time of boarding utilizing NEC software which relies of neural networks.

Chatbots

Chatbots are virtual travel agents that offer recommendations based on customer preferences and answers standard frequently asked questions and make bookings. Some of the leading popular chatbots are those from booking.com, Expedia and KLM (Marques 2018). The success of a chatbot, measured by loyalty, for travel is based on *trust, security, accuracy* and *experience* (Vinod 2016). Travelers have high expectations from chatbots, that the bot works as well as a human, and it is vital that the traveler trusts the chatbot. An offshoot of trust is security. The traveler must be assured of full security. Payment details, passport information and other personal data must be secure which means encryption is mandatory. A component of trust is accuracy. The chatbot must be accurate and relevant. It must provide the best results for any given search. For example, for corporate travelers, the chatbot experience might be quite valuable. Corporate travel policies and travel history can sit within the chatbot's rules. This drives more relevant, in-policy bookings to the front for each traveler. A Natural Language Processing (NLP) based chatbot, can assess the intent of the request by the traveler and respond based on the context, thereby improving trust and accuracy. And finally, the experience is what sets chatbots apart. This requires the display of only concise pertinent information which load quickly.

Corporate travel and multi-lingual real-time translation

There are several companies such as Pilot and Google Pixel buds that offer a real-time translation service, based on Artificial Neural Networks and Deep Learning. This is a boon for corporate business travelers. These apps operate through an earpiece and offers translation into multiple languages.

AI-enabled business

Contrary to common belief, a successful corporate AI program is not based on a secret sauce—specific algorithm or data science technique. A successful program requires an investment in data infrastructure that supports automated data collection, storage, ease of data access, a platform with analytical tools on which the input data can be pre-processed to build, train, store and deploy the machine learning model. The single biggest challenge is adoption at scale across the organization. This requires a commitment from the organization to make employees aware of the potential of AI through frequent communication of success stories and ensure that a consistent repeatable process be established to evaluate ideas, estimate potential and prioritization for investment to support a future product launch.

Overcoming complexity

The challenge in travel is that consumer expectations are influenced by their experiences outside of travel. Apple focused on a stripped-back focused design philosophy that overwhelmingly defined how consumers wanted to interact with technology. Amazon uses data from individual consumer preferences and purchases, browsing history and items that are related and purchased together to create a personalized list of products that customers want to buy using a technique called collaborative filtering. Amazon's AI/machine learning powered recommendation engine generates 35% of the company's revenues. Starbucks claims to deliver 87,000 drink combinations in its stores in the time people are willing to wait for a coffee. Netflix is an example of hyper-personalization and they have refined their recommendation engine over time from regional personas to global personas to personalization for a specific individual. The brand has also moved to artwork personalization to increase engagement.

The promise of AI in travel is to simplify and shape the future travel experience. The desired end state is the promise of frictionless door-to-door travel where all aspects of a trip including local transportation are managed.

Over the past few decades, complexity in travel has grown. Since airline deregulation, core complexity is associated with a larger number of scheduled flight options, fares, and airline alliances. Complexity has grown over the past decade with the launch of branded fares and la carte air ancillary pricing. Content fragmentation caused by Low Cost Carriers (LCC) that do not participated in the Global Distribution Systems (GDS) and IATA's New Distribution Capability pose a unique challenge. It is also anticipated that the IATA NDC adoption by airline will lead to a proliferation of time of day, date specific and routing specific fares. In addition, there is a new wave of complexity introduced by airlines and hotels as they become retailers to drive incremental revenues. Airlines what to optimize total revenue; base airfare and air ancillaries sold to customers and not just the base fare. Hotels have embarked on attribute-based pricing of rooms to both generate incremental revenues and improve customer satisfaction by selecting a room for a customer for precisely what they want to pay for.

Challenge of interpretability

Unlike statistical models, a machine-learning model is typically a black box. When a machine learning model is developed, it is important to understand the expected behavior of a trained model. However, frequently, a machine-learning model can produce results that are counter intuitive. Such scenarios should be debugged by data scientists who developed the model. It is important to understand "why" a model produces meaningful results and when it does not. An active area of research in academia and corporations is to develop methods to identify issues such as model bias and interpret the model response to gain insights.

Democratization of AI

With lower costs in cloud computing and data storage, leveraging AI is no longer limited to select large companies with multi-million-dollar product investment budgets, but to companies of all sizes to experiment and create new capabilities. Making the intelligence accessible for start-ups and small companies will accelerate the adoption of AI in travel (Boehmer, 2019). Besides adoption, travel entities such as GDS's, Online Travel Agencies (OTA's) and Travel Management Companies (TMC's) associate this with a future revenue stream. By giving developers and startups access to their API library, it is typically free during testing and in production up to a limit and then becomes a pay-as-you-go model as the company scales the application.

Quantum computing and AI

Over the past three decades, there has been much talk among physicists about the promise of Quantum Computing from technology heavyweights such as IBM, Microsoft, Google and Intel. China has invested billions in research to develop Quantum Computers. While traditional computer chips require that the data be encoded into binary digits represented by a zero or one (an "on" "off" switch), a quantum computer uses quantum–mechanical phenomena called quantum bits or "qubits" where they are both on and off at the same time. We attempted to solve the airline crew rostering problem (Hur 2018). To provide a form suitable for use on quantum computers, the traditional optimization problem needs to be converted into an equivalent QUBO (Quadratic Unconstrainted Binary Optimization) formulation. Formulating a QUBO is a pattern matching technique, common in machine-learning applications. It is the problem of minimizing a quadratic polynomial over binary variables. The conversion to QUBO for the crew rostering problem with 11,000 + variables and 1300 + constraints, resulted in 1 million terms that could not be directly imbedded on a quantum computer due to the high connectivity of the constraints. This experiment showed that we are a few years away to use quantum computers for large-scale optimization problems.

Google recently claimed (Martinis and Boixo 2019) it had achieved quantum supremacy, the inflexion point at which a quantum computer outperforms a classical computer, but the claim has been disputed by competitors such as IBM (Lichfield 2020) who believe that quantum supremacy is not the milestone we should care about.

Machine learning and especially training deep learning neural networks are compute intensive, require Graphical Processing Units (GPU's) and are candidates for quantum chips in the future. However, this is probably at least a decade away which means that conventional chips optimized for AI/ML algorithms will be the mainstay for the next decade.

Employees and AI

The adoption of AI in a company poses a fundamental challenge on how to raise employee awareness and take the necessary steps to scale participation across the organization. To raise awareness of the role of AI requires effective communication on where it can be used across the organization, encourage teams to be curious, learn and leverage the available data sources and bring forward new travel-related value propositions that adds value to customers.

There are two key challenges with leveraging AI/ML in any corporation. First is the skill to identify areas where AI can be leveraged to create a new solution or enhance an existing solution. Second, is the ability to scale the adoption of AI/ML across the organization. This begins with basic training with a toolkit before the analyst can advance into problem definition, model development, data access, calibration, and deployment. Raising awareness can be accomplished with small special interest groups focus groups that meet periodically to discuss use cases, success stories and pitfalls.

Many vendors provide "black-box" toolkits for AI/ML. Every organization should make an effort to evaluate toolkits, including cloud-based solutions, proprietary solutions, and open source solutions (in-house and cloud) to enable data science and development teams with AI tools and capabilities. This approach can also provide internal consulting support to teams who are developing AI/ML applications and have a preferred toolkit. The evaluation of AI toolkits should be based on criteria such as data ingestion and connectors, data wrangling and visualization, flexible modeling capabilities and deployment capabilities that support automation. Further, given the mix of employees a factor to be considered is tools for citizen data scientists, who prefer to drag and drop capabilities, versus technical data scientists who prefer to write code and build models using Python, R, Scala or other programming languages.

Machine-learning techniques can be grouped based on the types of problems the techniques are designed to solve.

Unsupervised Learning: Clustering (e.g. k Means, k Medians, Fuzzy cMeans, Hierarchical), Gaussian Mixture, Hidden Markov Model).

Supervised Learning: Classification (Support Vector Machines, Discriminant Analysis, Naïve Bayes, Nearest Neighbor) and Regression (Linear, Generalized Linear Model (GLM), Decision Trees, Ensemble Methods, Neural Networks).

Reinforcement learning: Genetic Algorithms, Multi-armed Bandit test and learn, Approximate Dynamic Programming, Markov Decision Processes.

Deep Learning: Multi-layer Neural Networks, Convolution Neural Networks (CNN), Recurrent Neural Networks (RNN).

In the literature, there are various nomenclatures to group AI/ML techniques, libraries and frameworks which can be overwhelming to an inexperienced analyst. For example, Nguyen et al. (Nguyen et al. 2019) categorize machine learning tools into Statistical, Mathematical (e.g. R, SPSS, etc.), General Purpose (e.g. Scikit-Learn), Narrowed Solutions (e.g. Vowpal Wabbit), Interactive Platforms (e.g. Jupyter), ML/DL with MapReduce (e.g. Spark ML/MLLib, etc.), Deep Learning with GPU (e.g. TensorFlow), and Deep Learning wrapper libraries (e.g. Keras).

Conclusions

Why is adoption of AI a necessity rather than a luxury? A question that needs to be understood is "why companies fail?" Companies fail when they can see what the technology is, but they struggle to adapt to the change because of their reluctance to give up existing capabilities they have perfected for decades and to fully integrate new ones. Although incumbents may imitate the new architecture, they have a hard time overcoming the way they have done things in the past and to match the superior performance of the new, purpose-built architecture. Thoughtful abandonment of legacy technology and architecture to embrace the future is a critical success factor for any organization. Ultimately, individuals that do not adapt will be detrimental to a company's success and will be left behind.

A successful organization must raise awareness and imbed a *sense and respond* mindset among employees to enable rapid adoption of AI technology. In travel, it is difficult to predict which companies will successfully capitalize on the AI-enabled landscape. A critical success factor will be the willingness of managers and individual contributors to be nimble, carefully define the scope and boundaries of the planned experiment, establish success criteria, run experiments, evaluate results quickly and move the application to the next stage with a fail fast attitude. We anticipate that these small fundamental steps will accelerate awareness and urgency for adoption of AI across the organization for competitive advantage in the marketplace. In travel this journey is just evolving.

Acknowledgements I would like to take this opportunity to thank Ross Darrow, former Sr. Principal in Sabre Research and currently Chief Scientist at Charter and Go, for his thoughtful feedback on this paper and his many contributions toward the Artificial Intelligence Special Interest Group (AISIG) Initiative – townhalls, big pitch presentations, AI training and quarterly newsletters at Sabre. I would also like to thank Paula Lippe, Cuneyd Kaya, Richard Ratliff, Christian Huff, Melvin Woodley, Rajeev Bellubbi and the many individuals who supported the launch of the AISIG initiative.

References

Bengio, Y., A. Lodi, A. Prouvost. 2020. Machine Learning for Combinatorial Optimization: A Methodological Tour d'Horizon. *European Journal of Operational Research*, March 12, 2020 (online version, publication forthcoming).

Boehmer, J. 2019. Amadeus 'Democratizing AI' for Travel Startups with Open API's. *The Beat*, December 06.

BreakingTravelNews. 2019. Robots to guide British Airways passengers through Heathrow. https://www.breakingtravelnews.com/news/article/robots-to-guide-british-airways-through-heathrow/, December 23.

Brownlee, J. 2017. A Gentle Introduction to Concept Drift in Machine Learning. In Machine Learning Algorithms, Machine Learning

Mastery. https://machinelearningmastery.com/gentle-introduction-concept-drift-machine-learning/. December 15.

Chui, M., J. Manyika, M. Miremadi, N. Henke, R. Chung, P. Nel, and S. Malhotra. 2018. *Notes from the AI Frontier: Applications and Value of Deep Learning*, McKinsey & Company, April 2018. https://www.mckinsey.com/featured-insights/artificial-intelligence/notes-from-the-ai-frontier-applications-and-value-of-deep-learning.

Columbus, L. 2019. 10 Ways AI and Machine Learning are Improving Endpoint Security. *Forbes*, September 25.

Cook, T.M. 1998. SABRE Soars. *OR/MS Today* (June) 26–31.

Davenport, T., and R. Ronanki. 2018. Artificial Intelligence for the Real World. *Harvard Business Review*, January–February 2018.

Goasduff, L. 2019. Top Trends on the Gartner Hype Cycle for Artificial Intelligence, 2019. https://www.gartner.com/smarterwithgartner/top-trends-on-the-gartner-hype-cycle-for-artificial-intelligence-2019/, September 12.

Hao, K. 2019a. We Analyzed 16,625 Papers to Figure Out Where AI is Headed Next. *MIT Review*, January 25.

Hao, K. 2019b. China has Started a Grand Experiment in AI Education. It could Reshape How the World Learns. MIT Technology Review, August 2. https://www.technologyreview.com/s/614057/china-squirrel-has-started-a-grand-experiment-in-ai-education-it-could-reshape-how-the/.

Hur, Y. 2018. Quantum Computing for Airline Problems. In: *AGIFORS 58-th Annual Symposium*, Tokyo, October 8–12.

IdeaWorks Company.com and Cartrawler. 2019. Cartrawler Worldwide Estimate of Ancillary Revenue for 2019. https://www.cartrawler.com/ct/ancillary-revenue/worldwide-ancillary-revenue-2019.

Khalil, E.B., P. Le Bodic, L. Song, G. Nemhauser, and B. Dilkina. 2016. Learning to Branch in Mixed Integer Programming. Proceedings of the *Thirtieth AAAI Conference on Artificial Intelligence*, pp. 724–731.

Kothari, A., M. Madireddy, and S. Ramasubramanian. 2016. Discovering Patterns in Traveler Behaviour Using Segmentation. *Journal of Revenue and Pricing Management* 15: 334–351.

Levesque, H. 2011. The Winograd Schema Challenge. *Commonsensereasoning.org*.

Lichfield, G. 2020. Inside the race to build the best quantum computer on earth. *MIT Review*, February 26.

Liu, B., Y. Xia, and P.S. Yu. 2000. Clustering via decision tree construction. In *Conference on Information & Knowledge Management*, ed. A. Agah, J. Callan, E. Rundensteiner, and S. Gauch. McLean, VA: ACM.

Marques, M. 2018. Top 3 Chatbots that are Changing the Travel Industry. *HiJiffy*, https://medium.com/hijiffy/top-3-chatbots-that-are-changing-the-travel-industry-d325082c50b8, March 14.

Martinis, J., and S. Boixo. 2019. Quantum Supremacy Using a Programmable Superconducting Processor. Google AI Blog, https://ai.googleblog.com/2019/10/quantum-supremacy-using-programmable.html, October 23.

Musser, G. 2019. Artificial Imagination: How Machines could Learn Creativity and Common Sense, Among Other Human Qualities. *Scientific American*, pp. 59–63, May.

Nguyen, G., S. Dlugolinsky, M. Bobák, V. Tran, Á. López Garcia, I. Heredia, P. Malik, and L. Hluchý. 2019. Machine Learning and Deep Learning Frameworks and Libraries for Large-Scale Data Mining: A Survey. *Artificial Intelligence Review* 52: 77–124.

Ratliff, R.M., J. Manjot, and B.R. Guntreddy. 2013. Applied O&D Revenue Opportunity Model for Dependent Demands. *AGIFORS Revenue Management Study Group*, May, Miami, FL.

Toyoglu, H. 2019. Revenue opportunity model (ROM) expert system. *Artificial Intelligence Special Interest Group (AISIG) Newsletter*, 1(3).

Turing, A. 1950. Computing Machinery and Intelligence. *Mind* 49: 433–460.

Vinod, B. 2014. Operations Research: Laying the Foundation for the Future of Airline Technology. *Ascend*, pp. 64–69.

Vinod, B. 2016. Chatbots in Travel: 4 Things the Industry must Get Right for Success. https://www.sabre.com/insights/chatbots-in-travel-4-things-the-travel-industry-must-get-right/, August 23.

Vinod, B. 2017. The Evolving Paradigm of Interactive Selling Based on Consumer Preferences. In Proceedings of the *"21st Century Airlines: Connecting the Dots"* by Nawal Taneja, ISBN: 1138093130, September, pp: 207–213.

Vinod, B., R.M. Ratliff, and V. Jayaram. 2018. An Approach to Offer Management: Maximizing Sales With Fare Products And Ancillaries. *Journal of Revenue & Pricing Management* 17: 91–101.

B. Vinod was Chief Scientist and Senior Vice President at Sabre (2008–2020) and is now a co-founder and Chief Operating Officer at Charter and Go. He is the author of *The Evolution of Yield Management in the Airline Industry: Origins to the Last Frontier*, that will be published by Springer in 2021.

Journal of Revenue and Pricing Management (2021) 20:376–380
https://doi.org/10.1057/s41272-021-00320-3

PRACTICE ARTICLE

The key to leveraging AI at scale

Deborah Leff[1] · Kenneth T. K. Lim[1]

Received: 12 April 2020 / Accepted: 8 August 2020 / Published online: 20 March 2021

Abstract
With the explosive growth of AI and ML-driven processes, companies are under more pressure than ever to drive innovation. For many, adding a Data Science capability into their organization is the easy part. Deploying models into a large, complex enterprise that is operating at scale is new, unchartered territory and quickly becoming the technology challenge of this decade. This article takes an in-depth look at the primary organizational barriers that have not only hindered success but often prevented organizations from moving beyond just experimentation. These obstacles include challenges with fragmented and siloed enterprise data, rigid legacy systems that cannot easily be infused with AI processes, and insufficient skills needed for data science, engineering and the emerging field of AI-ops. Operationalizing AI is hard, especially at the fast pace at which the business operates today. This paper uses real-world examples to guide a discussion around each of these hurdles and will equip industry leaders with a better understanding of how to overcome the challenges they will face as they navigate their data and AI journey.

Keywords Artificial Intelligence · AI · Machine Learning · ML · Digital transformation · Modernization · Organizational barriers

'Data is the new oil' (Clive Humby, 2006)

Introduction

A lot has changed since Clive Humby first alerted businesses of the importance of data almost 15 years ago.[1] There has been an explosion in the amounts of data and compute power available, and the number of machine learning algorithms developed. Technology companies, such as Google, Facebook, Amazon, Airbnb and Uber, have disrupted markets by successfully capitalizing on the technological advancements. In response, businesses have been making significant investments into modernizing their infrastructures and business models.

Almost every major business is investing in artificial intelligence (AI) and its applications now in a bid to emulate the successes of technological disrupters. There has been a surge in companies hiring data scientists, with demand for data scientists increasing by almost 1300%, and other advanced analysts by 220%, over 5 years.[2] Advanced analytics is no longer viewed as essential outside a few companies in certain industries, e.g. operational research in the travel and transportation industries.

It would be reasonable to expect businesses to have been successful at scaling their AI initiatives to drive results. Businesses have made significant investments over 15 years. It has also been several years since we emerged from the last 'AI winter'.[3] Businesses have more access to data scientists, data, compute power and state-of-the art algorithms.

However, only 8% of companies have reported being able to achieve AI at scale.[4] Most companies stuck experimenting with AI lack a clear AI strategy.[5] While almost every company expects business value from their AI investments,

✉ Deborah Leff
Deborah.leff@gmail.com

Kenneth T. K. Lim
kenneth.lim@ibm.com

1 76/78 Upper Ground, IBM United Kingdom Limited, South Bank, London SE1 9PZ, UK

[1] There is no direct source for this quote but, several reputable sources have credited the quote to Humby, e.g. https://hbr.org/2020/05/be-a-data-custodian-not-a-data-owner.

[2] https://www.burning-glass.com/wp-content/uploads/dynamics-of-data-science-skills-report.pdf.

[3] https://www.technologyreview.com/s/603062/ai-winter-isnt-coming/.

[4] https://hbr.org/2019/07/building-the-ai-powered-organization.

[5] https://sloanreview.mit.edu/projects/reshaping-business-with-artificial-intelligence/.

Chapter 14 was originally published as Leff, D. & Lim, K. T. K. Journal of Revenue and Pricing Management (2021) 20:376–380. https://doi.org/10.1057/s41272-021-00320-3.

almost 70% of companies have reported achieving minimal or no impact from their AI investments.[6]

It is worth reflecting on the reasons behind the lack of AI adoption and business success, despite the significant technological advancements and business investments over the last 15 years. Why are only 8% of companies able to achieve AI at scale and not the other way round? Why have companies not succeeded despite access to more data scientists, data and compute power? Companies are failing not because of technical challenges, but rather organizational barriers.

In this paper, we discuss the main organizational barriers to scaling AI. These barriers include not having the right data architecture, underestimating the data science lifecycle, not having a clear strategy to operationalize models, and a lack of business leader involvement and executive sponsorship. Then, we provide recommendations to overcome these barriers and successfully scale their AI initiatives. Finally, we conclude.

Organizational barriers to scaling AI

Not having the right data architecture

While every company has a data strategy, most businesses do not have the right data architecture in place to scale their AI initiatives. First, data tend to be stored in legacy systems across the organization and siloed within each line of business. Second, data are usually not organized or curated, making it hard to find or use. This hinders data scientists from extracting and understanding the right data needed.

Consider this simple example in the context of the airline industry. Although the concept of a 'passenger' seems easy to understand, it has multiple definitions depending on where the data are stored. The flight crew are included in fuel-related data but not in revenue-related data, which are likely to be stored in separate databases managed by different departments. Without documentation of the data fields and schema, the data scientist might end up analysing the wrong type of 'passengers', leading to inaccurate results. This mistake can go undetected and be especially costly considering the number of flights that happen over a year.

To compound the problem, most executives and managers are not aware of the limitations of their data architecture until they start an AI project. They usually expect data to be stored in a centralized data lake that is easily accessible with well-documented data dictionaries and simple schemas. The figure below shows an example of how executives' and managers' expectations differ from the more complex reality of their data architecture, in the context of a marketing campaign initiative.

Without a suitable data architecture, businesses might not even be able to start, let alone scale, their AI initiatives. Data scientists might not be able to get access to, or make sense of, the necessary data. As a result, projects suffer delays of weeks or months or ideas for the AI initiatives never get tested or explored (Fig. 1).

Underestimating data science lifecycle

Businesses have underestimated the data science lifecycle by focusing only on hiring data scientists to build machine learning models in open source tools and share insights from the results. This narrow approach might be useful for pilots and experiments but is insufficient for scaling AI. This approach is only one part of a much larger lifecycle (see Fig. 2).

The data science lifecycle starts with scoping the requirements for an AI initiative. This involves exploring ideas, assessing use cases, prioritizing the use cases by their feasibility and importance, and establishing clear success metrics. Then, time must be taken to gather, understand and share all the relevant data assets. Only once have the data assets been organized are the data scientists ready to build and evaluate their machine learning models. The trade-off between model accuracy and complexity needs to be considered as interpretable and simpler models are easier to implement, maintain and debug in production.

Once the machine learning models have been developed, they need to be deployed and integrated into a production environment for business use. The code written by the data scientist would need to be translated for production purposes, e.g. Python/R code used by the data scientist might need to be translated by data/software engineers into another language like C# for the production environment. This entire process requires clear communication and patience from all parties who will have different and sometimes competing considerations.

Finally, the deployed machine learning models need to be monitored and managed with a continuous integration and deployment (CI/CD) approach. The performance of the model will decrease with time and decisions must be made as to how often the model must be updated. The performance of the model extends beyond accuracy and includes other metrics like fairness in and explainability of the predictions, and business key performance indicators.

Unclear strategy to operationalize models

The data science lifecycle needs to be integrated, but is usually not compatible, with traditional IT systems. Most businesses have several legacy databases operated by different

[6] https://sloanreview.mit.edu/projects/winning-with-ai/.

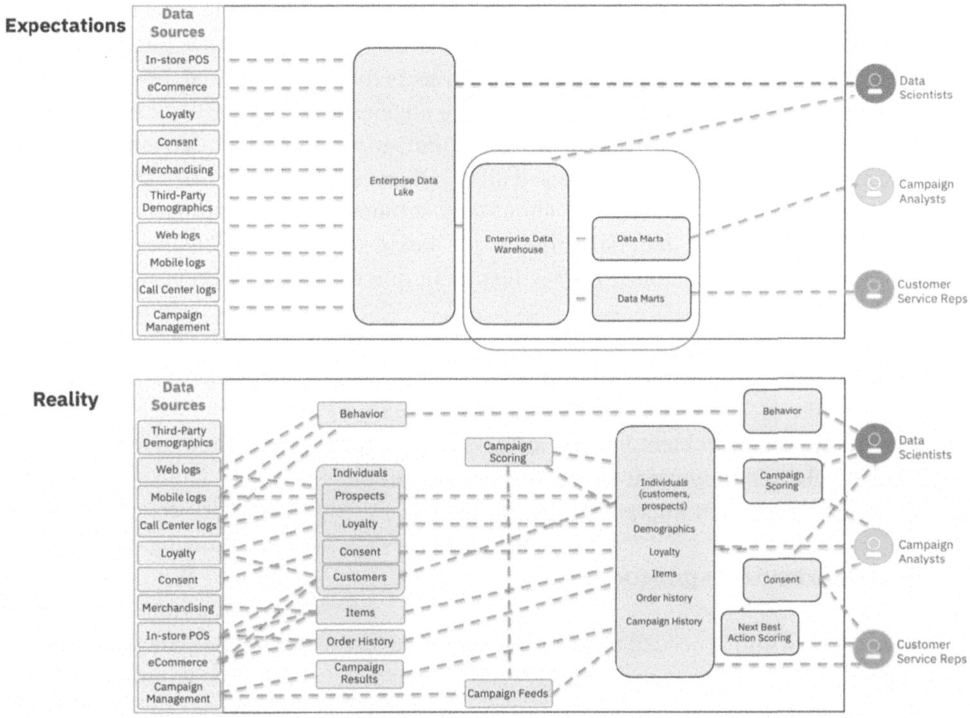

Fig. 1 Illustration of how executives' and managers' expectations (top) differ from the reality (bottom) of their data architecture, in the context of a marketing campaign initiative.

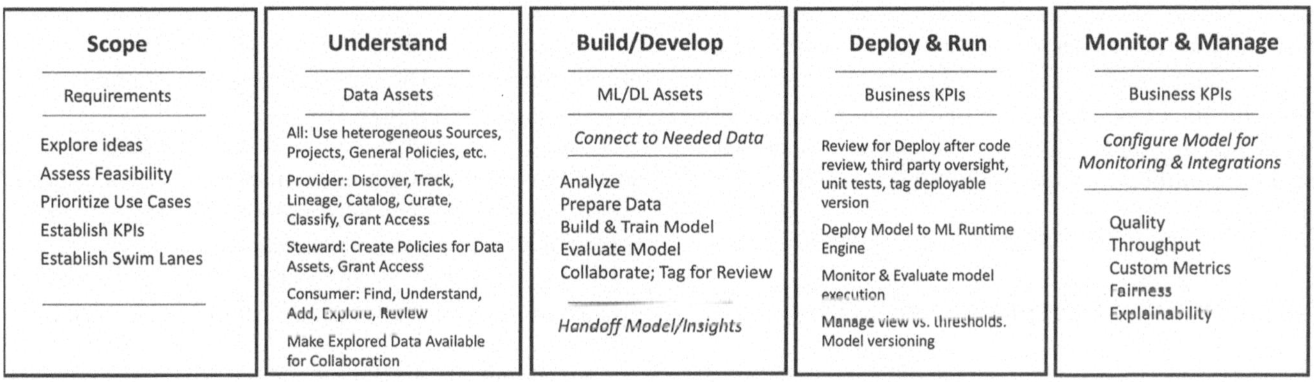

Fig. 2 An illustration of an end-to-end data science lifecycle.

vendors, making it difficult to efficiently extract the relevant data and integrate together for production purposes. There are also fundamental differences between IT and machine learning systems that need to be reconciled. For example, IT system behaviour is defined by code, while machine learning system behaviour is defined by data. Businesses need to modernize their IT systems as part of their AI initiative to succeed.

The work that IBM did with American Airlines is one example of how modernizing IT systems can help organizations in the travel industry gain competitive advantage with AI[7]. American Airlines wanted to give customers better self-service capabilities in the event of a forced rebooking due to a major weather event disrupting operations. IBM helped American Airlines modernize some of their key legacy customer-facing applications, build a weather forecasting model, develop a cloud-native application to rebook customers automatically and allow customers to choose alternatives on the application without needing to call or wait in line. This improved organizational efficiency and customer satisfaction. The application was deployed and scaled globally to all American Airline airports during Hurricane Irma in 2017.

[7] https://www.ibm.com/case-studies/american-airlines.

Insufficient business leader involvement

Too often, the data science and IT teams select projects that they think are great applications without involving business leaders enough. This happens because of a communication gap between business leaders and data scientists. Business leaders lack the technical knowledge about data science, while data scientists lack understanding of the business realities and nuance. This almost always results in one of three problems.

First, business leaders reject the AI results, instead relying on their own judgment and intuition because the AI results seem like a 'black box'. Second, the AI initiative solves an unimportant problem for the business. Third, the AI initiative solves an important business problem without domain context, making it impractical and unusable.

Insufficient executive sponsorship

Executive sponsorship is needed to ensure AI initiatives are implemented successfully as they usually take longer than expected due to a variety of challenges. For example, existing cultural norms in most organizations do not incentivize data owners to share access to the data with a wider team. This can slow or even derail AI initiatives. In another example, the pilot AI models are promising enough to be prepared for production, but legacy IT systems need to be modernized. Executive sponsorship can help the team responsible for the AI initiative to successfully navigate through the challenges.

Recommended best practices

The previous section listed five main organizational barriers that prevent businesses from scaling AI successfully: not having the right data infrastructure, underestimating the data science lifecycle, having an unclear strategy to operationalize AI models, and having insufficient business leader involvement and executive sponsorship in AI initiatives. This section provides our recommended best practices to overcome these organizational barriers.

First, get executive sponsorship for AI initiatives even before starting. Identify, and align the AI initiative with, what is most important to the executive sponsor. Understand and target important business problems to solve, with the view to operationalize the AI models. Companies tend to experiment with problems that are low risk but also not very important for the business if successful. This approach has two major shortcomings. First, failures get buried and successes do not make any meaningful impact. Second, initiatives that start as experiments tend to end as experiments, without progressing further. Executive sponsors can provide valuable insights into important business problems and support to see the initiative through.

Second, involve line of business leadership right from the onset. Business leaders provide important context to, and a deep understanding of, the problem. This will help ensure that the right problem is being solved with the most suitable approach, and that resulting solutions will be adopted by the business. To achieve this, business leaders and data scientists need to bridge the communication gap. Business leaders need to upskill and improve their understanding of AI, while data scientists need to take the time to understand the business function and realities. Both parties need to communicate regularly during the initiative duration and work together as a team.

Third, businesses need to start AI initiatives with an outcome-driven approach. This means having clear answers to the following questions. What process will the AI initiative change? Who will benefit? How will they benefit? Why is this important to the business? Where will it be executed? How will success be measured? What strategic initiative does the AI initiative align to?

Fourth, utilize enterprise solutions and develop a data science center of excellence. Open source technology is useful to build smaller scale proof of concepts but can become cumbersome to implement AI at scale, especially for businesses starting out. The right enterprise solutions can accelerate a company's journey to achieving AI at scale.

To successfully scale AI, a cultural change in mindset across the organization needs to happen, which a center of excellence can help achieve. Employees need to be comfortable with leveraging, and understand how to leverage, AI to make better informed decisions in their jobs. However, most organizations do not understand AI and employees resist adopting AI solutions.

Leaders must be transparent about when AI will begin to replace jobs and provide employees with opportunities to upskill. Middle managers must realize that AI will not replace them but that other managers who use AI will.[8] Front line employees need to be trained to leverage AI solutions in their daily job without being afraid that their jobs will be replaced by AI.

Finally, businesses should seek experience from others who have been successful to get started. It is difficult to scale AI, especially for the first time, and mistakes can be very costly. Once businesses have successfully scaled their AI initiatives for the first time, they will be able to repeat their success much more efficiently.

[8] https://hbr.org/cover-story/2017/07/the-business-of-artificial-intelligence.

Conclusion

It's been almost 15 years since the business world was alerted to data being the new oil. Businesses have made significant investments into advanced analytics, and have more access to data scientists, data, compute power and state-of-the art algorithms. However, only 8% of companies have reported being able to achieve AI at scale, with most companies achieving minimal or no impact from their AI investments. Organizational, and not technological, barriers have hindered businesses from successfully scaling and gaining value from AI. The main organizational barriers are not having the right data architecture, underestimating the data science lifecycle, not having a clear strategy to operationalize models, and a lack of business leader involvement and executive sponsorship.

There are five steps to overcome organizational barriers to successfully scaling and gaining value from AI. First, get executive sponsorship for AI initiatives even before starting to identify important business problems to solve and gain the necessary support. Second, involve line of business leadership right from the onset to ensure that the right problem is being solved with the most suitable approach, and that resulting solutions will be adopted. Third, start AI initiatives with an outcome-driven approach, focusing on the impact it will make to the business. Fourth, utilize enterprise solutions to accelerate the process of scaling AI and develop a data science center of excellence to bring about a cultural change in mindset of leveraging AI across the organization. Finally, seek experience from others who have been successful to get started.

Like oil, data have exponential value once refined. With these techniques, companies can unlock insights in their data to provide a better customer experience and optimized business process to drive growth and competitive advantage.

Deborah Leff is an independent Private Equity consultant and the former Global Industry CTO for IBM Data Science and AI. In that role, she worked with Senior Executives from leading companies around the world to identify, prioritize and successfully execute AI and ML projects to drive measurable business outcomes. She is the founder of Girls Who Solve, a STEM education club for high school girls that introduces High School girls to advantaged technologies and how they are used to solve real world problems. Ms. Leff is currently serving as an advisor to DG Capital and Censia and is a Board Member for Recruiter.com (RCRT).

Kenneth T. K. Lim is a data scientist with IBM's Data Science Elite team, where he works closely with clients across industries to improve business results with AI. He has a PhD from the University of Warwick, UK.

Journal of Revenue and Pricing Management (2021) 20:381–386
https://doi.org/10.1057/s41272-021-00321-2

THOUGHTS

The future of AI is the market

Ross M. Darrow[1]

Received: 28 May 2020 / Accepted: 24 September 2020 / Published online: 16 March 2021

Abstract
We've made great progress with the current generation of AI, with learning customer preferences embedded in recommender systems to better customize offerings. This is perfect for the New Distribution Capability that IATA is spearheading for the airline industry. To truly make this work, and not just for airlines but for all travel (hotels, rental cars, attractions, concerts, tours, and more), we need to move on from (a) supervised learning built upon masses of data to systems that learn on the fly with little data, and from (b) centralized (even if in the cloud) machine learning to distributed artificial intelligence, and from (c) recommender systems to marketplace approaches.

Keywords Customer Value Chain · Microeconomics · New distribution capability · Multi-armed bandits · Multi-agent systems · Marketplace · Distribution artificial intelligence

New distribution capability

Airlines are in the process of reinventing how they market and distribute their offerings with the new messaging standard for airlines to communicate with intermediaries called New Distribution Capability (NDC). Traditional airline revenue management is being reinvented. There are several references for this (Bacon 2019; Veinbergs 2019; Sabre, Inc 2019a).

A good way to look at NDC is to see that a new marketplace for travel is being created. At the same time, all the players in this new approach are applying machine learning to add value to their offers. This is great, but it misses that we are creating a marketplace. Microeconomics is meeting machine learning, and too little attention is being paid to how machine learning will be useful in this environment, and how marketplace ideas will inform the systems that developed and how consumers shop for and book travel in the future.

This paper highlights some of the main ideas of new thinking that is needed, especially emphasizing the ideas from Professor Michael Jordan. Many references are provided for those wishing to make an effective travel distribution future happen.

✉ Ross M. Darrow
ross_darrow@prodigy.net

1 Bedford, USA

Customer value chain

In his insightful book *Unlocking the Customer Value Chain*, Harvard professor Thales S. Teixeira (2019) shows that one important form of disruption takes place by changing the customer value chain, or the process by which consumers discover, buy, and use products and services. His key message is one of customer centricity. Most incumbent companies start with their resources ("we have this shopping system, this huge amount of data, so how do we leverage this?"), instead of truly understanding how customers buy and use products. He argues that an important form of disruption is decoupling. Some key examples:

- Car sharing companies decoupling the link between purchasing and driving a car
- Amazon decoupling the link between physically trying out a new TV (at Best Buy) and purchasing it
- Blue Apron decoupling the link between finding a recipe, shopping for ingredients, and cooking the meal

Decoupling disruptions occur because consumers like them. They either create value, offer value for less, or save time or effort in the process. The typical customer value chain consists of (a) evaluate, (b) choose, (c) purchase, and (d) consume. NDC is changing the first three, and it is a scramble to figure out who in the industry will occupy each part of the chain. (This is assuming that travel is still

Chapter 15 was originally published as Darrow, R. M. Journal of Revenue and Pricing Management (2021) 20:381–386. https://doi.org/10.1057/s41272-021-00321-2.

consumed by getting to the airport two hours in advance, checking your bag, going through security, waiting in the lounge, boarding, sitting in a tiny seat, reclaiming bag, finding transportation to hotel. The "consuming" part of the chain is ripe for disruption, but beyond the scope of this paper.)

It will take effective machine learning and data flows to improve the customer experience of evaluating options, especially since NDC will allow airlines to create custom bundles for each traveler.

This will be difficult, because often (usually?) buyers do not know their preferences, and sellers do not know which buyers they prefer. We have a system where noisy utilities are observed through repeated interactions. We need recommendation systems on both the consumer and supplier sides, with communication between them. And with all the noise, we need to make sure we do not think we have "learned" something that isn't so (called false discovery rate control in statistics, see Ramdas et al. 2019).

This is a great challenge, and current AI approaches will not be sufficient.

We need to move on

The last 15 years of AI (really, machine learning, not true general artificial intelligence) have shown dramatic improvements in creating recommendations for consumers, to the point that recommender systems have become common, with Python toolkits for developing these engines becoming more common (Deng 2019, and Note 1). Deep learning (really, new and improved neural networks) have become common for pattern recognition, including very complicated patterns that traditional regression analysis would not discover. Much excellent research has been done, with important implementations (see Chap 1 in Microsoft 2018). We need, however, to look beyond current AI:

- Current AI is (mostly) supervised learning, with much (deserved) celebration about deep learning. The work that has been done is important, but we need to look to the future. The future will be about learning with little data, exploring solutions, and learning on the fly.
- Current AI is (mostly) done with one algorithm (often an ensemble approach) in one place (even if that place is the cloud). We have all had great experience with this model, but now we need to move on. The future will be about distributed AI, learning on the fly with many little bits that are combined.
- Current AI is (mostly) about top-down pushing solutions to the users; that is, learning from data. This works, but we need to consider the limitations. Recommender systems should not just learn from user data, which only serves to make the user the product (Note 2). Instead, users should be empowered, broadcasting what they want (at that particular moment) and allowing vendors to reply.

Change is coming, and many strands of this change are coming together in economics, in the form of a marketplace. Leading the charge is Professor Michael Jordan, University of California at Berkeley, who has been discerning what is needed for a few years now. Highly recommended reading (it will change your AI mindset) includes:

(1) The most recent iteration is from (Jordan 2019 and Note 3)
(2) "ML Meets Economics: New Perspectives and Challenges" (Jordan 2018 and Note 4)
(3) "Clarity of thought on AI," (Jordan 2018)
(4) "On gradient-based methods for finding game-theoretic equilibria," (Note 5)

This is very different from what most AI researchers are pursuing. Consider the book *The Master Algorithm* (Domingos, 2015). "The author pictures a master algorithm, where machine learning algorithms grow to a perfect understanding of how the world and people in it work." (Note 6). Without a marketplace approach, this approach can be a little creepy.

Reinforcement learning

Much airline shopping is done anonymously (most consumers do not login to an online travel agency or do anything to reveal their identity until they have selected a flight or hotel and are ready to book and pay). And travel is often a big-ticket item (bigger than, say, a movie or a book or a concert). And different trips have different purposes for the same person (business trip, business/leisure hybrid, pure vacation, short-haul business trip where coach is fine, international travel where business class is a must for some people). And companies change their budget for travel across the year ("it's the fourth quarter, and we need to cut back on expenses").

Often, it makes more sense to learn about consumer preferences via online learning approaches, such as multi-armed bandit (Toyoglu and Ratliff 2019), contextual bandit (Surmenok 2017), and other reinforcement learning approaches. These are closer to human intelligence than supervised learning techniques that require massive amounts of data, though humans certainly don't learn according to bandit algorithms either.

Multi-armed bandits have an underlying model of picking the best arm among a series of slot machines (Surmenok 2017). Each arm produces an independent and identically distributed random reward independent of the past and other arms. The parameters of the reward distribution are fixed but unknown (some research has been done to allow the rewards to change over time). A reward can be observed only after playing an arm. The key idea is the explore-exploit tradeoff: *Exploiting* arms that yielded high rewards in the past for immediate rewards vs *Exploring* other arms to gather more information for long-term rewards. Unlike A/B testing, which fixes the sample size up front and provides statistical guarantees that you have found the best arm, bandits use online learning to minimize regret, where regret is the loss from using a sub-optimal arm. If two arms produce similar rewards (but one is slightly better than the other), but one arm is a real loser, bandits will drop the loser fast, but continue to use both high-performing arms. There are various algorithms for this, all well explained in (White 2013): (a) epsilon-greedy, where a varying percentage of attempts are tried on each arm, with online learning of what percentage of attempts to use for each arm, (b) upper confidence bounds, generating a statistical confidence interval, (c) Thompson sampling, using Bayesian techniques that are similar to sampling in proportion to the rewards seen so far.

Contextual bandits allow the incorporation of other variables (context) into the decision, such as type of trip (business vs leisure), type of shopping request (number of days to departure, length of stay, market, days-of-week from traveling, and so on). Michael Byrd has a paper about contextual bandits in this special issue.

Bandits are one form of reinforcement learning. An excellent introduction to the general reinforcement learning problem is (Dinculescu 2018). In general, reinforcement learning is a more difficult problem than supervised learning, but much research is ongoing. "Reinforcement learning has two fundamental difficulties not present in supervised learning—exploration and long-term credit assignment. They have always been there and will take more than a really nice function approximator to solve. We must find much better ways to explore, use samples from past exploration, transfer across tasks, learn from other agents including humans, act in different timescales and address the awkward problems with scalar rewards" (Sahni 2018).

Some research has applied evolution strategies instead of traditional reinforcement learning approaches (Salimans et al. 2017). This evolutionary approach "is invariant to action frequency and delayed rewards, tolerant of extremely long horizons, and does not need temporal discounting or value function approximation."

One of the issues with all of these approaches is that most research is still not considering a marketplace solution. The key point is that both sides need to be learning, suppliers and travelers. And Global Distribution Systems need to become the marketplace to bring them together and facilitate this learning. Or someone else will.

Distributed AI

If reinforcement learning approaches are the next big thing in AI, then distributed AI and multi-agent systems are the next, next big thing. Essentially, multi-agent systems will be the key to creating a marketplace that will bring suppliers and travelers together, allowing each to learn about the other through communication.

Multi-agent systems bring everything together:

- Artificial intelligence
- Intelligent agents
- Game theory (used so the agents can negotiate with each other)

Francesco Corea has written a widely distributed primer for multi-agent systems (Corea 2019). Michael Wooldridge has been researching this technology for over twenty years, and he's written an excellent reference (Wooldridge 2009). Wikipedia has a good quick overview, under "Multi-Agent System".

From (Corea 2019): "An agent is an entity with domain knowledge, goals and actions. Multi-agent systems are a set of agents which interact in a common environment. Multi-agent systems deal with the construction of complex systems involving multiple agents and their coordination. A multi-agent system (MAS) is a distributed computing system with autonomous interacting intelligent agents that coordinate their actions so as to achieve its goal(s) jointly or competitively."

Of course, we need to take these agents and put them into a multi-agent environment, which is where game theory and social science come in. (Shebalov 2019) looked at the idea of using multi-agent systems to take solutions from individual airline operations research optimizers (route & schedule planning, fleet assignment, aircraft routing, crew scheduling) and having agents negotiate an overall plan (e.g., "if you move this flight forward one hour, which reduces customer satisfaction slightly, we can save big on pilot & staff scheduling"). See also (Castro et al. 2010).

Chapter 1 (Wooldridge 2009) discusses how multi-agent systems are much more than:

- Distributed systems, because (a) agents are autonomous and (b) not all agents share a common goal
- Artificial intelligence, because while AI is important, agents need to integrate different machine learning components into an overall social ability that constitutes real intelligence

- Game theory, which is important for giving agents the ability to negotiate, but we need to add the ability to compute (in real-time) game theory solutions
- Social science, because while understanding psychology and social relationships are important, we may not want multi-agent systems to simply follow these human concepts

So, we need multi-agent systems using reinforcement learning approaches. Marketplaces will bring all of this together.

Marketplace

Interested in seeing real artificial intelligence? We have it, and have had it for a very long time. Professor Jordan gives the example that if an alien came to Earth looking for intelligence, the alien would find that it has existed ever since humans started trading. Trading networks = intelligence. Somehow, breakfast appears on your plate, sourced from all over the world, all coming together in a vast network that results in your cereal or your fruit or your coffee. That's intelligence, even though each component is not necessarily "intelligent."

The intelligence is in the network.

Professor Jordan's key message is to empower users, broadcast what they want (more or less, a publish and subscribe model), and let vendors reply. This is different from the top-down push approach that Facebook and social media platforms use.

Current AI only allows communication in the form of consumer shopping data. Current recommender systems scoop up vast amounts of data, apply an algorithm to it, and (ta-da!) a neural network or recommender system appears. Wouldn't it be fun to actually ask the consumer what they are thinking about, even during an anonymous shop? An example of this is Sabre's "Preference Driven Air Shopping," (Vinod et al. 2015; Sabre 2019b) which uses the TOPSIS algorithm (Opricovic and Tzeng 2004) to find the best options for what the consumer is searching for now.

The "fourth Generation of AI is not just one agent making a decision, but an interconnected web of data, agents, decisions." What we need is an "intelligent infrastructure" approach of "large-scale, distributed collections of data flows and loosely coupled decisions." The future of AI is about economics, which is studying the sharing of resources (Jordan 2019).

Still, recommendation systems need to be part of the market. Example: "Restaurant owners and diners, it would be nice if every evening when I walk out in Shanghai, I push a button on my phone and say I am interested in food tonight, the system knows where I am geolocated and I that like certain kinds of cuisines. And then it tells all the restaurant owners around there that I am available as a possible client. The restaurants learn a little bit about me, and my preferences and they bid on me, and maybe they give me a 10% discount, then I accept and we now have a pricing transaction. This system puts me in a market that I can accept or reject. It's not that complicated, but you have to do it at a scale with millions of entities on both sides, and the data changes every day." (Jordan 2018).

The future of AI is not about building something to play the role of a human being inside of a computer, but about networked systems where data flows are accompanied by value flows. For this value to be realized trust is critical, something few companies appreciate.

Conclusion

The next-generation AI ideas are mostly in a research phase now, but we all know how rapidly new ideas can explode onto the marketplace while we are buried in projects using current generation approaches. Exponential growth in any one topic can be huge, but here we will have converging exponentials (Diamandis and Kotler 2020). These will make a huge difference in how travelers' shop for, purchase, and consume travel products. Global Distribution Systems are in the best position to deliver this, but other companies (notably Google and Amazon) are very eager to enter the travel space. A recent book's title says it best: *The Future Is Faster Than You Think: How converging technologies are transforming business, industries, and our lives*, the final volume in the trilogy Exponential Learning Series. This book covers how travel will be consumed, and the authors have astonishing predictions (Diamandis and Kotler 2020).

Notes

(1) Deng (2019) provides a nice overview of the many flavors of recommendation engines. Some Python packages are:

- Surprise (http://surpriselib.com/)
- LightFM (https://github.com/lyst/lightfm)
- Case Recommender (https://github.com/caserec/CaseRecommender)
- Spotlight (https://github.com/maciejkula/spotlight)
- Rexy (https://github.com/kasramvd/Rexy)

For R, there is Recommender Lab (https://cran.r-project.org/web/packages/recommenderlab/index.html).

See https://github.com/grahamjenson/list_of_recommender_systems for a lengthy list.

(2) The quote "If you are not paying for it, you become the product" has been around a long time, with many attributions. Some links:

- https://www.forbes.com/sites/marketshare/2012/03/05/if-youre-not-paying-for-it-you-become-the-product
- https://quoteinvestigator.com/2017/07/16/product/, which traces the idea back to when network television was popular
- https://twitter.com/timoreilly/status/22823381903, where famous publisher Tim O'Reilly tweets the quote, attributing it to Bryce Roberts

(3) Prof. Jordan's original version appeared in an April 2018 *Medium* post. A newer version that also includes 11 discussion items by well-known researchers, plus a rejoinder, is in the inaugural issue of the *Harvard Data Science Review* (2019), "Artificial Intelligence—The Revolution Hasn't Happened Yet," July 2019. It's the rejoinder ("Dr. AI or: How I Learned to Stop Worrying and Love Economics") that actually has many of the ideas discussed here. See https://hdsr.mitpress.mit.edu/pub/wot7mkc1/release/8.

(4) From the Spark + AI Summit 2018, organized by Databricks, video at https://databricks.com/session/keynote-from-michael-i-jordan. Also see https://blogs.blackberry.com/2018/11/the-future-of-ai-its-not-about-a-computer-that-thinks like-a-human

(5) Presentation by Prof Jordan at the O'Reilly AI Conference, 2019, San Jose, California, USA. "On gradient-based methods for finding game-theoretic equilibria." The title sounds intimidating, but it's really a discussion about finding marketplace solutions. "The aim is to blend gradient-based methodology with game-theoretic goals as part of a large "microeconomics meets machine learning" program." A free highlight is at https://www.oreilly.com/radar/on-gradient-based-methods-for-finding-game-theoretic-equilibria/, and the Powerpoint slides are available (for free) at https://conferences.oreilly.com/artificial-intelligence/ai-ca-2019/public/schedule/detail/77921.

(6) The quote is from the Wikipedia review at https://en.wikipedia.org/wiki/The_Master_Algorithm. Slate has a detailed review at https://slate.com/technology/2015/09/pedro-domingos-master-algorithm-how-machine-learning-is-reshaping-how-we-live.html

References

Bacon, Tom. 2019. 4 ways that NDC is reshaping how airlines sell. https://www.eyefortravel.com/revenue-and-data-management/4-ways-ndc-reshaping-how-airlines-sell.

Castro, Antonio J.M., Ana Paula Rocha, Eugénio Oliveira. 2010. A negotiation based approach to airline operations recovery. https://www.academia.edu/1132120/A_Negotiation_Based_Approach_to_Airline_Operations_Recovery.

Corea, Franciso. 2019. Distributed artificial intelligence: a primer on multi-agent systems, agent-based modeling, and swarm intelligence. https://www.kdnuggets.com/2019/04/distributed-artificial-intelligence-multi-agent-systems-agent-based-modeling-swarm-intelligence.html. This paper has many good references.

Deng, Houtao. 2019. Recommender systems in practice: How companies make product recommendations. https://towardsdatascience.com/recommender-systems-in-practice-cef9033bb23a.

Diamandis, Peter H., and Steven Kotler. 2020. *The future is faster than you think: How converging technologies are transforming business, industries, and our lives (exponential technology series)*. New York: Simon & Schuster.

Dinculescu, Monica. 2018. An introduction to reinforcement learning (with otters). https://meowni.ca/posts/rl-with-otters/.

Domingos, Pedro. 2015. *The Master Algorithm*. New York: Basic Books.

Jordan, Michael I. 2018. Clarity of thought on AI, at https://medium.com/syncedreview/michael-i-jordan-interview-clarity-of-thought-on-ai-ed936d0dc421. This is from an interview he gave in China.

Jordan, Michael I. 2019. Artificial intelligence—The revolution hasn't happened yet. *Harvard Data Science Review*. https://doi.org/10.1162/99608f92.f06c6e61.

Microsoft. 2018. *The future computer: Artificial intelligence and its role in society*

Opricovic, Serafim, and Gwo-Hshung. Tzeng. 2004. Compromise solution by MCDM methods: A comparative analysis of VIKOR and TOPSIS. *European Journal of Operational Research* 156: 445–455.

Ramdas, Aaditya, Tijana Zrnic, Martin J. Wainwright, Michael I. Jordan. 2019. SAFFROM: An adaptive algorithm for online control, University of California at Berkeley. https://arxiv.org/pdf/1802.09098.pdf.

Sabre, Inc. 2019a. Explore Beyond NDC. https://www.sabre.com/ndc/.

Sabre, Inc. 2019b. Preference Driven Air Shopping - Airline Version. https://www.sabre.com/insights/innovation-hub/prototypes/preference-driven-air shopping-airline-version/.

Sahni, Himanshu. 2018. Reinforcement Learning never worked, and 'deep' only helped a bit. https://himanshusahni.github.io/2018/02/23/reinforcement-learning-never-worked.html.

Salimans, Tim, Jonathan Ho, Xi Chen, Szymon Sidor, Ilya Sutskever. 2017. Evolution Strategies as a Scalable Alternative to Reinforcement Learning. https://arxiv.org/abs/1703.03864.

Shebalov, Sergey. 2019. Integrating commercial planning and operations with multi-agent systems. AGIFORS 59th annual symposium proceedings

Surmenok, Pavel. 2017. Contextual bandits and reinforcement learning. https://towardsdatascience.com/contextual-bandits-and-reinforcement-learning-6bdfeaece72a.

Teixeria, Thales S. 2019. *Unlocking the customer value chain, how decoupling drives consumer disruption*. New York: Currency Publishers.

Toyoglu, Hunkar and Richard Ratliff. 2019. Product recommendations with multi-armed bandit. *AGIFORS Revenue Management Study Group Proceedings*.

Veinbergs, Jevgenijs. 2019. What shapes revenue management in the age of NDC?, at https://blog.yieldr.com/what-shapes-revenue-management-in-the-age-of-ndc/

Vinod, B., P. Xie, and R. Bellubbi. 2015. From shopper to customer: preference driven air shopping with targeted one-to-one shopping responses. *Ascend* 4: 11–13.

White, John Myles. 2013. *Bandit algorithms for website optimization*. Sebastopol: O'Reilly Books.

Wooldridge, Michael. 2009. *An introduction to multi-agent systems*, 2nd ed. Hoboken: Wiley.

Ross M. Darrow was Director at Sabre Research, responsible for advanced statistical modeling and decision support. After three decades at American Airlines and Sabre, he recently joined Charter and Go, an innovation start-up for charter airlines, as Chief Scientist.

The manufacturer's authorised representative in the EU is Springer
Nature Customer Service Centre GmbH, Europaplatz 3, 69115 Heidelberg,
Germany. If you have any concerns regarding our products, please
contact ProductSafety@springernature.com

Printed and bound by CPI Group (UK) Ltd, Croydon, CR0 4YY
28/09/2026
02224096-0001